Molecular Mechanisms
of Photosynthesis

Molecular Mechanisms of Photosynthesis

Third Edition

Robert E. Blankenship
Tempe, AZ, USA

Registered Offices
John Wiley & Sons, Inc., 111 River Street, Hoboken, NJ 07030, USA
John Wiley & Sons Ltd, The Atrium, Southern Gate, Chichester, West Sussex, PO19 8SQ, UK

Editorial Office
The Atrium, Southern Gate, Chichester, West Sussex, PO19 8SQ, UK
For details of our global editorial offices, customer services, and more information about Wiley products visit us at www.wiley.com.

Wiley also publishes its books in a variety of electronic formats and by print-on-demand. Some content that appears in standard print versions of this book may not be available in other formats.

Library of Congress Cataloging-in-Publication data is applied for

ISBN 9781119800019

Cover Design: Wiley
Cover Image: © Cover image produced by Melih Sener, John Stone, and Barry Isralewitz using VMD. Shown is the atomic-to-cell scale organization of a photosythetic purple bacterium and the energy conversion proteins therein.

Set in 10.5/12.5pt Minion by Straive, Pondicherry, India

10 9 8 7 6 5 4 3 2 1

Molecular Mechanisms of Photosynthesis

Third Edition

Robert E. Blankenship
Tempe, AZ, USA

Registered Offices
John Wiley & Sons, Inc., 111 River Street, Hoboken, NJ 07030, USA
John Wiley & Sons Ltd, The Atrium, Southern Gate, Chichester, West Sussex, PO19 8SQ, UK

Editorial Office
The Atrium, Southern Gate, Chichester, West Sussex, PO19 8SQ, UK
For details of our global editorial offices, customer services, and more information about Wiley products visit us at www.wiley.com.

Wiley also publishes its books in a variety of electronic formats and by print-on-demand. Some content that appears in standard print versions of this book may not be available in other formats.

Library of Congress Cataloging-in-Publication data is applied for

ISBN 9781119800019

Cover Design: Wiley
Cover Image: © Cover image produced by Melih Sener, John Stone, and Barry Isralewitz using VMD. Shown is the atomic-to-cell scale organization of a photosythetic purple bacterium and the energy conversion proteins therein.

Set in 10.5/12.5pt Minion by Straive, Pondicherry, India

10 9 8 7 6 5 4 3 2 1

I dedicate this book to the memory of my mother, whose early and constant encouragement started me down the road to a career in science.

Contents

Introduction to the third edition

It is now nearly 20 years since the first edition and more than 7 years since the second edition of *Molecular Mechanisms of Photosynthesis* were published. In that time, the scientific understanding of how photosynthesis works has continued to progress. The success of the first and second editions has prompted numerous requests for a third edition, which I am pleased to provide. I have tried to update the text to reflect this new understanding.

This book is an introduction to the basic concepts that underlie the process of photosynthesis as well as a description of the current understanding of the subject. Because it is such a complex process that requires some knowledge of many different fields of science to appreciate, it can be intimidating for a person who is not already conversant with the basics of all these fields. For this reason, a brief overview is provided in the first chapter, introducing and summarizing the main concepts. This chapter is then followed by a more in-depth treatment of each of the main themes in later chapters.

Photosynthesis is perhaps the best possible example of a scientific field that is intrinsically interdisciplinary. Our discussion of photosynthesis will span time scales from the cosmic to the unimaginably fast, from the origin of the Earth 4.5 billion years ago, to molecular processes that take only a few femtoseconds. This is a range of over 32 orders of magnitude. Appreciating this extraordinary scale will require us to learn a range of vocabularies and concepts that stretch from geology through physics and chemistry, to biochemistry, cell and molecular biology, and finally to evolutionary biology. Any person who wishes to appreciate the big picture of how photosynthesis works, and how it fits into the broad scope of scientific inquiry, needs to have at least a rudimentary understanding of all of these fields of science. This is an increasingly difficult task in this age of scientific specialization, because no one can truly be an expert in all areas. This book attempts to provide the starting point for a broadly based understanding of photosynthesis, incorporating key concepts from across the scientific spectrum. The emphasis throughout the book will be on molecular-scale mechanistic processes.

Many of the concepts that we will explore throughout the bulk of this book require an understanding of basic concepts of physical chemistry, including thermodynamics, kinetics, and quantum mechanics. It is beyond the scope of our broad, and therefore necessarily brief, treatment of photosynthesis to provide a comprehensive background in these areas that form the core of the mechanistic aspects of the subject. However, some modest understanding of these physical principles is essential to be able to appreciate the essence of the photosynthetic process. This is addressed in an appendix that introduces the physical basis of light and energy. This appendix can either be read as a preface to the bulk of the book or consulted as needed as a reference.

The book is aimed toward advanced undergraduates and beginning graduate students in a range of disciplines, including life sciences, chemistry, and physics, as well as more senior scientists seeking to learn about the remarkable process of photosynthesis. An understanding of basic principles of chemistry, physics, and biology is assumed.

Robert E. Blankenship

Acknowledgements

I thank my former advisors Ken Sauer and Bill Parson for initiating me into the fascinating world of photosynthesis. Their guidance, support, and friendship have been invaluable to me during the course of my career.

I thank the many friends and colleagues from around the world for reading and commenting on some material, for helpful discussions on specialized topics, and for kindly providing figures for publication. These include (in alphabetical order) Carl Bauer, Oded Béjà, Gary Brudvig, Don Bryant, Julian Eaton-Rye, Gyozo Garab, Govindjee, Beverley Green, Martin Hohmann-Marriott, Werner Kühlbrandt, Haijun Liu, Yuval Mazor, Johannes Messinger, Tom Moore, Jian-Ren Shen, and Junko Yano.

The beautiful graphic that adorns the cover of the book is courtesy of Melih Sener. This is an updated version of the graphics that were on the covers of the first and second editions.

Special thanks to my editor Rebecca Ralf and to Kerry Powell from Wiley-Blackwell for help with the production.

Finally, I thank my wonderful family, Liz, Larissa, and Sam, for their constant love and support.

About the companion website

This book is accompanied by a companion website:

https://www.wiley.com/go/blankenship/molecularphotosynthesis3e

The website includes:

- Powerpoints of all figures from the book for downloading
- PDFs of all tables from the book for downloading

Chapter 1

The basic principles of photosynthetic energy storage

1.1 What is photosynthesis?

Photosynthesis is a biological process whereby the Sun's energy is captured and stored by a series of events that convert the pure energy of light into the free energy needed to power life. This remarkable process provides the foundation for essentially all life and has over geologic time profoundly altered the Earth itself. It provides all our food and most of our energy resources.

Perhaps the best way to appreciate the importance of photosynthesis is to examine the consequences of its absence. The catastrophic event that caused the extinction of the dinosaurs and most other species 65 million years ago almost certainly exerted its major effect not from the force of the comet or asteroid impact itself, but from the massive quantities of dust ejected into the atmosphere. This dust blocked out the Sun and effectively shut down photosynthesis all over the Earth for a period of months or years. Even this relatively short interruption of photosynthesis, miniscule on the geological time scale, had catastrophic effects on the biosphere.

Photosynthesis literally means "synthesis with light." As such, it might be construed to include any process that involved synthesis of a new chemical compound under the action of light. However, that very broad definition might include a number of unrelated processes that we do not wish to include, so we will adopt a somewhat narrower definition of photosynthesis:

> Photosynthesis is a process in which light energy is captured and stored by a living organism, and the stored energy is used to drive energy-requiring cellular processes.

This definition is still relatively broad and includes the familiar chlorophyll-based form of photosynthesis that is the subject of this book, but also includes the very different form of photosynthesis carried out by some micro organisms using proteins related to rhodopsin, which contain retinal as their light-absorbing pigment. Light-driven signaling processes, such as vision or phytochrome action, where light conveys information instead of energy, are excluded from our definition of photosynthesis, as well as all processes that do not normally take place in living organisms.

Molecular Mechanisms of Photosynthesis, Third Edition. Robert E. Blankenship.
© 2021 Robert E. Blankenship 2021 by John Wiley & Sons Ltd.
Companion website: https://www.wiley.com/go/blankenship/molecularphotosynthesis3e

What constitutes a photosynthetic organism? Does the organism have to derive all its energy from light to be classified as photosynthetic? Here, we will adopt a relatively generous definition, including as photosynthetic any organism capable of deriving some of its cellular energy from light. Higher plants, the photosynthetic organisms that we are all most familiar with, derive essentially all their cellular energy from light. However, there are many organisms that use light as only part of their energy source and, under certain conditions, they may not derive any energy from light. Under other conditions, they may use light as a significant or sole source of cellular energy. We adopt this broad definition because our interest is primarily in understanding the energy storage process itself. Organisms that use photosynthesis only part of the time may still have important things to teach us about how the process works and therefore deserve our attention, even though a purist might not classify them as true photosynthetic organisms. We will also use both of the terms "photosynthetic" and "phototrophic" when describing organisms that can carry out photosynthesis. We will usually use photosynthetic to describe higher plants, algae, and cyanobacteria that derive most or all of their energy needs from light, and phototrophic to describe bacteria or archaea that can carry out photosynthesis but often derive much of their energy needs from other sources.

The most common form of photosynthesis involves chlorophyll-type pigments and operates using light-driven electron transfer processes. The organisms that we will discuss in detail in this book, including plants, algae, and cyanobacteria (collectively called **oxygenic** organisms because they produce oxygen during the course of doing photosynthesis) and several types of **anoxygenic** (non-oxygen-evolving) bacteria, all work in this same basic manner. All these organisms will be considered to carry out what we will term "chlorophyll-based photosynthesis." The retinal-based form of photosynthesis, while qualifying under our general definition, is mechanistically very different from chlorophyll-based photosynthesis, and will

not be discussed in detail. It operates using *cis–trans* isomerization that is directly coupled to ion transport across a membrane (Ernst *et al.*, 2014). The ions that are pumped as the result of the action of light can be either H^+, Na^+, or Cl^- ions, depending on the class of the retinal-containing protein. The H^+-pumping complexes are called bacteriorhodopsins, and the Cl^--pumping complexes are known as halorhodopsins. No light-driven electron transfer processes are known thus far in these systems.

For many years, the retinal-based type of photosynthesis was known only in extremely halophilic Archaea (formerly called archaebacteria), which are found in a restricted number of high-salt environments. Therefore, this form of photosynthesis seemed to be of minor importance in terms of global photosynthesis. However, in recent years, several new classes of microbial rhodopsins, known as proteorhodopsin, heliorhodopsin, and others, have been discovered (Béjà *et al.*, 2000; Pushkarev *et al.*, 2018; Inoue *et al.*, 2020). Proteorhodopsin pumps H^+ and has an amino acid sequence and protein secondary structure that are generally similar to bacteriorhodopsin. The proteobacteria that contain proteorhodopsin are widely distributed in the world's oceans, so the rhodopsin-based form of photosynthesis may be of considerable importance. Recent evidence suggests that the proteorhodopsins are responsible for a significant amount of primary productivity in the ocean (Gómez-Consarnau *et al.*, 2019).

As mankind pushes into space and searches for life on other worlds, we need to be able to recognize life that may be very different from what we know on Earth. Life always needs a source of energy, so it is reasonable to expect that some form of photosynthesis (using our general definition) will be found on most or possibly all worlds that harbor life. Photosynthesis on such a world need not necessarily contain chlorophylls and perform electron transfer. It might be based on isomerization such as bacteriorhodopsin, or possibly on some other light-driven process that we cannot yet imagine (Kiang *et al.*, 2007a, b; Schweiterman *et al.*, 2018).

1.2 Photosynthesis is a solar energy storage process

Photosynthesis uses light from the Sun to drive a series of chemical reactions. The Sun, like all stars, produces a broad spectrum of radiation output that ranges from gamma rays to radio waves. The solar output is shown in Fig. 1.1, along with absorption spectra of some photosynthetic organisms. Only some of the emitted solar radiation is visible to our eyes, consisting of light with wavelengths from about 400 to 700 nm. The entire visible range of light, and some wavelengths in the near infrared (700–1000 nm), are highly active in driving photosynthesis in certain organisms, although the most familiar chlorophyll *a*-containing organisms cannot use light with a wavelength longer than 700 nm.

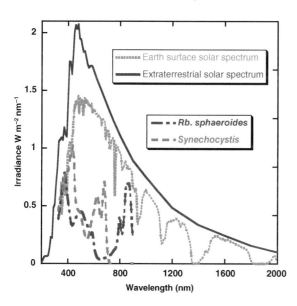

Figure 1.1 Solar irradiance spectra and absorption spectra of photosynthetic organisms. Solid curve: intensity profile of the extraterrestrial spectrum of the Sun; dotted line: intensity profile of the spectrum of sunlight at the surface of the Earth; dash-dot line: absorbance spectrum of *Rhodobacter sphaeroides*, an anoxygenic purple photosynthetic bacterium; dashed line: absorbance spectrum of *Synechocystis* PCC 6803, an oxygenic cyanobacterium. The spectra of the organisms are in absorbance units (scale not shown).

The spectral region from 400 to 700 nm is often called **photosynthetically active radiation** (PAR), although this is only strictly true for chlorophyll *a*-containing organisms. Recently, some oxygenic photosynthetic organisms that utilize radiation outside the PAR region have been discovered. These are discussed in detail in Chapters 2 and 7.

The sunlight that reaches the surface of the Earth is reduced by scattering and by the absorption of molecules in the atmosphere. Water vapor and other molecules such as carbon dioxide absorb strongly in the infrared region, and ozone absorbs in the ultraviolet region. The ultraviolet light is a relatively small fraction of the total solar output, but much of it is very damaging because of the high energy content of these photons (see Appendix for a discussion of photons and the relationship of wavelength and energy content of light). The most damaging ultraviolet light is screened out by the ozone layer in the upper atmosphere and does not reach the Earth's surface. Wavelengths less than 400 nm account for only about 8% of the total solar irradiance, while wavelengths less than 700 nm account for 47% of the solar irradiance (Thekaekara, 1973).

The infrared wavelength region includes a large amount of energy and would seem to be a good source of photons to drive photosynthesis. However, no organism is known that can utilize light of wavelength longer than about 1000 nm for photosynthesis (1000 nm and longer wavelength light comprises 30% of the solar irradiance). This is almost certainly because infrared light has a very low energy content in each photon, so that large numbers of these low-energy photons would have to be used to drive the chemical reactions of photosynthesis. No known organism has evolved such a mechanism, which would in essence be a living heat engine. Infrared light is also absorbed by water, so aquatic organisms do not receive much light in this spectral region.

The distribution of light in certain environments can be very different from that shown in Fig. 1.1. The differing spectral content, or color, of light in different environments represents differences in **light quality**. In later chapters, we will encounter some elegant control mechanisms that organisms use to adapt to changes in light quality. In a forest, the upper part of the canopy receives the full solar spectrum, but the forest floor receives only light that was not absorbed above. The spectral distribution of the filtered light that reaches the forest floor is therefore enriched in the green and far-red regions and is almost completely lacking in the red and blue wavelengths.

In aquatic systems, the intensity of light rapidly decreases as one goes deeper down the water column, owing to several factors. This decrease is not uniform for all wavelengths. Water weakly absorbs light in the red portion of the spectrum, so that the red photons that are most efficient in driving photosynthesis rapidly become depleted. Water also scatters light, mainly because of effects of suspended particles. This scattering effect is most prevalent in the blue region of the spectrum, because scattering by small particles is proportional to the frequency raised to the fourth power. The sky is blue because of this frequency-dependent scattering effect. At water depths of more than a few tens of meters, most of the available light is in the middle, greenish part of the spectrum, because the red light has been absorbed and the blue light scattered (Falkowski and Raven, 2007; Kirk, 2011). None of the types of chlorophylls absorb green light very well. However, other photosynthetic pigments, in particular some carotenoids (e.g. fucoxanthin, peridinin), have intense absorption in this region of the spectrum and are present in large quantities in many aquatic photosynthetic organisms. At water depths greater than about 100 m, the light intensity from the Sun is too weak to drive photosynthesis.

1.3   Where photosynthesis takes place

Photosynthesis is carried out by a wide variety of organisms. In all cases, lipid bilayer membranes are critical to the early stages of energy storage, such that photosynthesis must be viewed as a process that is at heart membrane-based. The early processes of photosynthesis are carried out by pigment-containing proteins that are integrally associated with the membrane. Later stages of the process that occur on a slower (e.g. millisecond)

time scale are mediated by proteins that are freely diffusible in the aqueous phase.

In eukaryotic photosynthetic cells, photosynthesis is localized in subcellular structures known as **chloroplasts** (Fig. 1.2). The chloroplast contains all the chlorophyll pigments and, in most organisms,

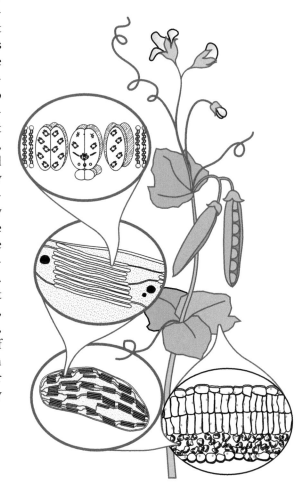

Figure 1.2 Exploding diagram of the photosynthetic apparatus of a typical higher plant. The first expansion bubble shows a cross-section of a leaf, with the different types of cells; the dark spots are the chloroplasts. The second bubble is a chloroplast; the thylakoid membranes are the dark lines; the stroma is the stippled area. The third bubble shows a grana stack of thylakoids. The fourth bubble shows a schematic picture of the molecular structure of the thylakoid membrane, with a reaction center flanked by antenna complexes. *Source:* Courtesy of Aileen Taguchi.

carries out all the main phases of the process of photosynthesis. Synthesis of sucrose and some other carbon metabolism reactions require extrachloroplastic enzymes. Chloroplasts are about the size of bacteria, a few micrometers in diameter. In fact, chloroplasts were derived long ago from symbiotic bacteria that became integrated into the cell and eventually lost their independence, a process known as **endosymbiosis** (see Chapter 12). Even today, they retain traces of their bacterial heritage, including their own DNA, although much of the genetic information needed to build the photosynthetic apparatus now resides in DNA located in the nucleus.

An extensive membrane system is found within the chloroplast, and all the chlorophylls and other pigments are found associated with these membranes, which are known as **thylakoids**, or sometimes called **lamellae**. In typical higher plant chloroplasts, most of the thylakoids are closely associated in stacks and are known as **grana thylakoid membranes**, while those that are not stacked are known as **stroma thylakoid membranes**. The thylakoid membranes are the sites of light absorption and the early or primary reactions that first transform light energy into chemical energy. The nonmembranous aqueous interior of the chloroplast is known as the **stroma**. The stroma contains soluble enzymes and is the site of the carbon metabolism reactions that ultimately give rise to products that can be exported from the chloroplast and used elsewhere in the plant to support other cellular processes.

In prokaryotic photosynthetic organisms, the early steps of photosynthesis take place on specialized membranes that are derived from the cell's cytoplasmic membrane. In these organisms, the carbon metabolism reactions take place in the cell cytoplasm, along with all the other reactions that make up the cell's metabolism.

1.4   The four phases of energy storage in photosynthesis

It is convenient to divide photosynthesis into four distinct phases, which together make up the complete process, beginning with photon absorption and ending with the export of stable carbon products from the chloroplast. The four phases are as follows: (1) light absorption and energy delivery by antenna systems, (2) primary electron transfer in reaction centers, (3) energy stabilization by secondary processes, and (4) synthesis and export of stable products.

The terms **light reactions** and **dark reactions** have traditionally been used to describe different phases of photosynthetic energy storage. The first three phases that we have identified make up the light reactions, and the fourth encompasses the dark reactions. However, this nomenclature is somewhat misleading, in that all the reactions are ultimately driven by light, yet the only strictly light-dependent step is photon absorption. In addition, several enzymes involved in carbon metabolism are regulated by compounds produced by light-driven processes. We will now briefly explore each of the phases of photosynthetic energy storage, with the emphasis on the basic principles. Much more detail is given in the later chapters dedicated to each topic.

1.4.1   Antennas and energy transfer processes

For light energy to be stored by photosynthesis, it must first be absorbed by one of the pigments associated with the photosynthetic apparatus. Photon absorption creates an excited state that eventually leads to charge separation in the reaction center. Not every pigment carries out photochemistry; the vast majority function as antennas, collecting light and then delivering energy to the reaction center where the photochemistry takes place. The antenna system is conceptually similar to a satellite dish, collecting energy and concentrating it in a receiver, where the signal is converted into a different form (Fig. 1.3). Energy transfer in antenna systems is discussed in more detail in Chapter 5.

The antenna system does not do any chemistry; it works by an energy transfer process that involves the migration of electronic excited states from one molecule to another. This is a purely physical process, which depends on a weak energetic coupling of the antenna pigments. In almost all cases, the pigments are bound to proteins in highly specific

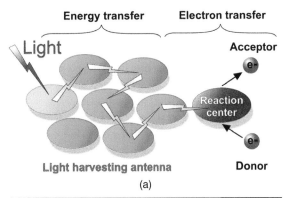

(b)

Figure 1.3   (a) Schematic diagram illustrating the concept of antennas in photosynthetic organisms. Light is absorbed by a pigment, which can be either a type of chlorophyll or an accessory pigment. Energy transfer takes place in which the excited state migrates throughout the antenna system and is eventually delivered to the reaction center where electron transfer takes place, creating an oxidized electron donor and a reduced electron acceptor. (b) Radio telescope as an analogy to the photosynthetic antenna. The dish serves as the antenna, collecting energy and delivering it to the receiver, which transduces it into a signal. *Source:* NASA.

associations. In addition to chlorophylls, common antenna pigments include carotenoids and open-chain tetrapyrrole bilin pigments found in phycobilisome antenna complexes.

Antenna systems often incorporate an energetic and spatial funneling mechanism, in which pigments that are on the periphery of the complex absorb at shorter wavelengths and therefore higher excitation energies than those at the core. As energy transfer takes place, the excitation energy moves from higher- to lower-energy pigments, at the same time heading toward the reaction center.

Antenna systems greatly increase the amount of energy that can be absorbed compared with a single pigment. Under most conditions, this is an advantage, because sunlight is a relatively dilute energy source. Under some conditions, however, especially if the organism is subject to some other form of stress, more light energy can be absorbed than can be used productively by the system. If unchecked, this can lead to severe damage in short order. Even under normal conditions, the system is rapidly inactivated if some sort of photoprotection mechanism is not present. Antenna systems (as well as reaction centers) therefore have extensive and multifunctional regulation, protection, and repair mechanisms.

The number of antenna pigments associated with each reaction center complex varies widely, from a minimum of a few tens of pigments in some organisms to a maximum of several thousand pigments in other types of organisms. The pigment number and type largely reflect the photic environment that the organism lives in. Smaller antennas are found in organisms that live in high intensity conditions, while the large antennas are found in environments where light intensity is low.

1.4.2   Primary electron transfer in reaction centers

The transformation from the pure energy of excited states to chemical changes in molecules takes place in the photosynthetic reaction center. The reaction center is a multisubunit protein complex that is embedded in the photosynthetic membrane. It is a pigment–protein complex, incorporating chlorophyll or bacteriochlorophyll and carotenoid pigments as well as other cofactors such as quinones or iron sulfur centers, which are involved in electron transfer processes. The cofactors are bound to extremely hydrophobic polypeptides that thread back and forth across the nonpolar portion of the membrane multiple times.

The reaction center contains a special dimer of pigments that in most or all cases is the **primary electron donor** for the electron transfer cascade. These pigments are chemically identical (or nearly so) to the chlorophylls that are antenna pigments, but their environment in the reaction center protein gives them unique properties. The final step in the antenna system is the transfer of energy into this dimer, creating an excited dimer that has been electronically excited to a higher energy level.

The basic process that takes place in all reaction centers is described schematically in Fig. 1.4a.

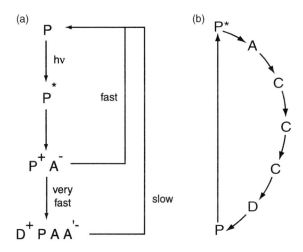

Figure 1.4   (a) General electron transfer scheme in photosynthetic reaction centers. Light excitation promotes a pigment (P) to an excited state (P*), where it loses an electron to an acceptor molecule (A) to form an ion-pair state P⁺A⁻. Secondary reactions separate the charges, by transfer of an electron from an electron donor (D) and from the initial acceptor A to a secondary acceptor (A'). This spatial separation prevents the recombination reaction. (b) Schematic diagram of cyclic electron transfer pathway found in many anoxygenic photosynthetic bacteria. The terms fast, slow, and very fast are relative to each other. The vertical arrow signifies photon absorption: P represents the primary electron donor: D, A, and C represent secondary electron donors, acceptors, and carriers.

A chlorophyll-like pigment (P) is promoted to an excited electronic state, either by direct photon absorption or, more commonly, by energy transfer from the antenna system. The excited state of the pigment is chemically an extremely strong reducing species. It rapidly loses an electron to a nearby electron acceptor molecule (A), generating an ion-pair state P⁺A⁻. This is the "primary reaction" of photosynthesis. The energy has been transformed from electronic excitation to chemical redox energy. The system is now in a very vulnerable position with respect to losing the stored energy. If the electron is simply transferred back to P⁺ from A⁻, a process called **recombination**, then the net result is that the energy is converted into heat and rapidly dissipated and is therefore unable to do any work. This pathway is possible because the highly oxidizing P⁺ species is physically positioned directly next to the highly reducing A⁻ species.

The system avoids the fate of recombination losses by having a series of extremely rapid **secondary reactions** that successfully compete with recombination. These reactions, which are most efficient on the acceptor side of the ion-pair, spatially separate the positive and negative charges. This physical separation reduces the recombination rate by orders of magnitude.

The final result is that within a very short time (less than a nanosecond) the oxidized and reduced species are separated by nearly the thickness of the biological membrane (~30 Å; 1 Å = 0.1 nm). Slower processes can then take over and further stabilize the energy storage and convert it into more easily utilized forms. The system is so finely tuned that in optimum conditions the photochemical **quantum yield** of products formed per photon absorbed is nearly 1.0 (see Appendix). Of course, some energy is sacrificed from each photon in order to accomplish this feat, but the result is no less impressive.

1.4.3   Stabilization by secondary reactions

The essence of photosynthetic energy storage is the transfer of an electron from an excited chlorophyll-type pigment to an acceptor molecule in a pigment–protein complex called the reaction center. The initial, or primary, electron transfer event is followed by separation of the positive and negative charges by a very rapid series of secondary chemical reactions. This basic principle applies to all photosynthetic reaction centers, although the

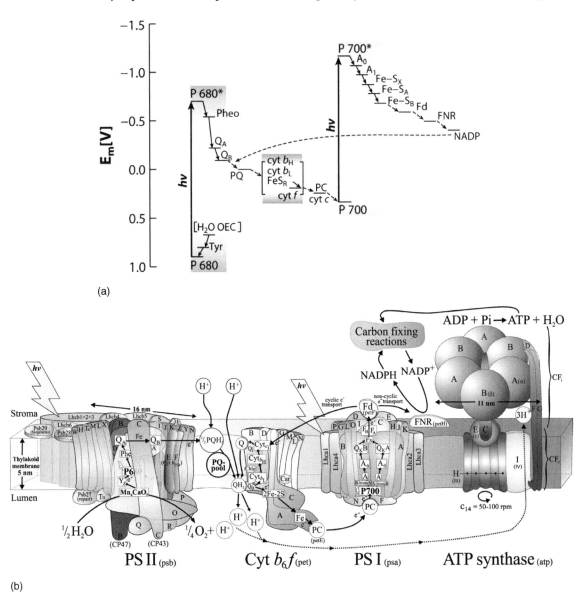

Figure 1.5 Schematic diagram of the noncyclic electron transfer pathway found in oxygenic photosynthetic organisms. The upper diagram (a) is an energetic picture of the electron transport pathway, incorporating the major reactions of photosynthesis into what is called the Z-scheme of photosynthesis. The lower diagram (b) is a spatial picture, showing the major protein complexes whose energetics are shown in the Z-scheme, and how they are arranged in the photosynthetic membrane. Neither view alone gives a complete picture, but together they summarize much information about photosynthetic energy storage. *Source:* (a) Hohmann-Marriott and Blankenship (2011) (p.532)/Annual Reviews. Reproduced with permission of *Annual Reviews*. (b) Courtesy of Dr. Jonathan Nield.

details of the process vary significantly from one system to the next.

In some organisms, one light-driven electron transfer and stabilization is sufficient to complete a **cyclic electron transfer chain**. This is shown schematically in Fig. 1.4b, in which the vertical arrow represents energy input to the system triggered by photon absorption, and the curved arrows represent spontaneous, or downhill, electron transfer processes that follow, eventually returning the electron to the primary electron donor. This cyclic electron transfer process is not in itself productive unless some of the energy of the photon can be stored. This takes place by the coupling of proton movement across the membrane with the electron transfer, so that the net result is a light-driven difference of pH and electrical potential, or electrochemical potential gradient across the two sides of the membrane. This electrochemical potential gradient, called a **protonmotive force**, is used to drive the synthesis of ATP.

The more familiar oxygen-evolving photosynthetic organisms have a different pattern of electron transfer. They have two photochemical reaction center complexes that work together in a **noncyclic electron transfer chain**, as shown in Fig. 1.5. The two reaction center complexes are known as Photosystems I and II. Electrons are removed from water by Photosystem II, oxidizing it to molecular oxygen, which is released as a waste product. The electrons extracted from water are transported via a quinone and the cytochrome b_6f complex to Photosystem I and, after a second light-driven electron transfer step, eventually reduce an intermediate electron acceptor, $NADP^+$ to form NADPH.

Protons are also transported across the membrane and into the thylakoid lumen during the process of the noncyclic electron transfer, creating a protonmotive force. The energy in this protonmotive force is used to make ATP (see Chapter 8).

Reaction centers and electron transfer processes in anoxygenic bacteria are discussed in more detail in Chapter 6, while these processes in oxygenic photosynthetic organisms are discussed in more detail in Chapter 7.

1.4.4   Synthesis and export of stable products

The final phase of photosynthetic energy storage involves the production of stable high-energy molecules and their utilization to power a variety of cellular processes. This phase uses the intermediate reduced compound, NADPH, generated by Photosystem 1, along with the phosphate bond energy of ATP to reduce carbon dioxide to sugars. In eukaryotic photosynthetic organisms, phosphorylated sugars are then exported from the chloroplast. The carbon assimilation and reduction reactions are enzyme-catalyzed processes that take place in the chloroplast stroma. These reactions are discussed in more detail in Chapter 9.

References

Béjà, O., Aravind, L., Koonin, E. V., Suzuki, M. T., Hadd, A., Nguyen, L. P., Jovanovich, S., Gates, C. M., Feldman, R. A., Spudich, J. L., Spudich, E. N., and DeLong, E. F. (2000) Bacterial rhodopsin: Evidence for a new type of phototrophy in the sea. *Science* 289: 1902–1906.

Ernst, O. P., Lodowski, D. T., Elstner, M., Hegemann, P., Brown, L. S., and Kandori, H. (2014) Microbial and animal rhodopsins: Structures, functions, and molecular mechanisms. *Chemical Reviews* 114: 126–163

Falkowski, P. and Raven, J. (2007) *Aquatic Photosynthesis*, 2nd Edn. Princeton, NJ: Princeton Univerity Press.

Gómez-Consarnau, L., Raven, J. A., Levine, N. M., Cutter, L. S., Wang, D., Seegers, B, Arístegui, J., Fuhrman, J. A., Gasol, J. M., and Sañudo-Wilhelmy, S. A. (2019) Microbial rhodopsins are major contributors to the solar energy captured in the sea. *Science Advances* 5: eaaw8855.

Hohmann-Marriott, M. F. and Blankenship, R. E. (2011) Evolution of photosynthesis. *Annual Review of Plant Biology* 62: 515–548.

Inoue, K., Tsunoda, S. P., Singh, M., Tomida, S., Hososhima, S., Konno, M., Nakamura, R., Watanabe, H., Bulzu, P.-A., Banciu, H., Andrei, A.-S., Uchihashi,

T., Ghai, R., Béjà, O., and Kandori, H. (2020) Schizorhodopsins: A family of rhodopsins from Asgard archaea that function as light-driven inward H^+ pumps. *Science Advances* 6: eaaz2441.

Kiang, N., Siefert, J., Govindjee, and Blankenship, R. E. (2007a) Spectral signatures of photosynthesis. I. Review of earth organisms. *Astrobiology* 7: 222–251.

Kiang, N., Segura, A., Tinetti, G., Govindjee, Blankenship, R. E., Cohen, M., Siefert, J., Crisp, D., and Meadows, V. S. (2007b) Spectral signatures of photosynthesis. II. Coevolution with other stars and the atmosphere on extrasolar worlds. *Astrobiology* 7: 252–274.

Kirk, J. T. O. (2011) *Light and Photosynthesis in Aquatic Ecosystems*, 3rd Edn. Cambridge: Cambridge University Press.

Pushkarev, A., Inoue, K., Larom, S., Flores-Uribe, J., Singh, M., Konno, M., Tomida, S., Ito, S., Nakamura, R., Tsunoda, S. P., Philosof, A., Sharon, I., Yutin, N., Koonin, E. V., Kandori, H., and Béjà, O. (2018) A distinct abundant group of microbial rhodopsins discovered using functional metagenomics. *Nature* 558: 595–599

Schweiterman, E. W., Kiang, N. Y., Parenteau, M. N., Harman, C. E., DasSarma, S., Fisher, T. M., Arney, G. N., Hartnett, H. E., Reinhard, C. T., Olson, S. L., Meadows, V. S., Cockell, C. S., Walker, S. I., Grenfell, J. L., Hegde, S., Rugheimer, S., Hu, R., and Lyons, T. W. (2018) Exoplanet biosignatures: A review of remotely detectable signs of life. *Astrobiology* 18: 663–708.

Thekaekara, M. P. (1973) Extratesrrrestrial solar spectral irradiance. In: A. J. Drummond and M. P. Thekaekara, (eds.) *The Extraterrestrial Solar Spectrum*. Mount Prospect, IL: Institution of Environmental Sciences, pp. 71–133.

Chapter 2
Photosynthetic organisms and organelles

2.1 Introduction

Green is all around us. The distinctive color of chlorophyll announces the presence of photosynthetic organisms, including trees, shrubs, grasses, mosses, cacti, ferns, and many other types of vegetation. But this is just the tip of the iceberg of photosynthetic life. In addition to these most visible organisms, there is a remarkable variety of microscopic life, including many types of algae and bacteria that carry out photosynthesis. This chapter will introduce the different types of photosynthetic and phototrophic organisms and will give some information about their cellular organization and structure.

All living things on Earth are related to each other. In some cases, the relationships are obviously close, such as between a dog and a coyote, or an orange tree and a lemon tree, while in other cases the relationships are apparent only upon close examination, such as between a bacterium and a human or an amoeba and a fish. To establish these less obvious relationships, it is necessary to look at a deeper level of analysis, down to the cellular and even the molecular levels (Alberts *et al.*, 2014; Nelson and Cox, 2017). At these levels of organization, the unity of life is readily apparent. All organisms are organized in the same fundamental way, with DNA serving as the master copy of the information needed to construct the organism, RNA as the intermediate working copy, and proteins as the workhorses of the cell, carrying out almost all the chemical reactions that make up metabolism. This basic pattern of information flow and metabolic responsibilities is known as the central dogma of molecular biology. Although some exceptions are known, such as viruses that use RNA for information storage or RNA molecules that act as enzymes, the basic pattern applies to all life. The chemical structures of the building blocks of DNA, RNA, and proteins are exactly the same in bacteria and humans. The process of copying DNA into RNA is called **transcription,** and the translation of the nucleic acid code into proteins is called **translation**. This latter process takes place on large protein-RNA complexes called **ribosomes**.

Cells are surrounded by membranes, which function as permeability barriers and also carry out many important functions. Membranes are composed of **lipids**, which are amphipathic molecules with a polar head group and nonpolar tail. The lipids are arranged in a bilayer structure, with the polar head groups toward the outside and inside of

Molecular Mechanisms of Photosynthesis, Third Edition. Robert E. Blankenship.
© 2021 Robert E. Blankenship 2021 by John Wiley & Sons Ltd.
Companion website: https://www.wiley.com/go/blankenship/molecularphotosynthesis3e

the cell, and the nonpolar tails pointed into the center of the bilayer. There are many types of lipids, which form several classes. Two of the most important are **phospholipids** and **glycolipids**. Phospholipids include a phosphate group that is esterified to a glycerol backbone. The most common type of lipids in chloroplasts are glycolipids, in which sugars are found in place of the phosphate groups. The nonpolar tails are long-chain fatty acids that are esterified to the glycerol groups. They almost always contain one or more double bonds, which increase the fluidity of the membranes of which they are the principal components. The cell membrane is often called the **cytoplasmic membrane**, while the space enclosed is called the **cytoplasm**. Additional membranes are found in photosynthetic organisms, in particular the thylakoid membrane, which is the site of photosynthesis in chloroplasts and cyanobacteria.

Membranes also contain proteins, either integral membrane proteins, which span the lipid bilayer, or peripheral membrane proteins, which are associated with one or other side of the membrane but do not cross the bilayer. Many of the proteins essential for photosynthesis are membrane proteins. All cells also contain a variety of carbohydrates, or sugars, as well as many other small molecules essential for proper cellular function. When viewed in this way, the similarities among the various classes of life far outweigh the differences.

Despite the fundamental similarities just pointed out, life nevertheless comes in a remarkable variety of shapes and sizes. These differences form the basis of our mechanisms of classification of living things.

2.2   Classification of life

There are many ways to organize and classify life. No one method is inherently superior to any other, as all such systematic organizations are ultimately only for our benefit. Classification methods are usually structured to accomplish a desired goal, such as rapid identification of organisms in the field or the laboratory. One of the most informative ways to classify organisms is based on evolutionary relationships. This evolutionary, or phylogenetic, approach has led to the recognition that there are three fundamental domains of living organisms: **bacteria**, **archaea** (formerly called archaebacteria), and **eukarya**. This division of the tree of life into the three domains is based on a classification of organisms according to the small subunit rRNA method introduced by Carl Woese (Fig. 2.1). This method relies on comparisons of the sequences of RNA molecules that are part of the ribosome, the protein-synthesizing particle. For bacteria and archaea, the RNA molecule that is used is known as 16S rRNA, while for eukaryotes it is a related molecule known as 18S rRNA. The S stands for Svedberg units, which derive from early methods of molecular weight determination using ultracentrifuges. The basis of this molecular evolution method is the fact that the positional order, or sequence, of building blocks of a biological macromolecule retains information about the evolutionary history of the organism. Two organisms that are closely related will have macromolecules (DNA, RNA, or proteins) whose sequences are highly similar, while distantly related organisms will have sequences that have diverged in the long time interval since their common ancestor. The selection of the rRNA molecule as the molecular chronometer is based on the fact that this molecule is universally present in all organisms, has the same function in all organisms, and has an excellent dynamic range. Parts of the molecule change slowly and are therefore useful for establishing distant evolutionary relationships, while other parts change more rapidly and are therefore more useful for fine distinctions among more closely related organisms.

The rRNA molecules are thought to be a proxy for the evolutionary relationships of the entire organisms that are being compared. This view is something of an oversimplification, because the method actually establishes only the evolutionary relationships of the rRNA molecule, which is part of the protein synthesis machinery of the cell. However, the rRNA molecules appear to be only very rarely transferred from one cell type to another, a process known as **horizontal gene transfer**. All these reasons make the rRNA molecules a good proxy for the evolutionary history of the organism as a whole. A tree of organismal evolutionary relationships is often called a **species tree**.

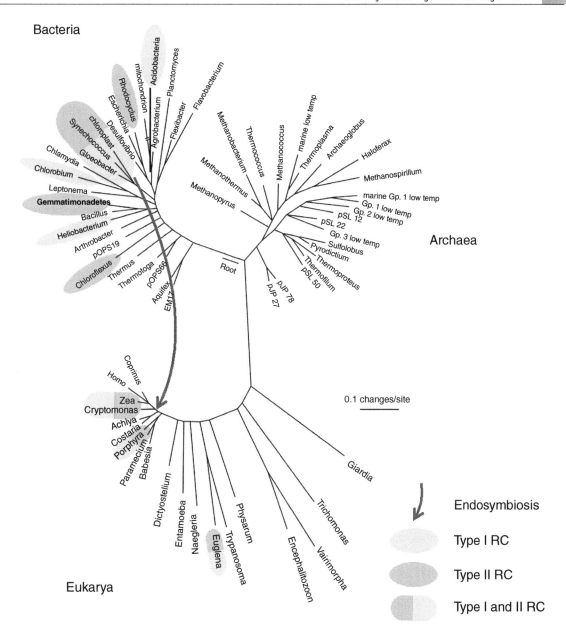

Figure 2.1   Small subunit rRNA phylogenetic tree of Life, with division into the three domains of bacteria, archaea, and eukarya. The highlighted taxa contain photosynthetic organisms. The color coding represents the type of reaction center complex found in the organism, with purple indicating Type II reaction centers and green indicating Type I reaction centers. The red arrow indicates the endosymbiotic origin of chloroplasts from cyanobacteria. *Source:* Blankenship (2010) (p. 435)/The American Society of Plant Biologists.

It is clear from analysis of complete genome sequences for many organisms that there has been significant horizontal transfer of genetic information among various bacteria and even between bacteria and eukaryotes (Soucy *et al.*, 2015). Therefore, the image of a single, branching evolutionary tree that applies to all organisms is increasingly being replaced by a more complex netlike arrangement, in which some parts of an organism bear different evolutionary relationships to other organisms (Doolittle, 1999). However, the process of horizontal gene transfer has apparently not been so extensive as to blur the essential distinctions among the different groups of organisms, as the groupings according to more traditional classification methods are reasonably consistent with those predicted by the RNA analysis, at least in broad outline.

The evolutionary history of any gene or gene family reflects the development of that gene, regardless of what organism has been its host during the course of evolution. An evolutionary tree of a particular gene is therefore called a **gene tree** and may be very different from the species tree of organisms. The origin and early evolution of life with special emphasis on photosynthesis are discussed in more detail in Chapter 12.

2.2.1   Nomenclature

Living organisms are classified according to the binomial nomenclature method, introduced by Linnaeus in the 1700s. The first name (always capitalized) is the genus name (plural, genera), while the second name (never capitalized) is the species name. Both names are italicized. The grouping of organisms into species, genera, and higher order taxa is based on a number of characteristics and represents a useful, but ultimately arbitrary, decision as to where to place the divisions along the continuous variations among related organisms. A genus is a group of organisms that share many but not all characteristics. Higher-order classifications that are intermediate between the genus and phylum, such as family and order, serve to classify groups of organisms into broader categories.

2.3   Prokaryotes and eukaryotes

An older, but still very useful, concept to distinguish among living things is the division into **prokaryote** and **eukaryote**. Prokaryotes are the structurally simplest life forms, including bacteria and Archaea. No Archaea that carry out chlorophyll-based photosynthesis have yet been found, so our discussion of them will be limited. Both these groups of organisms are nearly always single-celled and have a relatively simple cellular organization without a nucleus or other subcellular organelles. A bilayer lipid cytoplasmic membrane surrounds the cell and serves as the main permeability barrier. In **Gram-negative** bacteria, including most types of phototrophic bacteria, a second, more permeable, outer membrane is present, as well as a tough cell wall that provides mechanical stability (Madigan *et al.*, 2017). The space between the outer surface of the cytoplasmic membrane and the inner surface of the cell wall is called the **periplasm**. This region contains a number of soluble proteins, including some cytochromes and chemosensory binding proteins. These proteins are actually topologically outside the cell, but are prevented from being lost by the cell wall. The cell wall has several layers and a complex chemical structure consisting of lipids, proteins, and polysaccharides. Nutrients pass into the periplasm from outside the cell through pores, which are made of proteins called **porins**. A porin is an integral membrane protein that forms a small hole in the outer membrane. Ions and small molecules, such as sugars and amino acids, can easily pass through the pore, but larger molecules cannot. Bacteria are almost always submicroscopic cells, with typical dimensions on the order of one to a few micrometers. They usually divide by binary fission, producing two daughter cells.

Eukaryotes, which are identical to the eukarya domain discussed earlier, are more sophisticated cells than prokaryotes. They are usually much larger than bacteria, up to hundreds of microns in size. In many cases, eukaryotic cells make up multicellular organisms, in which different cells are highly specialized or differentiated. Eukaryotic cells also contain a number of internal structures called **organelles**. These organelles are surrounded by membranes and include the nucleus, the mitochondrion, the endoplasmic reticulum, and, most importantly for photosynthesis, the **chloroplast**.

At least two of these organelles, the mitochondrion and the chloroplast, have complex evolutionary histories, having been acquired by an early eukaryotic cell by a process called **endosymbiosis** (Margulis, 1993).

2.4   Metabolic patterns among living things

Organisms can also be classified according to their metabolic capabilities. While this method does not strictly follow evolutionary relationships, it is still very useful for understanding patterns of energy and metabolite flow, which is especially important in phototrophic organisms. A number of these patterns can be present simultaneously in a single organism, leading to names that are often intimidating. However, they are simply combinations of the individual metabolic patterns.

A fundamental metabolic distinction is between **autotrophs** and **heterotrophs**. The "troph" part is derived from a Greek word meaning "to feed" Autotrophs are "self-feeding" organisms that derive all their cellular carbon from CO_2, whereas heterotrophs are organisms that derive cellular carbon from organic carbon compounds. A second pattern relates to the source of energy for cellular processes. **Phototrophs** derive their energy from sunlight, whereas **chemotrophs** derive energy from various types of chemical compounds. If these compounds are organic chemicals, the organisms are **chemoorganotrophs**. If they are inorganic chemicals, they are called **chemolithotrophs**.

An organism that derives its energy from light and all its cellular carbon from CO_2 is known as a **photoautotroph**. Most photosynthetic organisms can grow in this manner. If the organism grows by using light as an energy source, but assimilates organic carbon, it is known as a **photoheterotroph**. Many phototrophic organisms can also grow in this way, and in a large number of cases, a single organism can grow either photoautotrophically or photoheterotrophically, depending on the availability of organic matter. Table 2.1 summarizes some of the metabolic relationships among living organisms.

Oxygen is central to the metabolism of most cells. If an organism is capable of growing in the presence of oxygen, it is classed as an **aerobe**. If it cannot grow in the presence of oxygen, it is called an **anaerobe**. Some organisms can switch back and forth from aerobic and anaerobic metabolisms and are call **facultative aerobes**. In most cases, aerobes utilize organic molecules as electron donors and O_2 as an electron acceptor in a process called **aerobic respiration**. Other inorganic compounds can sometimes serve as electron acceptors, a process known as **anaerobic respiration**. Finally, organisms that use organic compounds as both electron donors and acceptors in the absence of oxygen live by carrying out **fermentation**.

Table 2.1   Metabolic patterns in living organisms

Type of organism	Metabolic characteristics
Autotroph	Organism that is capable of living on CO_2 as sole carbon source
Heterotroph	Organism that uses organic carbon as carbon source
Phototroph	Organism that uses light as source of energy
Chemotroph	Organism that uses chemicals as energy sources
Photoautotroph	Organism that uses light as source of energy and CO_2 as sole carbon source
Photoheterotroph	Organism that uses light as source of energy and organic compounds as carbon source
Chemoorganotroph	Organism that obtains energetic needs from organic compounds
Chemolithotroph	Organism that obtains energetic needs from inorganic compounds
Chemolithoheterotroph	Organism that obtains energetic needs from inorganic compounds and uses organic compounds as carbon source
Chemolithoautotroph	Organism that obtains energetic needs from inorganic compounds and uses CO_2 as sole carbon source

2.5 Phototrophic prokaryotes

There are seven distinct major groups, or phyla, of bacteria that are capable of photosynthesis (Hohmann-Marriott and Blankenship, 2012; Thiel *et al.*, 2018). Six of these are **anoxygenic**, in that they do not produce molecular oxygen as a by-product of photosynthesis (Blankenship *et al.*, 1995). These anoxygenic phototrophic bacteria include the **purple bacteria**, the **green sulfur bacteria**, the **filamentous anoxygenic phototrophs (FAP)**, formerly known as **green nonsulfur bacteria**, the **heliobacteria**, the **chloroacidobacteria,** and the **gemmatimonadetes**. Each will be introduced briefly below. The single oxygen-evolving, or **oxygenic**, group of photosynthetic bacteria is the **cyanobacteria** (Bryant, 1994). Each of these groups is a relatively coherent collection of

organisms with many similar properties in the pigments they contain and the way they carry out photosynthesis. The groups have been established over many years and are based on a number of characteristics, including pigment composition and metabolic capabilities. Fortunately, these groups are also generally similar to those revealed by the rRNA classification method discussed above. Figure 2.2 illustrates schematically the molecular complexes and metabolisms found in each of the phyla of anoxygenic phototrophic prokaryotes (left panel) and oxygenic photosynthetic organisms (right panel).

2.5.1 Purple bacteria

Purple bacteria are anoxygenic phototrophs that are extremely versatile metabolically (Hunter *et al.*, 2009 and refs. therein). Many species can

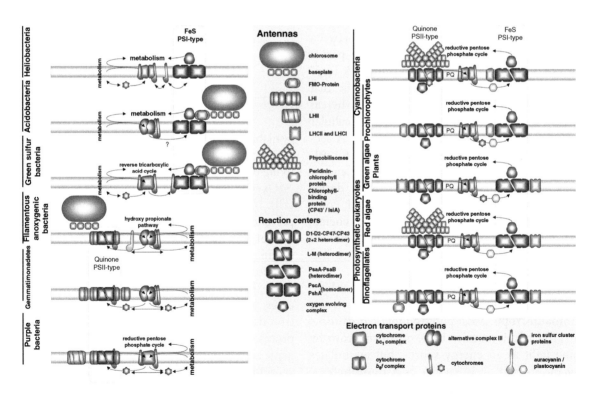

Figure 2.2   Photosynthetic machinery and electron transport of photosynthetic organisms, including a description of photosynthetic complexes. Left panel: anoxygenic phototrophs; right panel: oxygenic photosynthetic organisms. *Source:* Courtesy of Martin Hohmann-Marriott.

grow photoautotrophically, photoheterotrophically, fermentatively, or by either aerobic or anaerobic respiration. Other species are less versatile. Interestingly, many nonphototrophic bacteria are found in the bacterial phylum that includes the purple bacteria, which is known as the **proteobacteria**. The nonphototrophic proteobacteria include many familiar bacteria such as *E. coli*.

In nature, purple bacteria are very widely distributed, especially in anaerobic environments such as sewage treatment ponds. Purple bacteria can use a wide variety of reductants as electron donors, including H_2S or other sulfur-containing compounds, a variety of organic compounds, or even H_2.

An electron micrograph of the purple bacterium *Rhodobacter capsulatus* is shown in Fig. 2.3. The cover of this book shows a model of the intracytoplasmic membrane of the purple bacterium *Rhodobacter sphaeroides*.

Purple bacteria have been the subject of detailed structural and spectroscopic studies, making them the best understood of all phototrophic organisms in terms of energy collection and primary electron transfer processes. Almost all species contain

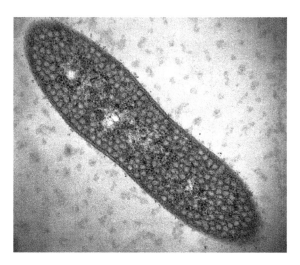

Figure 2.3 Thin section transmission electron micrograph of the purple bacterium *Rhodobacter capsulatus*. The diameter of the cell is about 1 µm. *Source:* Courtesy of Steven J. Schmitt and Michael T. Madigan.

bacteriochlorophyll *a*, while a few instead contain bacteriochlorophyll *b*.

The purple bacteria are subdivided into sulfur and nonsulfur groups, depending on the range of ability to metabolize reduced sulfur compounds (Frigaard and Dahl, 2009). However, the terms sulfur and nonsulfur are somewhat misleading because all purple bacteria have the capability to carry out extensive sulfur metabolism. Most purple bacteria use the Calvin–Benson cycle, also known as the reductive pentose phosphate cycle for CO_2 fixation.

The "purple" name comes from the color found in many of the common species, which results from the combination of bacteriochlorophyll and carotenoids. Representative species include *Rhodobacter sphaeroides*, *Rhodospirillum rubrum*, *Allochromatium vinosum*, and *Blastochloris viridis* (formerly known as *Rhodopseudomonas viridis*). Some of these organisms are not in fact purple in color, such as *Blc. viridis*, which is a greenish color, as its name suggests.

When grown aerobically, most species of purple bacteria derive their energy from aerobic respiration and completely repress pigment synthesis and expression of the structural proteins involved in photosynthetic energy conversion. Photosynthesis is therefore only observed if the cells are grown under anaerobic conditions. Under these conditions, the ultrastructure of the cytoplasmic membrane changes dramatically and invaginates in toward the cell cytoplasm in vesicles, tubes, or lamellae, which are then called **intracytoplasmic membranes**. The photosynthetic apparatus is localized in these intracytoplasmic membranes.

One group of purple bacteria, known as aerobic anoxygenic phototrophs (AAP), has the opposite pattern, in that they make pigments and carry out photosynthesis only under aerobic conditions (Yurkov and Beatty, 1998). This is counterintuitive, in that they seem to perform photosynthesis only when they don't really need to. It is not yet clear what advantage this pattern of metabolism gives these organisms. They are widely distributed and have been found throughout the open ocean (Kolber *et al.*, 2001). This group of purple bacteria is not capable of autotrophic metabolism and cannot assimilate CO_2 using the Calvin–Benson cycle. Instead, they grow using photoheterotrophic metabolism whereby organic matter from the environment is assimilated with the help of light energy.

Most purple phototrophic bacteria are capable of N_2 fixation. In fact, certain classes of *Rhizobia*, the bacteria that live symbiotically in nodules of leguminous plants, contain bacteriochlorophyll and may actually use photosynthesis to supplement the energy requirements of nitrogen fixation (Fleischman and Kramer, 1998). Other *Rhizobia* do not express photosynthetic characteristics, but since they are proteobacteria, based on 16S rRNA analysis, they are therefore relatives of purple phototrophic bacteria.

2.5.2   Green sulfur bacteria

In contrast to the versatile purple bacteria, the green sulfur bacteria are metabolic specialists (Overmann, 2006). They are almost always obligate anoxygenic photoautotrophs, unable to grow with only organic carbon as a carbon source (Tang and Blankenship, 2010). The green sulfur bacteria do not fix carbon using the Calvin–Benson cycle; instead, they use the reverse tricarboxylic acid cycle to fix CO_2 (Fuchs, 2011). They are also strict anaerobes, are incapable of any form of respiration, and are active nitrogen fixers. Green sulfur bacteria can be found in the anaerobic zone at the bottom of lakes or below the chemocline (the transition from aerobic to anaerobic conditions) in a stratified lake. These organisms preferentially utilize H_2S as an electron donor, which is abundant in these environments, although they can also use a variety of other donors such as thiosulfate or elemental sulfur (Frigaard and Dahl, 2009). The green sulfur bacteria can be found living in the lowest light intensities of any known phototrophic organisms (Beatty *et al.*, 2005; Overmann, 2006) and contain highly specialized antenna structures known as **chlorosomes**. These antenna complexes contain bacteriochlorophyll *c*, *d*, or *e* as principal pigments. The chlorosome is attached to the cytoplasmic side of the cell membrane, which does not invaginate as in purple bacteria. The green sulfur bacteria also contain bacteriochlorophyll *a*, which functions in both antennas and reaction centers and small amounts of chlorophyll *a*, which functions in reaction centers.

2.5.3   Filamentous anoxygenic phototrophs

The FAP, sometimes called the green nonsulfur bacteria (Hanada and Pierson, 2006), have metabolic characteristics that are very different from those of the green sulfur bacteria, and in most respects, the two groups of organisms are not closely related. This is in contrast to the purple sulfur and purple nonsulfur bacteria, which are very close relatives by comparison. The FAPs are the earliest branching group of bacterial phototrophs according to the 16S rRNA analysis discussed earlier. They are in most cases capable of photoautotrophic, photoheterotrophic, and aerobic respiratory growth. When grown aerobically, the FAPs suppress pigment synthesis and do not express the structural proteins involved in photosynthesis, similar to many purple bacteria. Most use a unique carbon fixation pathway known as the **hydroxypropionate pathway** for autotrophic growth (Fuchs, 2011), although they grow best by assimilating organic carbon during photoheterotrophic growth. The FAP bacteria contain bacteriochlorophyll *a* localized in reaction centers and integral membrane antenna complexes. These complexes are generally similar to those found in the purple bacteria. Most FAP bacteria also contain bacteriochlorophyll *c*, which is located in chlorosome antenna complexes that are generally similar to those found in green sulfur bacteria and is the major characteristic shared with them. The FAP bacteria are often found in microbial mats, where they live in association with cyanobacteria. They are especially widespread in thermophilic environments.

Surprisingly, a newly discovered anoxygenic bacterium that is a member of the FAP phylum contains a reaction center complex that is unlike the complex found in all other FAPs and is more similar to that found in the green sulfur bacteria (Tsuji *et al.*, 2020). This organism, *Candidatus* Chlorohelix allophototropha also appears to fix carbon using the Calvin–Benson cycle.

2.5.4 Heliobacteria

The heliobacteria are the third most recently discovered of the groups of anoxygenic phototrophs (Gest, 1994; Madigan, 2006). They are also the only group of phototrophs that belong to the Gram-positive group of bacteria. Heliobacteria are the only group of phototrophic bacteria known to form endospores, a characteristic of many other Gram-positive bacteria. They are strict anaerobes and are rapidly killed by exposure to oxygen. The heliobacteria do not appear to be capable of photoautotrophic growth and require organic carbon compounds such as pyruvate in the growth medium. They contain bacteriochlorophyll *g* as their main photopigment, and also small amounts of chlorophyll *a*. They are active N_2 fixers and are often isolated from rice paddies, where they may make a contribution to the nitrogen economy of those environments.

By any measure, the heliobacteria have the least sophisticated photosynthetic metabolism of all known phototrophic bacteria. They have a homodimeric reaction center complex (discussed more in Chapters 6 and 12) with very few protein subunits, no antenna complexes and no autotrophic carbon fixation pathway. They may be important in understanding the origin and early evolution of photosynthesis.

2.5.5 Chloracidobacteria

The chloracidobacteria are the second most recently discovered group of phototrophic bacteria, found in hot spring microbial mats (Bryant *et al.*, 2007). They have a homodimeric reaction center and chlorosome peripheral antenna complexes. They are aerobic organisms and are not capable of autotrophic carbon fixation. They contain bacteriochlorophyll *a*, *c* chlorophyll *a*, and small amounts of Zn bacteriochlorophyll *a* (Tsukatani *et al.*, 2012).

2.5.6 Gemmatimonadetes

The most recently discovered group of phototrophic bacteria are the Gemmatimonadetes (Zeng *et al.*, 2014). These organisms contain bacteriochlorophyll *a*, are semiaerobic, and are incapable of assimilating inorganic carbon. Their photosynthetic apparatus is remarkably similar to that found in purple bacteria, and they have almost certainly acquired the ability to carry out photosynthesis by large-scale horizontal gene transfer.

2.5.7 Cyanobacteria

The cyanobacteria are a large and diverse group of photosynthetic prokaryotes (Bryant, 1994; Flores and Herrero, 2014). They are the only group of photosynthetic bacteria that have oxygenic metabolism, producing molecular oxygen as a byproduct of photosynthesis. Cyanobacteria were previously known as blue-green algae, although this name is misleading as they are not true algae, which are all eukaryotes. As we will explore in more detail in later chapters, the mechanism of photosynthesis in cyanobacteria is remarkably similar to that in photosynthetic eukaryotes, which therefore makes study of them of special interest. Cyanobacteria are remarkably tough and resilient organisms that inhabit almost any environment where light is available, ranging from freshwater, marine, and terrestrial environments to extreme environments such as hot springs and even the surfaces and subsurface regions of rocks in both Antarctica and scorching hot deserts. Nearly all cyanobacteria are photoautotrophs, although some species can also grow photoheterotrophically. An electron micrograph of a representative cyanobacterium is shown in Fig. 2.4.

The cyanobacterial group includes all phototrophic bacteria that produce oxygen as a byproduct of photosynthesis. Most species contain chlorophyll *a* and phycobiliproteins, which function as antenna complexes. However, some types of oxygenic photosynthetic prokaryotes deviate from this pattern. These include the chlorophyll *b*-containing prochlorophytes and the chlorophyll *d*-containing

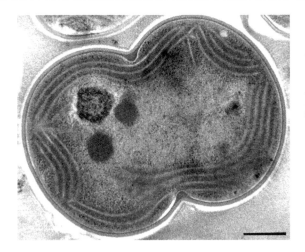

Figure 2.4   Thin section transmission electron micrograph of the cyanobacterium *Synechocystis* PCC 6803 prepared by high-pressure cryofixation. Scale bar 0.4 μm. *Source:* Courtesy of Robert Roberson.

organisms. Phylogenetic studies have shown that all these organisms form a single large phylum according to 16S rRNA (Flores and Herrero, 2014). This phylum is dominated by traditional cyanobacteria, so all prokaryotic oxygenic phototrophs are now usually called cyanobacteria for simplicity.

Many species of cyanobacteria can fix nitrogen from N_2 to NH_3, although, to do this, they face a special challenge. The enzyme system that fixes N_2, called **nitrogenase**, is very sensitive to O_2. The O_2 produced by Photosystem II in cyanobacteria is therefore incompatible with N_2 fixation. Cyanobacteria solve this problem in one of several different ways. In some filamentous forms, which grow as strings of cells, approximately every tenth cell will change its characteristics and become a special N_2-fixing cell called a **heterocyst** (Wolk *et al.*, 1994). In these cells, Photosystem II is absent, an exceptionally thick cell wall inhibits diffusion of O_2 into the cell, and O_2 scavenging systems keep these cells anaerobic to protect the nitrogenase. The other major strategy employed is to carry out N_2 fixation only when it is dark, when the cells are not producing O_2. An unusual group of nitrogen fixing cyanobacteria has lost all genes that code for Photosystem II and is an obligate symbiont with a eukaryotic alga (Thompson *et al.*, 2012).

A few groups of cyanobacteria can switch from using H_2O as an electron donor to using H_2S, with elemental sulfur as the product (Padan, 1979; Liu *et al.*, 2020). They are thus capable of true anoxygenic photosynthesis, although if H_2S is absent they produce O_2 in much the same way as other cyanobacteria. The anoxygenic metabolism therefore represents an additional capability in these organisms, and they thus differ significantly from the other anoxygenic phototrophic prokaryotes, which cannot produce O_2 under any environmental conditions.

Most cyanobacteria contain an extensive internal system of membranes called thylakoids. These membranes contain the photosynthetic apparatus (van de Meene *et al.*, 2006; Liberton *et al.*, 2011). All cyanobacteria contain chlorophyll *a*. Most species lack chlorophyll *b* and contain bilin pigments that are organized into large antenna complexes called phycobilisomes (Chapter 5). A group of cyanobacteria, called **prochlorophytes**, contain chlorophyll *b* in addition to chlorophyll *a* (Matthijs *et al.*, 1994). This chlorophyll *b*-containing group might logically be assumed to be closely related to the organisms that became the chloroplasts of green algae and higher plants, which contain chlorophyll *b*. However, this relationship is not supported by analyses of some genetic markers (see below and Chapter 12). The chlorophyll *b* in these organisms is contained in antenna complexes that are structurally quite different from those of plant and algal chloroplasts. The prochlorophytes do not contain organized phycobilisomes, although some of them do contain genetic information for certain phycobiliproteins.

An important group of prochlorophytes is the genus *Prochlorococcus* (Partensky *et al.*, 1999). These cells are found in the deeper regions of the photic zone in the oceans. They were overlooked for many years because they are extremely small organisms, less than 1 μm in diameter, and passed through the holes in standard collection filters. Because of their small size, they are sometimes called picoplankton, along with other small marine cyanobacteria. They are also unusual in that the chlorophyll *a* present, divinyl chlorophyll *a*, is chemically slightly different from the chlorophyll *a* found in all other oxygenic phototrophs (Chapter 4), a change that adapts them better to the photic environment where they are found. Other prochlorophytes include *Prochloron*, which was the first to be discovered. It grows as a symbiont with a marine animal known as an ascidian in the Great Barrier Reef off Australia and in other places in the South Pacific. Evidence strongly suggests that the prochlorophytes are polyphyletic in origin (Palenik and Haselkorn, 1992; Urbach *et al.*, 1992).

Two recently discovered groups of cyanobacteria are of particular interest. They contain the long-wavelength-absorbing pigments chlorophyll *d* and chlorophyll *f*, which absorb out to nearly 750 nm in the near infrared (Miyashita *et al.*, 1996; Chen *et al.*, 2010). These organisms live primarily in filtered light environments where other organisms above them absorb most of the visible light, so that only the near infrared radiation penetrates more deeply where these organisms live.

2.6   Photosynthetic eukaryotes

Eukaryotic photosynthetic organisms all contain the subcellular organelle called the chloroplast. An overview diagram of the photosynthetic complexes in the chloroplasts of a variety of photosynthetic eukaryotes is shown in Fig. 2.2 (right panel). Chloroplasts are one of a larger group of organelles known as **plastids**, some of which carry out other functions, such as starch or pigment storage in flowers and fruits. As discussed above, a variety of evidence clearly shows that chloroplasts originated by a process called endosymbiosis, in which a cyanobacterial-like cell was initially a symbiont with a protoeukaryotic cell and then eventually became a semiautonomous but essential part of the host cell (Margulis, 1993). The chloroplast contains DNA, which is organized and regulated in a manner typical of bacteria, not eukaryotes. This DNA encodes a number of chloroplast proteins that function in photosynthesis and chloroplast-localized ribosomal protein synthesis machinery.

After the initial endosymbiotic event, a significant degree of genetic transfer to the nucleus took place, so the chloroplast no longer contains enough information to be completely free of the nucleus. The majority of chloroplast proteins are therefore coded for by nuclear DNA, which is transcribed into RNA, the proteins synthesized on cytoplasmic ribosomes and then imported into the chloroplast. In addition, the plastid is the site of the early steps in lipid biosynthesis for the entire cell, so essential cellular components are also exported from the chloroplast. This division of labor requires a sophisticated control and regulation mechanism, which is discussed in more detail in Chapter 10. Mitochondria also originated by endosymbiosis, but in this case the symbiont was a proteobacterium instead of a cyanobacterium.

In addition to the **primary endosymbiosis**, which formed the first photosynthetic eukaryote, there is abundant evidence that there have been several **secondary endosymbioses**, in which a eukaryotic photosynthetic organism underwent a second endosymbiosis and in some cases even tertiary endosymbiosis (Keeling, 2013). Some of the classes of algae discussed below originated via this mechanism. The evolutionary relationships among all the types of photosynthetic organisms and the complex history of the various groups of eukaryotic photosynthetic organisms are discussed in more detail in Chapter 12.

An electron micrograph of a typical higher plant chloroplast is shown in Fig. 2.5. A schematic diagram of the chloroplast is shown in Fig. 2.6. The chloroplast has dimensions of a few microns, slightly larger than the size of a typical bacterium. It is surrounded by a **chloroplast envelope**, made up of a double membrane with two complete bilayers separated by an **intermembrane space**. The region inside the inner chloroplast envelope membrane is called the **stroma**. The stroma is like the cytoplasm of the chloroplast and contains numerous soluble enzymes, in particular the enzymes involved in carbon fixation. An extensive internal membrane system inside the chloroplast, called the thylakoid membrane, contains chlorophyll and the electron transport system that carries out the initial light energy capture and

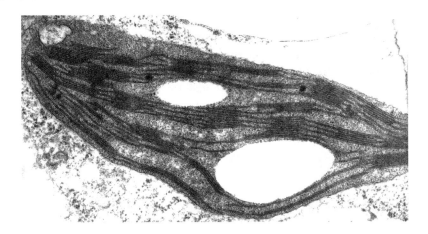

Figure 2.5   Electron micrograph of chloroplast from tobacco. *Source:* Courtesy of Kenneth Hoober.

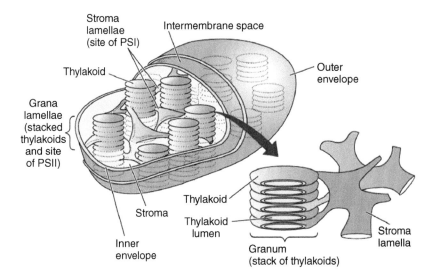

Figure 2.6   Schematic diagram of a chloroplast, showing the inner and outer envelope membranes, the thylakoid membranes – which are divided into grana and stroma lamellae – and the nonmembraneous stroma, containing soluble enzymes. *Source:* Taiz *et al.* (2018)/Oxford University Press.

storage. In higher plants, these thylakoids are pressed together in multiple places to form collections of very densely packed membranes, called **grana**, which are in turn connected by other membranes that are not pressed together. These membranes are known as **stroma lamellae**. In cyanobacteria and many algae, the thylakoid membranes are not found together in densely stacked grana, but are instead associated in stacks of two or a few membranes. As we will learn in more detail in Chapter 7, the components of the photosynthetic apparatus in algae and plants are not uniformly distributed in the thylakoid membranes. Photosystem II is localized primarily in the grana membranes, whereas Photosystem I is found mostly in the stroma lamellae. The thylakoid membranes appear in many pictures to be arranged like a stack of coins. However, in reality, they are highly interconnected, and actually form one or a few interconnected membranes, as shown in Fig. 2.7 (Daum and Kühlbrandt, 2011). Like all biological membranes, the thylakoid is intrinsically asymmetric, with the components arranged with a particular vectorial orientation in the membrane. This results in an overall sidedness to the thylakoid membrane system. The side of the thylakoid that is toward the stroma is called the stromal side, whereas the enclosed space that is in contact with the opposite side of the thylakoid is called the **lumen**. This distinction between the two sides of the thylakoid membrane is a crucial point, as many of the functions of the chloroplast components rely on the presence of a membrane system that is osmotically intact and impermeable to ions.

2.6.1 Algae

Algae are a large group of eukaryotic organisms (Graham *et al.*, 2008). Most of them are pigmented and carry out oxygenic photosynthesis. They are either unicellular, and therefore usually microscopic in size, or colonial, containing many cells. The colonial algae are macroscopic in size and can sometimes form huge structures that may look like plants but are quite distinct. There are many different groups of algae, which are usually distinguished by their pigment compositions and morphological features. In aquatic habitats, algae are the dominant photosynthetic life forms, although they are also found on land, including habitats as seemingly unlikely as the surface of snowfields and the hairs of polar bears. Many competing systems of algal classification are in use. We will not attempt to enumerate all the myriad types, but will instead just list some of the most commonly studied types in terms of their photosynthetic properties. In general, the algae all have rather similar electron transport chains, but widely variable antenna complexes from one group to another.

The **green algae** (**chlorophytes**) are the most widely studied, because their properties are the closest to higher plants. They contain both chlorophyll

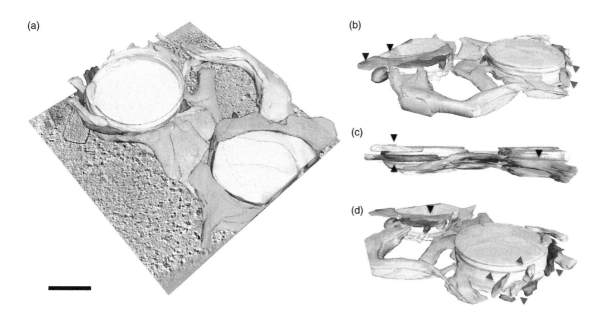

(a)

(b)

(c)

(d)

Figure 2.7   Electron microscopic tomographic surface representation of the thylakoid network within a ruptured chloroplast. The different views are of the same thylakoid network from different angles. *Source:* Daum and Kühlbrandt (2011). Reproduced with permission of Oxford University Press.

a and chlorophyll *b* as photopigments. They are certainly the evolutionary precursors to plants. The **red algae** (**rhodophytes**) are mostly marine organisms that contain chlorophyll *a* and phycobilisomes, antenna complexes similar to those found in most cyanobacteria. They often have a complex life cycle. The green and red algae, plus one other group (the **glaucophytes**), are thought to be primary endosymbionts, in that they arose from a single endosymbiotic event. All other algal groups are the result of additional endosymbiotic events, in which a eukaryotic alga was itself incorporated into an organism to form a new type of chimeric cell that in many cases retained the photosynthetic capability of the endosymbiont. Most of these secondary and in some cases tertiary endosymbiotic events involved the red algal line and the complex history of this group includes a dizzying array of gain, loss, and regain of photosynthesis. Many of these organisms contain chlorophyll *c* as an accessory pigment. The **chromoalveolate** hypothesis proposes that most of the non-green eukaryotic algae have been derived from secondary endosymbiosis of red algae and subsequent events (Cavalier-Smith, 1999).

2.6.2 Plants

Plants are the most complex of all photosynthetic organisms (Taiz *et al.*, 2018). The simplest plants are the **bryophytes**, including the **mosses**, **liverworts**, and **hornworts**. They are in many ways like algae and do not have true roots or leaves, or a vascular (liquid-transporting) system. They do not produce hard tissues for support. The **vascular plants** include the **ferns** and the **seed plants**. The ferns reproduce by means of spores, and have roots, leaves, and vascular tissues, as well as woody tissues for mechanical support. The seed plants reproduce by means of seeds and also contain roots and leaves, as well as vascular and woody tissues. The seed plants are divided into two groups: **gymnosperms** and **angiosperms**. Gymnosperms are the more primitive group and include coniferous trees. Angiosperms, also known as flowering plants, make up the vast majority of the species of plants around us.

Remarkably, the basic structure of the photosynthetic membrane and the mechanism of photosynthesis are generally similar in all oxygenic photosynthetic organisms. Some cells include novel antenna complexes, and certain regulatory mechanisms are clearly more sophisticated as one moves from cyanobacteria to algae to higher plants. However, the same basic principles and complexes are found throughout this wide range of organisms. It seems that nature perfected the ability to carry out photosynthesis several billion years ago and has not made major alterations since then. Even the anoxygenic photosynthetic bacteria, while clearly much more primitive than oxygenic forms, carry out photosynthesis using the same basic physical principles. As we proceed, we will examine each of these groups, comparing and contrasting them, trying to find common principles, and pointing out significant differences where they occur.

References

Alberts, B., Johnson, A. D., Lewis, J., Morgan, D., Raff, M., Roberts, K., and Walter, P. (2014) *The Molecular Biology of the Cell*, 6th Edn. New York: W.W. Norton.

Beatty, J. T., Overmann, J., Lince, M. T., Manske, A. K., Lang, A. S., Blankenship, R. E., Van Dover, C. L., Martinson, T. A., and Plumley, F. G. (2005) An obligately photosynthetic bacterial anaerobe from a deep-sea hydrothermal vent. *Proceedings of the National Academy of Sciences USA* 102: 9306–9310.

Blankenship, R. E. (2010) Early evolution of photosynthesis. *Plant Physiology* 154: 434–438.

Blankenship, R. E., Madigan, M. T., and Bauer, C. E., (eds.) (1995) *Anoxygenic Photosynthetic Bacteria*. Dordrecht: Kluwer Academic Press.

Bryant, D. A., (ed.) (1994) *The Molecular Biology of Cyanobacteria*. Dordrecht: Kluwer Academic Press.

Bryant, D. A., Costas, A. M., Maresca, J. A., Chew, A. G. M., Klatt, C., Bateson, M. M., Tallon, L. J., Hostetler, J., Nelson, W. C., Heidelberg, J. F., and Ward, D. M. (2007) *Candidatus* Chloracidobacterium thermophilum: An aerobic phototrophic acidobacterium. *Science* 317: 523–526.

Cavalier-Smith, T. (1999) Principles of protein and lipid targeting in secondary symbiogenesis: Euglenoid, dinoflagellate, and sporozoan plastid origins and the eukaryote family tree. *Journal of Eukaryotic Microbiology* 46: 347–366.

Chen, M., Schliep, M., Willows, R. D., Cai, Z.-L., Neilan, B. A., and Scheer, H. (2010) A red-shifted chlorophyll. *Science* 329: 1318–1319.

Daum, B. and Kühlbrandt, W. (2011) Electron tomography of plant thylakoid membranes. *Journal of Experimenatal Botany* 62: 2393–2402.

Doolittle, W. F. (1999) Phylogenetic classification and the universal tree. *Science* 284: 2124–2128.

Fleischman, D. and Kramer, D. (1998) Photosynthetic rhizobia. *Biochimica et Biophysica Acta* 1364: 17–36.

Flores, E. and Herrero, A., (eds.) (2014) *The Cell Biology of Cyanobacteria*. Poole: Caister.

Frigaard, N. U. and Dahl, C. (2009) Sulfur metabolism in phototrophic sulfur bacteria. *Advances in Microbial Physiology* 54: 103–200.

Fuchs, G. (2011) Alternative pathways of carbon dioxide fixation: Insights into the early evolution of life? *Annual Review of Microbiology* 65: 631–658.

Gest, H. (1994) Discovery of the heliobacteria. *Photosynthesis Research* 41: 17–21.

Graham, J. E., Wilcox, L. W., and Graham, L. E. (2008) *Algae*, 2nd Edn. San Francisco: Benjamin Cummings.

Hanada, S. and Pierson, B. K. (2006) The family Chloroflexaceae. In: M. Dworkin, S. Falkow, E. Rosenberg, K.-H. Schliefer, and E. Stackebrandt, (eds.) *The Prokaryotes*, 3rd Edn. Berlin: Springer-Verlag, pp. 815–842.

Hohmann-Marriott, M. F. and Blankenship, R. E. (2012) The photosynthetic world. In: J. J. Eaton-Rye, B. C. Tripathy, and T. D. Sharkey, (eds.) *Photosynthesis: Plastid Biology, Energy Conversion and Carbon Assimilation*. Dordrecht: Springer.

Hunter, C. N., Daldal, F., Thurnauer, M. C., and Beatty, J. T., (eds.) (2009) *The Purple Phototrophic Bacteria*. Dordrecht: Springer.

Keeling, P. J. (2013) The number, speed, and impact of plastid Endosymbioses in eukaryotic evolution. *Annual Review of Plant Biology* 64: 583–607.

Kolber, Z. S., Plumley, F. G., Lang, A. S., Beatty, J. T., Blankenship, R. E., Van Dover, C. L., Vetriani, C., Koblizek, M., Rathgeber, C., and Falkowski, P. G. (2001) Contribution of aerobic photoheterotrophic bacteria to the carbon cycle in the ocean. *Science* 292: 2492–2495.

Liberton, M., Austin, J. R., Berg, R. H., and Pakrasi, H. B. (2011) Unique thylakoid membrane architecture of a unicellular N-2-fixing cyanobacterium revealed by electron tomography. *Plant Physiology* 155: 1656–1666.

Liu, D., Zhang, J., Lu, C., Xia, Y., Liu, H., Jiao, N., Xun, L., and Liu, J. (2020) *Synechococcus* sp. strain PCC7002 uses sulfide: Quinone oxidoreductase to detoxify exogenous sulfide and to convert endogenous sulfide to cellular sulfane sulfur. *MBio* 11: e03420-19.

Madigan, M. T. (2006) The family Heliobacteriaceae. In: M. Dworkin, S. Falkow, E. Rosenberg, K.-H. Schliefer and E. Stackebrandt, (eds.) *The Prokaryotes*, 3rd Edn. Berlin: Springer-Verlag, pp. 951–964.

Madigan, M. T., Bender, K. S., Buckley, D. H., Sattley, W. M., and Stahl, D. (2017) *Brock Biology of Microorganisms*, 15th Edn. San Francisco: Pearson Benjamin Cummings.

Margulis, L. (1993) *Symbiosis in Cell Evolution: Microbial Communities in the Archean and Proterozoic Eons*. San Francisco: W. H. Freeman.

Matthijs, H. C. P., van der Staay, G. W. M., and Mur, L. R. (1994) Prochorophytes: The 'other' cyanobacteria. In: D. Bryant, (ed.) *The Molecular Biology of Cyanobacteria*. Dordrecht, The Netherlands: Kluwer Academic Publishers, pp. 49–64.

van de Meene, A. M. L., Hohmann-Marriott, M. F., Vermaas, W. F. J., and Roberson, R. W. (2006) The three-dimensional structure of the cyanobacterium *Synechocystis* sp. PCC 6803. *Archives of Microbiology* 184: 259–270.

Miyashita, H., Ikemoto, H., Kurano, N., Ikemoto, H., Chihara, M., and Miyachi, S. (1996) Chlorophyll *d* as a major pigment. *Nature* 383: 402.

Nelson, D. L. and Cox, M. M. (2017) *Lehninger Principles of Biochemistry*, 7th Edn. New York: W. H. Freeman.

Overmann, J. (2006) The family Chlorobiaceae. In: M. Dworkin, S. Falkow, E. Rosenberg, K.-H. Schliefer, and E. Stackebrandt, (eds.) *The Prokaryotes*, 3rd Edn. Berlin: Springer-Verlag, pp. 359–378.

Padan, E. (1979) Facultative anoxygenic photosynthesis in cyanobacteria. *Annual Review of Plant Physiology* 30: 27–40.

Palenik, B. and Haselkorn, R. (1992) Multiple evolutionary origins of prochlorophytes, the chlorophyll *b*-containing prokaryotes. *Nature* 355: 265–267.

Partensky, F., Hess, W. R., and Vaulot, D. (1999) Prochlorococcus: A marine photosynthetic prokaryote of global significance. *Microbiology and Molecular Biology Reviews* 63: 106–127.

Soucy, S. M., Huang, J., and Gogarten, J. P. (2015) Horizontal gene transfer: Building the web of life. *Nature Reviews Genetics* 16: 472–482.

Taiz, L., Zeiger, E., Møller, I. M., and Murphy, A. (2018) *Fundamentals of Plant Physiology Sunderland.* Sunderland, MA: Sinauer Associates.

Tang, K.-H. and Blankenship, R. E. (2010) Both forward and reverse TCA cycles operate in green sulfur bacteria. *Journal of Biological Chemistry* 285: 35848–35854.

Thiel, V., Tank, M., and Bryant, D. A. (2018) Diversity of chlorophototrophic bacteria revealed in the ohmics era. *Annual Review of Plant Biology* 69: 21–49.

Thompson, A. W., Foster, R. A., Krupke, A., Carter, B. J., Musat, N., Vaulot, D., Kuypers, M. M. M., and Zehr, J. P. (2012) Unicellular cyanobacterium symbiotic with a single celled eukaryotic alga. *Science* 337: 1546–1550.

Tsuji, J. M., Shaw, N. A., Nagashima, S., Venkiteswaran, J. J., Schiff, S. L., Hanada, S., Tank, M., and Neufeld, J. D. (2020) Anoxygenic phototrophic *Chloroflexota* member uses a Type I reaction center. https://doi.org/10.1101/2020.07.07.190934

Tsukatani, Y., Romberger, S. P., Golbeck, J. H., and Bryant, D. A. (2012) Isolation and characterization of homodimeric type-I reaction center complex from *Candidatus* Chloracidobacterium thermophilum, an aerobic chlorophototroph. *Journal of Biological Chemistry* 287: 5720–5732.

Urbach, E., Robertson, D. L., and Chisholm, S. W. (1992) Multiple evolutionary origins of prochlorophytes within the cyanobacterial radiation. *Nature* 355: 267–270.

Wolk, C. P., Ernst, A., and Elhai, J. (1994) Heterocyst metabolism and development. In: D. A. Bryant, (ed.) *The Molecular Biology of Cyanobacteria.* Dordrecht: Kluwer Academic Publishers, pp. 769–823.

Yurkov, V. V. and Beatty, J. T. (1998) Aerobic anoxygenic phototrophic bacteria. *Microbiology and Molecular Biology Reviews* 62: 695–724.

Zeng, Y., Feng, F., Medová, H., Dean, J., and Koblizek, M. (2014) Functional type 2 photosynthetic reaction centers found in the rare bacterial phylum Gemmatimonadetes. *Proceedings of the National Academy of Sciences USA* 111: 7795–7800.

Chapter 3

History and early development of photosynthesis

Our understanding of the complex process of photosynthesis early in the twenty-first century is the product of several centuries of effort on the part of countless dedicated scientists all over the world. In this chapter, we will discuss some of the landmark developments that form the underpinnings of our current picture. We will first discuss the developments that led to the determination of the chemical equation of photosynthesis, and will then examine certain key experiments that have given rise to the current mechanistic understanding of photosynthesis (Rabinowitch, 1945; Govindjee *et al.*, 2005; Hill, 2012; Nickelsen, 2015). This historical treatment is eclectic, rather than exhaustive, and is designed to give some sense of the path that the field has taken to the present state and to highlight a few of the personalities who have brought us here, rather than to enumerate all the developments that have taken place and all the individuals who have contributed.

3.1 Van Helmont and the willow tree

In the 1640s, a Flemish physician named Jan Baptista van Helmont (1577–1644) planted a willow tree in a tub of earth, which weighed 200 lbs. For five years, he watered the tree and weighed the leaves that fell off each autumn. At the end of that time, he pulled the tree from the tub and weighed both the tree and the tub of earth. The tub of the earth was essentially unchanged in weight, having lost only two ounces, but the tree and its leaves weighed 169 lbs. He concluded that the tree had come from the water that he had given the tree, rather than from the "humus" of the soil, which was the ancient view derived from Aristotle (384–322 BC). Van Helmont's conclusion was only partially correct, as the mass of a plant derives largely from both water and carbon dioxide, the latter of which was completely unknown at the time. However, van

Molecular Mechanisms of Photosynthesis, Third Edition. Robert E. Blankenship.
© 2021 Robert E. Blankenship 2021 by John Wiley & Sons Ltd.
Companion website: https://www.wiley.com/go/blankenship/molecularphotosynthesis3e

Helmont's emphasis on analysis by weighing was considerably ahead of its time, as the law of conservation of matter in chemistry, which is based on a careful weighing of reactants and products of chemical reactions, was not formulated by the French chemist Antoine Lavoisier (1743–1794) for over another 100 years.

3.2 Carl Scheele, Joseph Priestley, and the discovery of oxygen

Carl Scheele (1742–1786) was a Swedish chemist who was almost certainly the first person to isolate oxygen, producing it chemically in about 1771. Unfortunately, Scheele did not publish his results until 1777, well after both Priestley and Lavoisier had published their work. Their findings were at least partially based on Scheele's results, which he had communicated to them. Scheele also died at an early age of 44, almost certainly poisoned by some of the chemicals that he worked with, which he liked to taste and sniff.

Joseph Priestley (1733–1804) made many important discoveries, especially in relation to the properties and handling of gases. Priestley was an eighteenth-century English country minister who had a passion for science (Jaffe, 1976). His formal education was limited primarily to theology and languages, whereas his scientific knowledge was mostly self-taught. He was deeply influenced by a meeting with Benjamin Franklin, whom he met on a trip to London. He lived next to a brewery in Leeds, which provided a constant supply of carbon dioxide, so initially, he studied that gas's properties. One of his first results was the invention of seltzer water, for which the British Royal Society awarded him a gold medal.

We remember Priestley especially for his discovery of oxygen, first in an indirect way, by observing the action of plants in 1771, and then in pure form in 1774 by heating mercuric oxide and collecting the gas given off. He described the 1771 experiments thus:

Finding that candles would burn very well in air in which plants had grown a long time . . . I thought it was possible that plants might also restore the air which has been injured by the burning of candles. Accordingly, on the 17th of August, 1771, I put a sprig of mint into a quantity of air in which a wax candle had burned out and found that on the 27th of the same month another candle burnt perfectly well in it.

Priestley's interpretation of this and related experiments (Fig. 3.1) was that the candle (or mouse – he had a plentiful supply of mice and often used them as test subjects) produced large amounts of phlogiston, which was the basis for interpreting all chemical processes at that time. Phlogiston, which is derived from the Greek word for "to set on fire," was thought to be a flammable substance possessed by all substances that can burn. Upon combustion or respiration, the phlogiston was released into the air and contaminated it. Plants had the unique ability to recapture the phlogiston that had been released by burning. In his later experiments, Priestley was able to prepare and analyze in some detail significant quantities of pure oxygen, which he called "vital air." Priestley was a brilliant experimentalist, whose abilities to generate and manipulate gases were far better than those of most of his contemporaries. However, his skills at interpreting the observations he made were not as strong. He interpreted all his observations in terms of the phlogiston theory, even after most other scientists had abandoned it. He was usually content to simply describe his results with little interpretation. Priestley traveled to France in 1774 and met with Lavoisier, to whom he openly

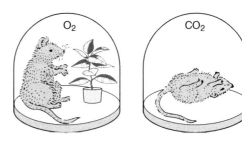

Figure 3.1   Joseph Priestley and the discovery of oxygen.

described his many experiments. Lavoisier used this information, as well as many of his own experiments, to overturn the phlogiston theory and establish the modern study of chemistry.

Priestley was both a religious and a political nonconformist and vocally supported both the American and the French revolutions. He was often in trouble for his unorthodox views, was attacked by the press and denounced in Parliament, and the scientific establishment increasingly shunned him. In 1791, an angry mob burned his house to the ground. In 1794, he fled to the USA and settled in Pennsylvania, where he was warmly welcomed by both religious and scientific leaders. He lived there until he died in 1804.

3.3   Ingenhousz and the role of light in photosynthesis

Jan Ingenhousz (1730–1799) was a Dutch physician and man of the world; he was at one-time court physician to the Austrian empress Maria Theresa and had also visited England on several occasions (Magiels, 2010). On one of these trips in 1773, he heard a lecture by John Pringle, the President of the Royal Society, in which he presented a medal to Priestley. In his lecture, Pringle lauded Priestley's accomplishments. Ingenhousz thus became interested in the subject, but took no action until several years later when, in 1779, he rented a house in the English countryside and in the short space of three months conducted more than 500 experiments on the properties of plants and their effects on the air. During this frenzied summer, Ingenhousz discovered the essential role of light in the process of photosynthesis. By October of the same year, he had already written up his results in the form of a book entitled *Experiments upon Vegetables, Discovering Their Great Power of Purifying the Common Air in Sunshine and Injuring It in the Shade and at Night*. In addition to discovering the effect of sunlight, Ingenhousz discovered

plant respiration, although he greatly overemphasized the effects of it, as can be seen from the title of his book. Ingenhousz was a brilliant, but not very careful, scientist and often jumped to conclusions with little evidence. Nevertheless, his contributions to the understanding of photosynthesis were extremely important, including a book published in 1796, *Food of Plants and Renovation of the Soil*. In this book, he summarized and presented for the first time the process of photosynthesis in terms of the new chemistry proposed by Lavoisier.

3.4   Senebier and the role of carbon dioxide

Jean Senebier (1742–1809) was a Swiss minister, contemporary, and bitter rival of Ingenhousz. His most important contribution to the developing understanding of photosynthesis was his finding in 1783 of the essential role that carbon dioxide, or "fixed air," has in the process. Senebier and Ingenhousz engaged in a long-standing and acrimonious priority dispute. This was not helped by Senebier's poor writing style, such that he would take entire volumes to describe his results in tedious detail, with no summaries provided. For his part, Ingenhousz tried to claim in his 1796 book that he had actually discovered the role of carbon dioxide in his 1779 experiments, although he had clearly denied that it was important in writings in 1784.

3.5   De Saussure and the participation of water

The final contribution to the overall equation of photosynthesis was made by Nicolas de Saussure (1767–1845), a Swiss scientist. He confirmed the observations of Ingenhousz and Senebier and added careful measurements of the relative amounts of carbon dioxide taken up and oxygen given off by plants. He proved by careful weighing experiments, published in 1804, that the increase in dry weight of the plant is considerably greater than the weight of carbon in the carbon dioxide taken up. He correctly surmised that the balance of the weight came from water, although he mistakenly thought that the assimilation of water was a process separate from the incorporation of carbon dioxide.

3.6   The equation of photosynthesis

The period between 1771 and 1804 was exciting in the history of photosynthesis. In this short time, the basic chemical equation of photosynthesis was established, which in 1804 could be written as:

$$\text{carbon dioxide} + \text{water} + \text{light} \rightarrow \text{organic matter} + \text{oxygen} \tag{3.1}$$

This was not yet a balanced equation, as the nature of the organic matter had not been established, and even the proper chemical formula for water had not yet been established! The concept of atoms combining to form molecules was just being formulated by John Dalton (1766–1844) near the end of this period, but unambiguous chemical formulas were not established for another 50 years.

It should be clear that the elucidation of the equation of photosynthesis was an amazing feat, which required contributions from many brilliant and hardworking scientists of differing backgrounds and points of view. It should also be clear from this example that science is very much an enterprise carried out by human beings, who are often blinded by preconceived ideas, ego, and jealousy.

3.6.1   The balanced equation for photosynthesis

It was another 60 years before the chemical equation of photosynthesis could be written in modern chemical symbols and properly balanced. A key aspect of this development was the determination of the redox state of the organic matter produced during photosynthesis. This involved careful measurements of the photosynthetic quotient, the ratio between the carbon dioxide assimilated and the oxygen produced. The first accurate measurements of the photosynthetic quotient were made in 1864 by a Frenchman, Jean Baptiste Boussingault (1802–1887). This determination consisted of measuring the volumes of CO_2 taken up and O_2 produced during photosynthesis. Boussingault found that the photosynthetic quotient was close to 1 for a number of plants. This established that the fixed carbon is at the redox level of carbohydrate (where the ratio of H to O is 2:1). Bolstering this view, Julius von Sachs, a German plant physiologist, found in the same year that the carbohydrate, starch, accumulates in leaves only when they are illuminated, and only in those parts of the leaf that are directly illuminated. This effect can be illustrated dramatically by actually printing photographs on leaves! The process is accomplished by taping a photographic negative over a leaf, illuminating it to form starch, extracting the pigments, and then developing the image by treating it with iodine, which forms a dark-colored complex with starch. Remarkably high-quality images can be obtained in this manner (Walker, 1992).

The information that the photosynthetic quotient is 1, and that the organic matter is a carbohydrate such as starch or sugar, allows us to write a minimally balanced equation for photosynthesis:

$$CO_2 + H_2O \rightarrow (CH_2O) + O_2 \tag{3.2}$$

where (CH_2O) is representative of a carbohydrate. One example of a carbohydrate is glucose, $C_6H_{12}O_6$, which makes the overall balanced photosynthetic equation:

$$6CO_2 + 6H_2O \rightarrow C_6H_{12}O_6 + 6O_2 \tag{3.3}$$

As we will see in Chapter 9, glucose is not the carbohydrate directly formed in photosynthesis, but it has almost the same energy content, so this is adequate for our present needs.

Julius Robert Mayer (1814–1878), a German physician and physicist who first enunciated the law of conservation of energy, proposed in 1845 that in photosynthesis light energy is converted to chemical energy, thus completing the formulation of the equation of photosynthesis.

3.7   Early mechanistic ideas of photosynthesis

When the overall reaction of photosynthesis had been established, attention turned to elucidating the details of the mechanism of the process. The early ideas in this regard were erroneous and much too simplistic. Richard Willstätter (1872–1942) and Arthur Stoll (1887–1971) proposed in 1918 that the product was actually formed directly as a molecular species, formaldehyde (CH_2O), in a direct, concerted process involving chlorophyll, CO_2, and H_2O. This view was revived later by Otto Warburg (1883–1970) in support of his unorthodox formulation of photosynthesis, which we will discuss a little later. We now know that these mechanistic ideas, which were certainly reasonable at the time, are not valid, because there are literally dozens of intermediate states that have been identified, and the reduction in CO_2 can be separated from the production of oxygen, and vice versa. The key to understanding in more detail how the mechanism of photosynthesis really works came from the analysis of simple photosynthetic organisms by van Niel and by Hill's experiments showing that CO_2 reduction and O_2 evolution can be decoupled.

3.7.1   Van Niel and the redox nature of photosynthesis

The cornerstone of our current understanding of photosynthesis is that it is a light-induced reduction–oxidation (redox) chemical process. This principle was first clearly set forth in the 1930s by the Dutch microbiologist Cornelis van Niel (1897–1985), working at Stanford University. Van Niel carried out a series of experiments on the metabolic characteristics of non-oxygen-evolving (anoxygenic) photosynthetic bacteria (van Niel, 1941). These organisms contain bacteriochlorophylls, pigments related to, but distinct from, the chlorophylls contained in cyanobacteria, algae, and plants. They assimilate CO_2 into organic matter, but do not produce molecular oxygen. In order for these bacteria to assimilate CO_2, they must be supplied with a reducing compound. Many different compounds will suffice, most notably H_2S, which is first oxidized to elemental sulfur and then further oxidized. In place of H_2S, a variety of organic compounds can also be utilized, or even molecular hydrogen. Van Niel's seminal contribution was the recognition that these compounds could all be represented by the general formula H_2A, and that the overall equation of photosynthesis could be reformulated in a more general way as follows:

$$CO_2 + 2H_2A \rightarrow (CH_2O) + 2A + H_2O \quad (3.4)$$

The oxygen-evolving form of photosynthesis can then be seen as a special case of this more general formulation, in which H_2O is H_2A and O_2 is 2A. When presented in this manner, the redox nature of photosynthesis is much more obvious. In fact, it is a simple further step to separate the oxidation and reduction into two chemical equations, one for the oxidation and the other for the reduction:

$$2H_2A \rightarrow 2A + 4e^- + 4H^+ \quad (3.5a)$$

$$CO_2 + 4e^- + 4H^+ \rightarrow (CH_2O) + H_2O \quad (3.5b)$$

This separation into oxidation and reduction reactions leads to a number of important predictions. First, it suggests that the two processes might possibly be physically or temporally separated. Indeed, this was found to be the case by Hill in his classic experiments on the use of artificial electron acceptors, which will be discussed in the next paragraph. A second prediction is that the oxygen produced by oxygen-evolving photosynthesis comes from H_2O and not from CO_2. This is indeed the case, although it took many years for this fact to be established unequivocally, because of the rapid exchange of oxygen between CO_2 and H_2O via carbonic acid, H_2CO_3, which is constantly forming and breaking down. Van Niel correctly imagined that the oxidation and reduction reactions of Eqs. (3.5a) and (3.5b) were not the primary processes carried out by light, but were, rather, the result of a primary oxidant and reductant generated in the light, which went on to react with the substrates to form the products. However, some of van Niel's detailed ideas about the nature of the primary oxidant and reductant were incorrect, and he was overly rigid in insisting that all assimilated carbon had first to be oxidized all the way to CO_2 before it could be subsequently reduced. Nevertheless, the truth of the basic principle of photosynthesis as a light-induced redox process was unequivocally established by van Niel's work, along with that of other scientists of the time, notably Hans Gaffron (1902–1979).

3.7.2   The Hill reaction: separation of oxidation and reduction reactions

In the 1930s, Robert (Robin) Hill (1899–1991), working at Cambridge University, was able to separate and investigate individually the oxidation and reduction reactions of photosynthesis. This taking apart of a complex system into individual parts, and investigation of them in detail in the absence of the complexities of a living cell is a cornerstone of biochemistry, often called reductionism. Hill established that it was possible to restore high rates of oxygen evolution to chloroplast suspensions if the latter were supplied with any of a number of artificial electron acceptors (Hill, 1939). Initially, he used ferric salts, and the Fe^{3+} was reduced to Fe^{2+} through the action of light, at the same time producing O_2. The reduction in artificial acceptors with concomitant O_2 production is today known as the Hill reaction. An example of the Hill reaction is given in Eq. (3.6):

$$2\,H_2O + 4\,Fe^{3+} \rightarrow O_2 + 4\,Fe^{2+} + 4H^+ \qquad (3.6)$$

Hill measured the O_2 in an ingenious way, which is worth relating if only to give an idea of the remarkable advances made by many of the pioneers of the field despite their primitive instrumentation. Hill obtained whole blood from a slaughterhouse, which

has a dark blue color when deoxygenated and bright red color when oxygenated. He combined this with his chloroplast preparation and illuminated the mixture, monitoring the degree of oxygenation of the blood using a hand-held spectroscope. At first, the results were disappointing, because the sample produced little oxygen. It is now clear that this was because the outer chloroplast envelope membranes were broken during the preparation, and the enzymes needed for CO_2 assimilation were lost. In searching for the factors needed to restore the lost activity, Hill made a fundamental discovery: namely, that it was possible to replace the reduction of CO_2 with the reduction of artificial electron acceptors, thereby restoring high rates of O_2 production. The physiological compound that acts as the light-driven electron acceptor facilitating CO_2 production is $NADP^+$, the oxidized form of nicotinamide adenine dinucleotide phosphate. The reduced form of this compound, NADPH, then serves as the reductant for CO_2 assimilation.

Hill did not set out to discover the reaction that bears his name. Instead, he was trying to establish whether an isolated chloroplast was capable of the complete process of photosynthesis, which was an important issue at the time. In fact, it is quite difficult to isolate chloroplasts with the envelope membranes still intact, and this was not routinely achieved until the mid-1960s. This is a good example of Louis Pasteur's famous saying that "fortune favors the prepared mind."

3.8   The Emerson and Arnold experiments

In 1932, Robert Emerson (1903–1959) and his undergraduate research student at the California Institute of Technology, William Arnold (1904–2001), published two seminal papers. They were the first scientists to exploit the use of very short flashes of light to probe photosynthesis. At the time, this was extremely difficult technically; one did not just order a flash instrument from a scientific supply house. Emerson and Arnold built their own apparatus from automobile ignition points, capacitors, neon lights, and even a hot plate (which served as a ballast resistor) (Myers, 1994). Using this device, they were able to obtain flash durations as short as 10 µs. Emerson and Arnold utilized the green alga *Chlorella pyrenoidosa* as their experimental organism and measured oxygen evolution using manometers, U-shaped tubes containing liquids in which the level was used to determine the volume of gas produced or consumed. Emerson was a master in the techniques of manometry, which he had learned from his former mentor, Otto Warburg, and could measure minute changes of O_2 with great accuracy.

In the first series of experiments (Emerson and Arnold, 1932a), they varied the time between flashes and found that if there was a long time between flashes, then the yield of O_2 per flash was independent of the time between flashes and did not depend on temperature between 1 and 25 °C (Fig. 3.2). With shorter times between flashes, the yield fell off dramatically at the lower temperature, but did not change at the higher temperature. This result was elegantly interpreted as evidence that

photosynthesis involved both a light stage and a dark stage. The light stage, which we now refer to as a photochemical reaction, can happen extremely rapidly and is not dependent on temperature. The dark stage, which we now know to be a series of enzymatic reactions, is slower and, like most chemical reactions, depends on temperature activation to proceed. These results, which were partially anticipated by other scientists using continuous and crude intermittent light methods, were readily interpretable on the basis of the understanding of photosynthesis available at the time. However, they did no more than set the stage for the remarkable result that Emerson and Arnold found in their second set of experiments, in which they examined the light-intensity dependence of the photochemical reaction.

For the second series of experiments, Emerson and Arnold (1932b) varied the light intensity of the flashes, using a flash spacing that they knew from the first series of experiments was long enough for the dark enzymatic reactions to proceed to completion. This experimental protocol enabled them to

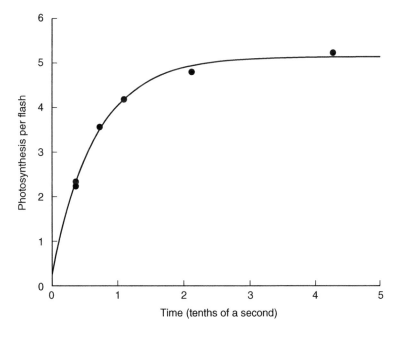

Figure 3.2   Emerson and Arnold's experiment establishing a light stage and a dark stage of photosynthesis. Time refers to the time between flashes. *Source:* Emerson and Arnold (1932a)/Rockefeller University Press.

isolate the photochemical reaction and study it without interference from the later steps. At very low intensities the yield of O_2 per flash was low and depended linearly on the flash energy. However, at higher intensities, the curve saturated, so additional flash energy gave no additional O_2 (Fig. 3.3).

Most experimenters would have been satisfied with just establishing that the curve saturated at a high light intensity. After all, if at very high light intensity every chlorophyll molecule absorbs a photon and produces photoproducts, then one expects that additional light will give no further products until the slow enzymatic reaction restores the chlorophyll to an active state. The beauty of the experiment lies in the fact that Emerson and Arnold took great pains to obtain a quantitative measure of how much O_2 was produced per chlorophyll in the sample. This may sound like a simple matter, but at the time the quantitative absorption properties of chlorophyll were not well known, so Emerson and Arnold had to determine this in order to know how many chlorophyll molecules were in their sample. The measurement of the amount of O_2 produced

was easier, utilizing the resulting volume and the known properties of gases.

The final result was a huge surprise. Only one O_2 was produced for every 2500 chlorophyll molecules, far less than the one per chlorophyll that was expected! Their findings were difficult to reconcile with the common idea at the time that each chlorophyll molecule directly reduced CO_2. It was many years before the true significance of this experiment was appreciated. We now know that the vast majority of chlorophyll molecules act as antennas and function only to collect light, transferring the energy to a special chlorophyll molecule (which is part of a protein complex known as the reaction center) that actually does the photochemistry. This is like a satellite dish, which collects radio waves and sends them to a receiver, where the signal is detected (Fig. 1.3). However, in 1932, Emerson and Arnold were only able to propose that a large number of chlorophyll molecules acted as a group to carry out photosynthesis, although it was not clear how this cooperation came about. The collection of chlorophyll molecules and associated enzymes

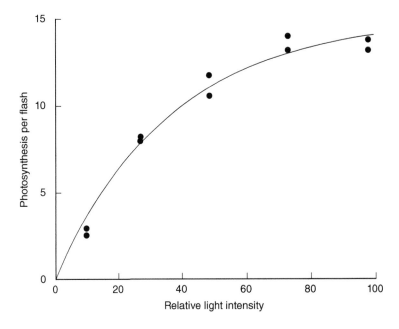

Figure 3.3   Emerson and Arnold's experiment establishing the light saturation curve for photosynthesis in flashing light. *Source:* Emerson and Arnold (1932b) (p. 1940)/Rockefeller University Press.

came to be known as a **photosynthetic unit,** a term proposed by Gaffron and Wohl (1936). We will explore the details of the structure and function of antennas, reaction centers, and other components of the photosynthetic unit in later chapters.

3.9 The controversy over the quantum requirement of photosynthesis

In the 1940s and 1950s, a controversy raged in the field of photosynthesis over the minimum quantum requirement for the process (Nickelsen and Govindjee, 2011). The quantum requirement is the number of photons that need to be absorbed for a photochemical process to take place. It is the reciprocal of the quantum yield. Otto Warburg, the Nobel prize-winning German biochemist who had developed the manometric techniques that were standard for measurement in these experiments, steadfastly maintained that the minimal quantum requirement for photosynthesis was three to four photons per O_2 evolved. Essentially everyone else obtained much higher values, in the range of 8–10 photons per O_2 produced. Foremost among these researchers was Warburg's former student, Emerson, who had earlier carried out the experiments with Arnold described above. The argument raged on for many years and was really only settled after Emerson's premature death in an airplane crash in 1959, followed by Warburg's death in 1970.

This disagreement may seem to be only an academic issue, but the outcome was essential to the development of a deeper understanding of the underlying chemical mechanism of photosynthesis. The discussion really boils down to energetics. The energy content of the three photons that Warburg thought were all that was needed is just barely enough to account for the free energy difference between the reactants and the products (see Chapter 13). Warburg was pleased with this result, which coincided with his nineteenth-century romantic view of nature, summarized by the

comment often attributed to him: "In a perfect world photosynthesis must be perfect." Emerson's view was more practical, and thousands of subsequent measurements in many laboratories have supported his higher numbers for the quantum requirement for photosynthesis.

Exactly why Warburg obtained the results he did is still not entirely clear, but it is thought to have to do with interactions of photosynthesis and respiration, including transient "gushes" and "gulps." The measurement shows only net oxygen production; to get the rate of photosynthesis, it is necessary to correct for the rate of respiration. If the rate of respiration is unchanged between light and dark, this correction will be accurate; but if photosynthesis inhibits respiration (as some modern evidence suggests), the correction will lead to erroneously low values for the quantum requirement. In retrospect, it is clear that Warburg, despite being a brilliant experimenter and very experienced professional scientist of the highest rank, fell into the very human trap of thinking that he knew what the answer should be and then not being sufficiently objective in evaluating his own experiments.

Unfortunately, the quantum requirement controversy took up the enormous time and effort of many of the foremost scientists of the day and didn't directly lead to a new understanding of the mechanism of photosynthesis. However, in the process of thoroughly examining the conditions required for the measurement of the quantum requirement for photosynthesis, some important new discoveries were made, which ultimately did lead to a much deeper understanding. Chief among these were the phenomena known as the "red drop" and "enhancement."

3.10 The red drop and the Emerson enhancement effect

As part of his attempt to settle the controversy with Warburg, Emerson and coworkers made careful measurements in the 1940s of the quantum requirements for photosynthesis as a function of wavelength and obtained a most remarkable result. As the wavelength of light utilized for the experiment approached the red edge of the absorption of the chlorophyll, the quantum requirement went up dramatically. The action spectrum for photosynthesis is remarkably congruent with the absorption spectrum throughout much of the visible wavelength range, but drops off more quickly in this far-red region (Fig. 3.4). An action spectrum is a plot of the effectiveness of light to cause a given effect, in this case oxygen evolution, versus the wavelength of light. This decrease in quantum yield (the reciprocal of the quantum requirement) at long wavelengths came to be known as the "red drop." The interpretation of the red drop is that those chlorophyll molecules that absorb light at the extreme red edge of the absorption band do not do photosynthesis as efficiently as the chlorophylls that absorb light of shorter wavelengths. The long-wavelength chlorophylls somehow behaved differently. Other measurements of photosynthesis and chlorophyll fluorescence in red algae, organisms that contain an antenna complex known as a phycobilisome, also suggested that the long-wavelength chlorophylls were somehow inactive in photosynthesis. The red drop result was easily reproduced, but the significance of it was not understood until later.

However, the result of another experiment by Emerson and coworkers was even more bizarre (Emerson *et al.*, 1957). He found that if the ineffective long-wavelength light was supplemented with shorter-wavelength light, it suddenly became capable of driving photosynthesis at good rates. A sample of algae was illuminated with red light, and the intensity adjusted to give a particular rate of O_2 production, measured as always using a manometer. This light was then turned off, and a second light source, this time the inefficient far-red light, was directed on the sample. The intensity of this light was adjusted to give a rate of O_2 production comparable to that of the red light. This required that the intensity of the far-red light be increased significantly, as expected from the earlier experiments that had shown its weak effect. The remarkable result was that, when both beams of light were directed on the sample at the same time, the rate of O_2 production was greatly increased and was much higher than the sum of the two individual rates! This result came to be known as the **enhancement effect**, because of the enhancing effect of the short-wavelength light. Additional experiments by Jack Myers and Stacey French (1960) showed that enhancement worked even when the two beams of light were not present at exactly the same time. These results made no sense in the context of the 1950s understanding of the mechanism of photosynthesis. Several years went by before a reasonable explanation was proposed for these and other puzzling results.

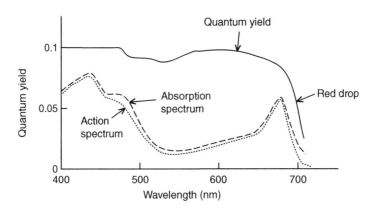

Figure 3.4 Absorption spectrum of chloroplasts (dashed line) and action spectrum for photosynthesis (dotted line). The red drop in the quantum yield of photosynthesis (solid line).

3.11   Antagonistic effects

A final experiment that pointed the way to the existence of two distinct photochemical systems working in series in photosynthetic organisms was carried out by Louis Duysens and coworkers from the Netherlands (Duysens *et al.*, 1961). Duysens was a pioneer in developing sensitive spectrophotometric methods to monitor photosynthetic systems. The experiment was to measure the oxidation–reduction state of cytochrome *f* in the sample upon illumination using various wavelengths of light. When the cytochrome is reduced, the absorbance spectrum changes, permitting quantitative measurements of its redox state (Fig. 3.5). Duysens found that far-red light caused the cytochrome to become oxidized, whereas shorter-wavelength light caused it to become reduced. The two colors of light had opposite, or antagonistic, effects. A particularly clear effect was observed using the red alga *Porphyridium cruentum*, which has phycobilisome antenna complexes. The effects are easily observed with this organism, because, as we now know, the phycobilisome antenna complex preferentially directs excitations mostly to one of the two photosystems. As is often the case, the choice of the experimental system in which an effect is emphasized was important to the initial understanding of the effect. Subsequent measurements have shown that the effect is, of course, a general one, observed in all oxygen-evolving photosynthetic organisms.

3.12   Early formulations of the Z scheme for photosynthesis

All these experiments (and some others not discussed here) suddenly crystallized into a consistent formulation for photosynthesis about 1960. Robin Hill and Fay Bendall published a short paper in *Nature* in 1960 outlining the concept of two sequential photochemical systems arranged in tandem, so that the products of one system became the substrates of the other system (Hill and Bendall, 1960). Their formulation was based primarily on the observation that the redox potentials for two cytochromes found in chloroplasts were intermediate between the potentials of both the reductant (H_2O) and the oxidant ($NADP^+$) involved in photosynthesis. In order for them to be participants in the light-driven electron flow, it was necessary to propose two photochemical processes and an energetically downhill intermediate step. Their original scheme is shown in Fig. 3.6. It has come to be known as the Z (for zigzag) scheme of photosynthesis (Govindjee and Björn, 2017).

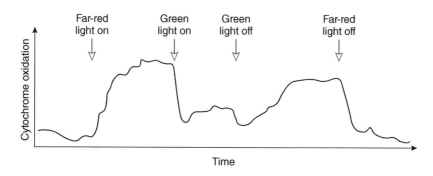

Figure 3.5   Antagonistic effects on cytochrome oxidation. Irradiation with the light of one color causes the cytochrome to become more oxidized, while irradiation with light of a different color causes it to become more reduced. This experiment was the clearest early evidence for two photochemical systems connected in series in oxygenic photosynthetic organisms. *Source:* Duysens *et al.* (1961)/Springer Nature.

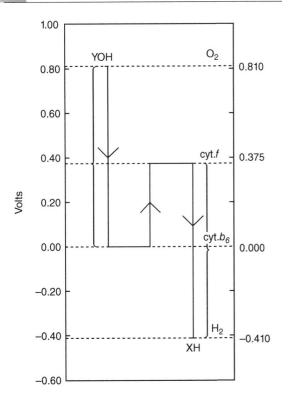

Figure 3.6   Hill and Bendall's (1960) original formulation of the Z scheme for photosynthesis. Modern Z schemes are plotted with the *y*-axis reversed in sign, so that more reducing species are at the top of the diagram and more oxidizing species are at the bottom. The action of light is shown as a vertical arrow in which a relatively highly oxidizing species is converted to a highly reducing one by the action of light. *Source:* Hill and Bendall (1960) (p. 137)/Springer Nature.

The slightly later publication by Duysens and coworkers provided the crucial experimental evidence for this proposal, in the form of the antagonistic effects described above. One photochemical reaction, now known as Photosystem II, oxidizes water and reduces cytochrome *f*, while the other, now known as Photosystem I, oxidizes cytochrome *f* and reduces $NADP^+$ (Duysens *et al.*, 1961; Duysens, 1989).

The puzzling results of the red drop and enhancement effects are easily explained by this formulation of two photochemical reactions connected in series, if the absorption spectra of the pigments that feed

energy to them are not quite the same. The shorter-wavelength chlorophylls (and the phycobilisome antennas in those organisms that contain them) preferentially drive Photosystem II, whereas the longer-wavelength pigments preferentially drive Photosystem I. Optimum rates of photosynthesis are observed when both short and long wavelengths are present, as found in the enhancement experiments. The light absorbed by the long-wavelength pigments does not have enough energy to drive Photosystem II, so the entire system grinds to a halt if only far-red light is used, thus explaining the red drop phenomenon. In retrospect, it was fortunate that the two photosystems had somewhat different wavelength optima, as the key early experiments demonstrating their existence all relied on preferential excitation of one or the other photosystem by carefully selecting the illumination regime.

The series formulation for oxygen-evolving photosynthesis has been tested and questioned many times since 1960 and has withstood all challenges. There is now no doubt that this basic framework for photosynthetic electron flow in oxygen-evolving organisms is correct. As we will see in later chapters, the two photosystems have been separated, biochemically purified, and their structures determined. The proteins that make them up to have been identified, and the genes that code for them identified and sequenced. While many questions remain about how the two photosystems interact and how both energy input to them from the antennas and electron flow between them is regulated, the Z scheme is an essential feature of the modern understanding of photosynthesis.

Of course, there are some types of photosynthetic organisms that contain chlorophyll-like pigments and exhibit light-dependent growth, but do not evolve oxygen. We encountered them earlier in our discussion of the van Niel formulation for the redox nature of photosynthesis. These anoxygenic (non-oxygen evolving) photosynthetic organisms do not display any of the enhancement or red drop effects described above that led to the discovery of two photosystems in oxygen-evolving organisms, which suggests that they contain only a single type of photosystem. Biochemical and genetic studies confirm that these organisms, which are all bacteria, indeed contain only a single type of photosystem. These simpler photosynthetic organisms have been essential for learning about the basic chemical mechanisms of photosynthesis. These organisms are discussed in detail in Chapter 6. They also provide clues to how the complex two-photosystem version of photosynthesis found in more advanced organisms may have arisen and evolved. This is discussed in Chapter 12.

3.13   ATP formation

So far, we have focused first on the overall process of photosynthesis and then on the discoveries leading up to the discovery of the series formulation of electron transport in oxygen-evolving photosynthetic organisms, resulting in the reduction of $NADP^+$ to NADPH. There is another critical part to the story, however, involving the light-dependent formation of ATP and the subsequent utilization of these two products to reduce CO_2 to carbohydrates. The discoveries of these processes paralleled the discoveries of the electron transfer processes.

The discovery that chloroplasts could make ATP in a light-dependent manner was made in 1954, by Daniel Arnon and coworkers at the University of California, Berkeley (Arnon *et al.*, 1954). The idea that chloroplasts could make ATP, in a process called photophosphorylation, initially met with considerable resistance, because it was well known that mitochondria produced large amounts of ATP and, since chloroplasts in many ways drive the mitochondrial reaction in the opposite direction, this initially seemed backward (Arnon, 1984). An analogous discovery of light-driven ATP formation in non-oxygen-evolving purple bacteria was made by Howard Gest and Martin Kamen (1948). The chemiosmotic hypothesis, the theoretical framework for the mechanism of how photon energy is stored in ATP, was provided by the incisive analysis of Peter Mitchell in the 1960s and 1970s, for which he received the Nobel prize in 1978 (Mitchell, 1979). We will discuss the details of the ATP synthesis process in Chapter 8.

3.14   Carbon fixation

At approximately the same time as Arnon was demonstrating photophosphorylation on one side of the Berkeley campus of the University of California, Melvin Calvin, Andrew Benson, and coworkers were working to understand the details of the carbon assimilation process itself on the other side of the campus (Calvin, 1989; Benson, 2002). They elucidated the chemical reactions that convert CO_2 and assimilatory power into carbohydrates. These reactions have become known as the Calvin–Benson cycle, and Calvin was awarded the Nobel Prize for chemistry in 1961 in recognition of the brilliant elucidation of this complex set of reactions. He and his coworkers used the newly developed method of radioactive tracers, injecting algae with $^{14}CO_2$ and then following the path of the radioactivity in the products (Creager, 2013). We will discuss the details of the Calvin–Benson cycle and other aspects of carbon metabolism in Chapter 9.

References

Arnon, D. I. (1984) The discovery of photosynthetic phosphorylation. *Trends in Biochemical Sciences* 9: 1–5.

Arnon, D. I., Allen, M. B., and Whatley, F. R. (1954) Photosynthesis by isolated chloroplasts. *Nature* 174: 394–396.

Benson, A.A. (2002) Following the path of carbon in photosynthesis: A personal story. *Photosynthesis Research* 73: 29–49.

Calvin, M. (1989) 40 years of photosynthesis and related activities. *Photosynthesis Research* 21: 3–16.

Creager, A. N. H. (2013) *Life Atomic: Radioisotopes in Science and Medicine.* Chicago: University of Chicago Press.

Duysens, L. M. N. (1989) The discovery of the 2 photosynthetic systems – a personal account. *Photosynthesis Research* 21: 61–79.

Duysens, L. M. N., Amesz, J., and Kamp, B. M. (1961) Two photochemical systems in photosynthesis. *Nature* 190: 510–511.

Emerson, R. and Arnold, W. (1932a) A separation of the reactions in photosynthesis by means of intermittent light. *Journal of General Physiology* 15: 391–420.

Emerson, R. and Arnold, W. (1932b) The photochemical reaction in photosynthesis. *Journal of General Physiology* 16:191–205.

Emerson, R., Chalmers, R., and Cederstrand, C. (1957) Some factors influencing the longwave limit of photosynthesis. *Proceedings of the National Academy of Sciences USA* 43: 133–143.

Gaffron, H. and Wohl, K. (1936) Zur theorie der assimilation. *Naturwissenschaften* 24: 81–90; 103–107.

Gest, H. and Kamen, M. D. (1948) Studies on the phosphorus metabolism of green algae and purple bacteria in relation to photosynthesis. *Journal of Biological Chemistry* 176: 299–318.

Govindjee, S. D. and Björn, L. O. (2017) Evolution of the Z-scheme of photosynthesis: A perspective. *Photosynthesis Research* 133: 5–15.

Govindjee, S. D., Beatty, J. T., Gest, H., and Allen, J. F., (eds.) (2005) *Discoveries in Photosynthesis*. Dordrecht: Springer.

Hill, R. (1939) Oxygen produced by isolated chloroplasts. *Proceedings of the Royal Society of London. Series B* 127: 192–210.

Hill, J. F. (2012) Early pioneers of photosynthesis research. In: J. J. Eaton-Rye, B. C. Tripathy, and T. D. Sharkey, (eds.) *Photosynthesis*. Dordrecht: Springer.

Hill, R. and Bendall, F. (1960) Function of the two cytochrome components in chloroplasts: A working hypothesis. *Nature* 186: 136–137.

Jaffe, B. (1976) *Crucibles: The Story of Chemistry*, 4th Edn. New York: Dover.

Magiels, G. (2010) *From Sunlight to Insight: Jan Ingen-Housz, the Discovery of Photosynthesis & Science in the Light of Ecology*. Brussels: Academic & Scientific Publishers.

Mitchell, P. (1979) Keilin's respiratory chain concept and its chemiosmotic consequences. *Science* 206: 1148–1159.

Myers, J. (1994) The 1932 experiments. *Photosynthesis Research* 40: 303–310.

Myers, J. and French, C. S. (1960) Relationships between time course, chromatic transient, and enhancement phenomena of photosynthesis. *Plant Physiology* 35: 963–969.

Nickelsen, K. (2015) *Explaining Photosynthesis: Models of Biochemical Mechanisms, 1840–1960*. Dordrecht: Springer.

Nickelsen, K. and Govindjee, S. D. (2011) *The Maximum Quantum Yield Controversy: Otto Warburg and the "Midwest-Gang"*. Bern: Bern Studies in the History and Philosophy of Science.

van Niel, C. B. (1941) The bacterial photosyntheses and their importance for the general problem of photosynthesis. *Advances in Enzymology* 1: 263–328.

Rabinowitch, E. I. (1945) *Photosynthesis and Related Processes*, Vol. 1. New York: Interscience Publishers.

Walker, D. A. (1992) *Energy, Plants and Man*, 2nd Edn. Brighton: Oxygraphics.

Chapter 4

Photosynthetic pigments: structure and spectroscopy

The lifeblood of a photosynthetic organism is its pigments. Without them, light cannot be absorbed, and therefore energy cannot be stored. There are a remarkable number of pigments found in different photosynthetic organisms, and they serve a variety of functional roles. In this chapter, we will learn about the different types of pigments, with an emphasis on how the chemical and spectroscopic properties of the pigments are determined by their structures and the functions that they perform in the photosynthetic process. The chlorophylls are named *a–f*, and the bacteriochlorophylls *a–g*, in order of their discovery. In addition, we will consider carotenoids and bilins, the two other major classes of photosynthetic pigments.

Chlorophylls have long been investigated (Scheer, 1991; Grimm *et al.*, 2006). The word chlorophyll was first used by Pelletier and Caventou in 1818 to describe the green pigments that are involved in photosynthesis in higher plants. Three Nobel prizes have been given at least in part for studies on the structural determination of chlorophyll. Richard Wilstätter was honored in 1915 for his work that established the major features of the chlorophyll structure, including the empirical formula and the presence of magnesium (Mg). Hans Fischer was awarded the 1930 Nobel prize in part because he determined the complete structure, and Robert Woodward received the 1965 prize in part for his work that culminated in the total synthesis of chlorophyll.

4.1 Chemical structures and distribution of chlorophylls and bacteriochlorophylls

The empirical chemical formula for chlorophyll *a* is $C_{55}H_{72}N_4O_5Mg$. This simple representation is entirely inadequate to convey the essential properties of this extraordinary molecule. The structural formula for chlorophyll *a* is shown in Fig. 4.1. It is a squarish planar molecule, about $10\,\text{Å}$ on a side. A space-filling model of chlorophyll *a* is shown in Fig. 4.2. The Mg atom in the center of the planar portion is coordinated to four nitrogen atoms. The nitrogens are each part of a substructural element of the molecule that is derived from pyrrole, a cyclic organic compound with a nitrogen atom in a five-membered ring with four carbons. For this reason, chlorophylls

Molecular Mechanisms of Photosynthesis, Third Edition. Robert E. Blankenship.
© 2021 Robert E. Blankenship 2021 by John Wiley & Sons Ltd.
Companion website: https://www.wiley.com/go/blankenship/molecularphotosynthesis3e

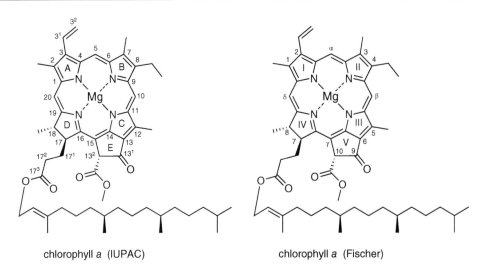

chlorophyll *a* (IUPAC)   chlorophyll *a* (Fischer)

Figure 4.1   Numbering schemes for chlorophylls and bacteriochlorophylls. Chlorophyll *a* is shown, although the same basic numbering scheme applies to all chlorophyll-type pigments. Left: the current IUPAC standard system. Right: the older Fischer numbering system. Hydrogen atoms are not shown.

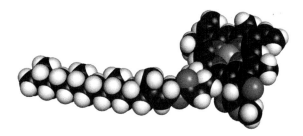

Figure 4.2   Space filling model of chlorophyll *a*. Carbon is shown as black, hydrogen as white, oxygen as red, nitrogen as blue, and magnesium as orange.

and related compounds are often referred to as tetrapyrroles. A fifth ring is formed in the lower right corner, and a long hydrocarbon tail is attached to the lower left (in the standard representation). Chemically, the chlorophylls are related to the porphyrins, which are also tetrapyrroles, but the porphyrins are generally more symmetric molecules.

The five rings in chlorophylls are lettered A through E, and the substituent positions on the macrocycle are numbered clockwise, beginning in ring A, as shown in Fig. 4.1, according to the officially recognized International Union of Pure and Applied Chemistry (IUPAC) nomenclature. An older nomenclature known as the Fischer system is

also shown in Fig. 4.1. All of the older literature uses the Fischer nomenclature, so it is necessary to be conversant with both systems. In this book, the IUPAC system will be used exclusively.

By convention, the *y* molecular axis of all chlorophylls is defined as passing through the N atoms of rings A and C, with the *x* axis passing through the N atoms in rings B and D. The *z* axis is perpendicular to the plane of the macrocycle. An extensive delocalized π electron system extends over most of the molecule, with the exception of ring D, in which the C-17–C-18 double bond is reduced to a single bond. The tail is formed by condensation of four five-carbon isoprene units and is then esterified to ring D. It is often called the phytyl tail, after the polyisoprenoid alcohol precursor phytol that is attached during biosynthesis. It is also sometimes called the isoprenoid tail.

Most of the chlorophylls are classified chemically as chlorins rather than porphyrins, by virtue of the reduced ring D. Most of the bacteriochlorophylls are similarly called bacteriochlorins, because of the reduction of both rings B and D. All chlorophylls and bacteriochlorophylls contain the extra ring E, which is called the isocyclic ring.

Most chlorophyll-type pigments contain three chiral carbon atoms, C-13^2, C-17, and C-18.

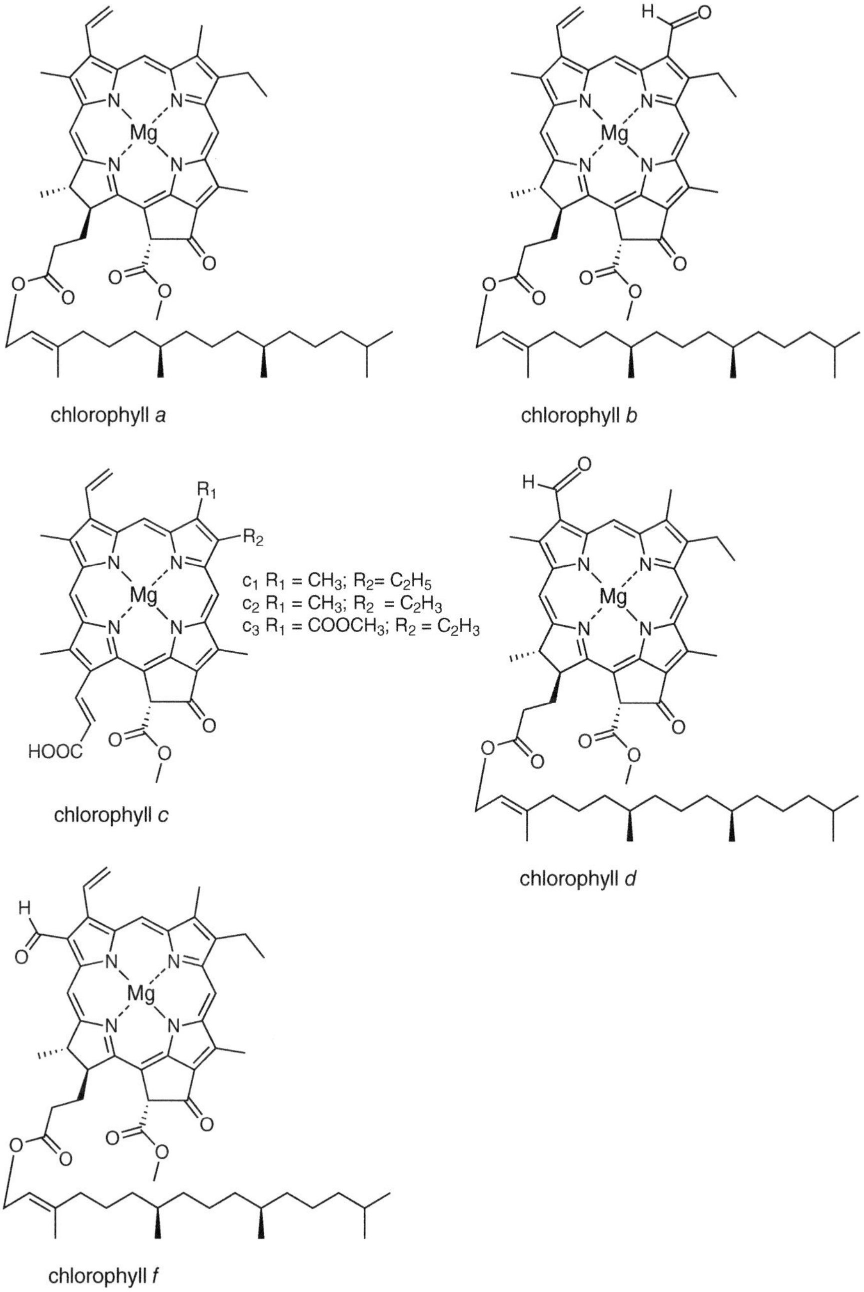

Figure 4.3 Chemical structures of chlorophylls *a, b, c, d,* and *f*. R_1, R_2, etc. refer to ring substituents. In some cases, more than one possible group can be found at some positions.

Bacteriochlorophyll *a* contains two additional chiral centers, C-7 and C-8. In all cases, the stereochemical fidelity of the biosynthetic enzymes is extremely high, so the compounds found in cells are a single species and not mixtures of diastereomers (except as noted below).

The structures of all major chlorophylls and bacteriochlorophylls are shown in Figs. 4.3 and 4.4. The distribution of photosynthetic pigments in different classes of photosynthetic organisms is given in Table 4.1.

4.1.1 *Chlorophyll* a

Chlorophyll *a* is found in all known eukaryotic photosynthetic organisms. Among prokaryotes, it is found in large quantities only in the cyanobacteria (including the prochlorophytes), although traces of chlorophyll *a* or minor variants are found in some anoxygenic bacteria, where it is thought to have an important function as an intermediate in the electron transport chain. Some prochlorophytes contain divinyl chlorophyll *a*, in which the substituent at the C-8 position on ring B is vinyl instead of ethyl.

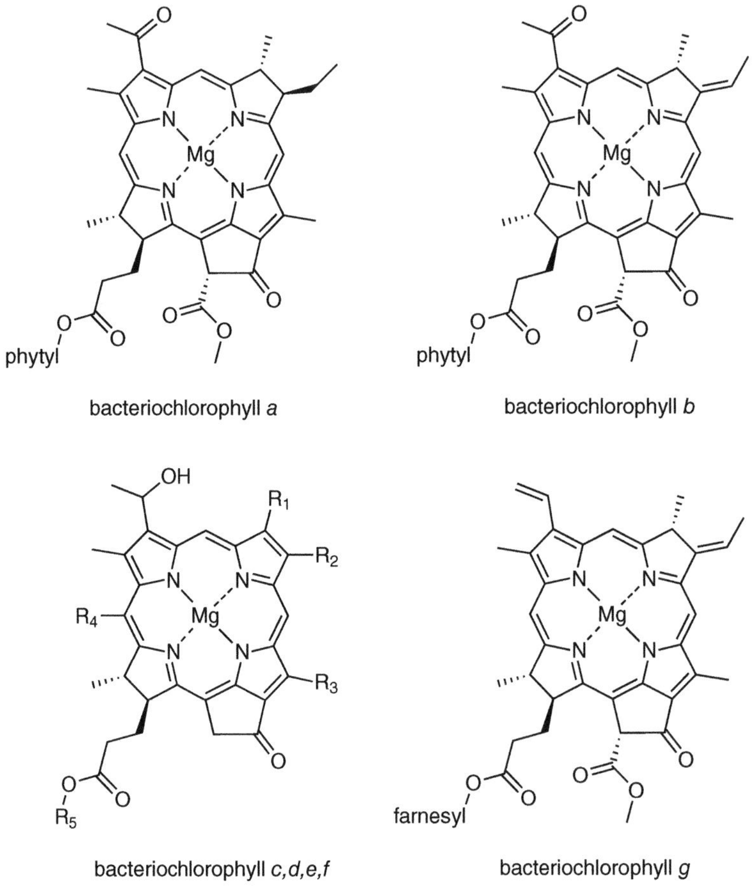

phytyl

bacteriochlorophyll *a*

phytyl

bacteriochlorophyll *b*

OH

R_1

R_2

R_4

R_3

R_5

bacteriochlorophyll *c,d,e,f*

farnesyl

bacteriochlorophyll *g*

Bchl *c* R_1=Me; R_2=Et, Pr, Bu, R_3=Me, Et; R_4=Me; R_5=stearyl, farnesyl, others
Bchl *d* R_1=Me; R_2=Et, Pr, Bu, neoPent; R_3=Me, Et; R_4=H; R_5=farnesyl, others
Bchl *e* R_1=CHO; R_2=Et, Pr, Bu, neoPent; R_3=Et; R_4=Me; R_5=farnesyl, others
Bchl *f* R_1=CHO; R_2=Et, Pr, Bu, neoPent; R_3=Et; R_4=H; R_5=farnesyl, others

Figure 4.4   Chemical structures of bacteriochlorophylls *a, b, c, d, e, f,* and *g*.

Table 4.1   Distribution of chlorophylls and bacteriochlorophylls

Type of organism	Chl *a*	Chl *b*	Chl *c*	Chl *d,f*	BChl *a*	BChl *b*	BChl *c,d,e*	BChl *g*	Carotenoids	Bilins
Purple bacteria					+[a]	+[a]			+	
Green sulfur bacteria					+		+		+	
Filamentous anoxygenic phototrophs					+		+		+	
Heliobacteria								+	+	
Cyanobacteria	+	+[b]	+[b]	+[b]					+	+
Green algae	+	+							+	
Diatoms	+		+						+	
Brown algae	+		+						+	
Dinoflagellates	+		+						+	
Cryptomonads	+		+						+	+
Red algae	+								+	+
Plants	+	+							+	

[a] Purple *a* or *b*, but not both in the same species.
[b] Most cyanobacteria contain Chl *a* as their only chlorophyll-type pigment. Prochlorophytes contain in addition Chl *b*. Some types also contain Chl *c*, *d*, or *f*.

An important variant of chlorophyll *a* is chlorophyll *a'*. This pigment differs from chlorophyll *a* only in the stereochemistry at the C-13^2 position. It is found in small but reproducible amounts in photosystem I complexes, where one molecule forms half of P700, the special pair of pigments that is the primary electron donor (see Chapter 7). The spectral and redox properties of chlorophyll *a'* are very similar to those of chlorophyll *a*. Current evidence suggests that chlorophyll *a'* is made from pre-existing chlorophyll *a*, although the putative C-13^2 invertase enzyme has not been identified.

4.1.2 Chlorophyll b

Chlorophyll *b* is identical to chlorophyll *a* except at the C-7 position, where a formyl group replaces the methyl group. This change shifts the maximum absorption to shorter wavelengths. Chlorophyll *b* is the major accessory light-absorbing pigment in light-harvesting complexes in the majority of eukaryotic photosynthetic organisms, including plants and green algae, and is not found in reaction center complexes. In photosynthetic prokaryotes, it is found only in the prochlorophytes, in some cases as divinyl chlorophyll *b*.

4.1.3 Chlorophyll c

Chlorophyll *c* is perhaps the most unusual of all the chlorophylls, in that it does not have an isoprenoid tail and also does not have ring D reduced. It is therefore chemically classified as a porphyrin, and not as a chlorin. Chlorophyll *c* is found exclusively in various groups of marine algae, such as diatoms and dinoflagellates. It functions as an accessory light-harvesting pigment in pigment–protein complexes similar to those involving chlorophyll *b* in plants and green algae. There are several structural variants of chlorophyll *c*, which vary in some of the peripheral ring substituents, as shown in Fig. 4.3.

4.1.4 Chlorophyll d

Chlorophyll *d* is different from chlorophyll *a* in only one respect: the substituent at the C-3 position is a formyl group in chlorophyll *d*, instead of the vinyl group found in chlorophyll *a*. For many years, chlorophyll *d* was known only as a trace constituent of certain red algae and was suspected to be an experimental artifact. However, in 1996, a cyanobacterium, *Acaryochloris marina,* was discovered as a symbiont in a marine animal called an ascidian. This organism contains chlorophyll *d* as the major pigment, although it also contains small amounts of chlorophyll *a*. It now appears that chlorophyll *d* is only found in cyanobacteria, and that earlier reports of chlorophyll *d* in red algae result from contamination of the algal surface by epiphytic cyanobacteria that contain chlorophyll *d* (Larkum and Kuhl, 2005). The electron withdrawing formyl group at the C-3 position adjacent to the *y* molecular axis of the pigment has the effect of shifting the absorption maximum of chlorophyll *d* to longer wavelengths compared to chlorophyll *a*. It has therefore received attention as a possible pigment to expand the solar spectrum in bioenergy applications (Chen and Blankenship, 2011).

4.1.5 Chlorophyll e

Chlorophyll *e* was isolated and named provisionally in the 1940s, but the structure has never been determined and so its structure and function remain uncertain.

4.1.6 Chlorophyll f

Chlorophyll *f* is the most recently discovered of all the chlorophyll-like pigments. It contains a formyl group at the C-2 position. It was found in cyanobacterial cultures isolated from microbial structures called stromatolites (Chen *et al.*, 2010; Chen *et al.*, 2012), but is now known to be present in a wide range of cyanobacteria that have been grown using far-red light (Gan et al., 2014). Similar to the case in chlorophyll *d*, the formyl group at the C-2 position shifts the absorbance maximum of this pigment substantially to longer wavelengths compared to chlorophyll *a*.

4.1.7 Bacteriochlorophyll a

The chemical structure of bacteriochlorophyll *a* is shown in Fig. 4.4. It is the principal chlorophyll-type pigment in the majority of anoxygenic photosynthetic bacteria. The chemical differences between the structures of chlorophyll *a* and bacteriochlorophyll *a* are the acetyl group at the C-3 position and the single bond in ring B between C-7 and C-8, instead of the double bond found in chlorophylls. This reduces the degree of conjugation in the macrocycle and also reduces the symmetry of the molecule compared with chlorophylls. These structural changes exert major effects on the spectral properties, which are discussed below.

A few species of the purple photosynthetic bacteria have been found that use zinc (Zn) as the central metal instead of Mg in bacteriochlorophyll *a* (Wakao *et al.*, 1996). These organisms are found in highly acidic environments where Mg^{2+} is readily displaced by H^+, whereas Zn is more stable as a central metal. These are the only two metal ions that have been found incorporated into natural chlorophylls, although many other metals can be inserted synthetically into the metal-free pigments. The reason for this specificity is probably that Mg is very readily available, whereas Zn is a trace element in almost all environments and would therefore often be a limiting nutrient. As discussed below, most other metals are unsuitable for photosynthesis, because pigments with these metals incorporated have a very short excited state lifetime. Zn bacteriochlorophyll has also been found in reaction centers of chloroacidobacteria (Tsukatani *et al.*, 2012).

4.1.8 Bacteriochlorophyll b

Bacteriochlorophyll *b* is found only in a few species of purple bacteria. It differs from bacteriochlorophyll *a* only by the presence of an exocyclic double bond at C-8 in ring B, which is called an ethylidine substituent. Its chemical structure is shown in Fig. 4.4. Bacteriochlorophyll *b* has the longest-wavelength absorbance band of any known chlorophyll-type pigment. *in vivo*, its absorbance maximum is at 960–1050 nm.

4.1.9 Bacteriochlorophylls c, d, e, and f

Bacteriochlorophylls c, d, e, and f will be considered as a group, because they are found only in green photosynthetic bacteria, organisms that contain the antenna complex known as a chlorosome. They are also unusual among chlorophylls in that they are invariably found as complex mixtures of closely related compounds instead of as a single compound of unique structure. Several distinct structural features are found in these pigments, whose structures are shown in Fig. 4.4. Ring B contains a C-7–C-8 double bond, as in chlorophylls, making these pigments chlorins instead of bacteriochlorins. They also have a hydroxyethyl substituent at the C-3^1 position in ring A. This functional group is essential to the aggregation of these pigments in the chlorosome, which will be discussed in Chapter 5. The C-3^1 carbon is chiral, and both R and S diastereomers are found in cells (the C-17 and C-18 chiral carbons are stereochemically pure). These pigments also have hydrogens at the C-13^2 position, instead of the bulky carboxymethyl substituent found in all other chlorophylls. This change allows the chlorin rings to pack together more closely. These pigments are structurally programmed for aggregation, and indeed, in the chlorosome, they are found as large oligomeric complexes with little protein.

The differences among the bacteriochlorophylls c, d, e, and f occur primarily in the C-20 methine bridge position, where bacteriochlorophylls c and e have a methyl substituent, and at the C-7 position, where bacteriochlorophylls e and f have a formyl substituent, like chlorophyll b. These changes tune the light absorption properties of these pigments, with the wavelength of maximum absorption decreasing as one goes from bacteriochlorophyll c to f. Other differences are found at the C-8 and C-12 positions, where a complex variety of substituents can occur, even in a single organism. The tails of these bacteriochlorophylls are also different from those of most other chlorophylls. The bacteriochlorophylls c, d, and e found in the green sulfur bacteria contain a farnesol tail instead of a phytol.

This is one isoprene unit shorter than phytol. The filamentous anoxygenic phototrophic bacteria, which contain only bacteriochlorophyll c (as well as bacteriochlorophyll a) primarily utilize the 18-carbon straight-chain stearol substituent, although a variety of other tails are found in varying amounts.

Bacteriochlorophyll f has never been found in nature. The compound that is known as bacteriochlorophyll f has the C-7 formyl substituent of bacteriochlorophyll e, as well as the C-20 H of bacteriochlorophyll d. It is thus the logical completion of this set of pigments. However, organisms that contain bacteriochlorophyll f have been created by inactivating the methylase enzyme that adds the methyl group to C-20 (Vogl et al., 2012).

4.1.10 Bacteriochlorophyll g

Bacteriochlorophyll g is found only in the anoxygenic heliobacteria. Its structure is shown in Fig. 4.4. It is essentially a molecular hybrid of chlorophyll a and bacteriochlorophyll b, in that it contains the C-3 vinyl substituent of chlorophyll a and the C-8 exocyclic ethylidine substituent of bacteriochlorophyll b. It also contains a farnesyl tail instead of phytyl. Bacteriochlorophyll g is very unstable and isomerizes into chlorophyll a or closely related compounds.

4.2   Pheophytins and bacteriopheophytins

The metal-free chlorophylls are known as pheophytins. The structures of pheophytin a and bacteriopheophytin a are shown in Fig. 4.5. In these compounds, two hydrogen ions replace the central Mg^{2+}. Acidic conditions promote the displacement of the metal. Pheophytins are formed during the degradation of chlorophylls, so are often viewed as primarily breakdown products resulting from loss of the central metal. However, small amounts of pheophytin are an essential component of some reaction center complexes and are almost certainly

pheophytin *a*   bacteriopheophytin *a*

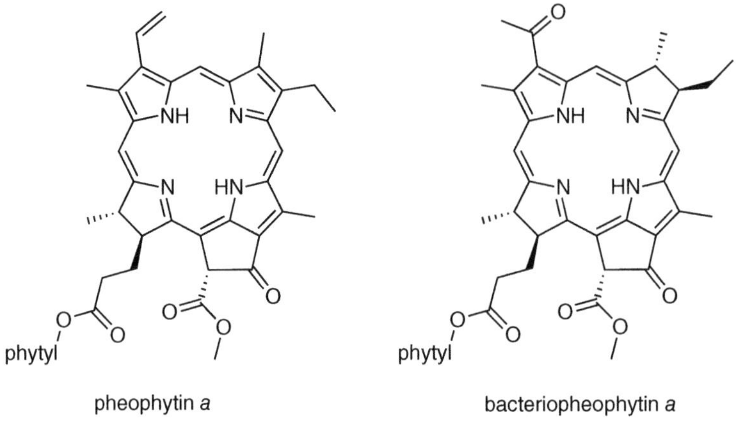

Figure 4.5   Chemical structures of pheophytin *a* and bacteriopheophytin *a*.

made in a specific pathway, although it is not known. Pheophytin is somewhat easier to reduce than the corresponding chlorophyll, so it can function as an electron acceptor at a place in the sequence of electron carriers where chlorophyll will not work.

4.3   Chlorophyll biosynthesis

The chlorophyll biosynthetic pathway contains 17 enzymatic steps, as shown in Fig. 4.6. It begins with the formation of δ-aminolevulinic acid (ALA) (Suzuki *et al.*, 1997; Beale, 1999; Chew and Bryant, 2007; Masuda, 2008; Chen, 2014; Bryant *et al.*, 2020). Eight molecules of ALA are condensed, eventually forming the symmetric metal-free porphyrin: protoporphyrin IX. At this point the pathway branches, with one branch leading to heme and the other to chlorophyll, depending on whether Fe or Mg is incorporated. The chlorophyll branch includes additional steps in which the fifth ring is fashioned and a double bond in ring D is reduced to form chlorophyllide, which still lacks the isoprenoid tail. The isoprenoid tail is attached as the last step. Biosynthesis of bacteriochlorophyll *a* is identical to chlorophyll synthesis through the chlorophyllide *a*

intermediate, but contains two additional steps: the reduction of ring B and the conversion of the vinyl group at C-3 to an acetyl moiety. The order of the biosynthetic steps and the identities of most of the enzymes that carry out the transformations have been well established for both chlorophyll *a* and bacteriochlorophyll *a* synthesis by a combination of biochemical and genetic experiments. The biosynthetic pathway of chlorophylls is shown in Fig. 4.6. The enzymes that catalyze the various steps are identified in the figure legend.

The first few steps in the pathway produce ALA. Most photosynthetic organisms use a unique pathway that involves the ATP-dependent charging of the amino acid glutamic acid to a glutamyl tRNA, the same reaction that takes place when glutamic acid is incorporated into a growing peptide during protein synthesis. This reaction is one of a very small number of reactions known in biology in which a tRNA molecule is used in a biochemical step other than protein synthesis. Step 2 is the NADPH-dependent reductive cleavage of the acid to an aldehyde to form L-glutamic acid 1-semialdehyde. The final step in ALA synthesis is the transamination rearrangement to form ALA.

Surprisingly, one group of photosynthetic organisms makes ALA by a completely different route. This is the one-step condensation of the amino acid glycine with succinyl-CoA, followed by decarboxylation, catalyzed by the enzyme ALA synthase. This enzyme is found in some (but not all) purple photosynthetic bacteria. It is also found in the mitochondrion of all eukaryotic cells and is the pathway used by mitochondria to make ALA for heme biosynthesis.

The next series of biosynthetic steps lead to protoporphyrin IX. Step 4 is the condensation of two molecules of ALA to form the cyclic compound porphobilinogen, which represents the pyrrole monomer. Step 5 is the further condensation of four molecules of porphobilinogen to form the open-chain tetrapyrrole hydroxymethylbilane. Steps 6–8 involve ring closure and successive oxidative decarboxylation steps. Note the remarkable "flipping" of pyrrole ring D during step 6, so that the positions of the acetate and propionate substituents are interchanged. This proceeds by way of a spiro intermediate at C-16.

The final step in this phase of the biosynthetic pathway is the oxidation of protoporphyrinogen IX to protoporphyrin IX, step 9 in the sequence shown in Fig. 4.6. This is an important step, because the former compound looks superficially like a porphyrin, but is not fully conjugated. The pyrrole rings are essentially independent of each other, and the compound is colorless. The product, protoporphyrin IX, is a fully conjugated porphyrin and is highly colored. Excited protoporphyrin IX reacts readily with molecular oxygen to form the highly damaging species singlet oxygen. The system is highly regulated so that high concentrations of free protoporphyrin IX and other photosensitive intermediates do not build up.

Protoporphyrin IX is metallated by insertion of Mg at step 10. The heme biosynthesis pathway branches at this point (Bryant *et al.*, 2020). Those molecules destined to become heme have Fe inserted instead of Mg. The next phase of the biosynthesis of chlorophylls involves the construction of the isocyclic ring E by cyclizing the propionic acid attached to C-13. This reaction proceeds by first esterifying the carboxylic acid moiety and then undergoing a stereospecific oxidative cyclase reaction, steps 12–14. The intermediate at this step, divinyl protochlorophyllide, is then acted on by two separate enzymes on opposite sides of the ring. The two enzymes are not sensitive to whether the other change has been made in the substrate, so the pathway branches, and a given molecule can first have step 15 and then step 16 take place, or vice versa. Step 15 is the reduction of the vinyl group at C-8 to an ethyl group, and step 16 is the reduction of ring D.

The reduction of ring D is one of the most interesting steps in the biosynthesis of chlorophylls. In all oxygenic photosynthetic organisms, a light-driven enzyme, protochlorophyllide reductase, (often called POR), carries out this step. This enzyme uses NADPH as a source of reductant but also has an absolute requirement for light, making it one of a very small number of light-driven enzymes known in all of biology (another is the DNA repair enzyme photolyase). The structure of the light-driven POR from a cyanobacterium has been determined (Dong *et al.*, 2020). In anoxygenic photosynthetic bacteria, an entirely different enzyme complex that is not light-driven carries out this reduction. In this case, the reductant is reduced ferredoxin instead of NADPH. All oxygenic organisms except higher plants have both the light-dependent and the light-independent enzymes. The evolutionary implications of this pattern are discussed in Chapter 12.

The final step in chlorophyll *a* biosynthesis is the attachment of the tail, catalyzed by the enzyme chlorophyll synthase and using phytol pyrophosphate as the substrate. Phytol pyrophosphate is made by reduction of the unsaturated polyisoprenoid compound geranylgeranyl pyrophosphate. Depending on the state of growth of the organism, the reduction can take place either before or after attachment of the tail to the macrocycle.

The same basic pathway is used for the synthesis of all other chlorophylls and bacteriochlorophylls. Most of the reactions are the same, except that some steps are O_2-dependent in aerobes but use a different enzyme and oxidant in anaerobes (Ouchane *et al.*, 2004; Raymond and Blankenship, 2004). The major differences are found at the end of the pathway. Bacteriochlorophyll *a* is made from chlorophyllide *a* using two additional steps. The first is the

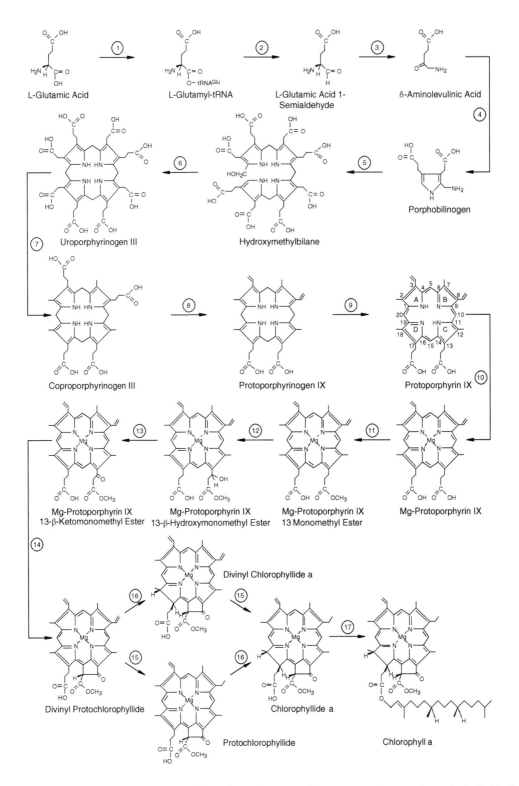

Figure 4.6 Outline of chlorophyll *a* biosynthesis from glutamate. The enzymes that catalyze the individual numbered reactions are (1) glutamyl-tRNA synthetase; (2) glutamyl-tRNA reductase; (3) glutamate 1-semialdehyde aminotransferase; (4) porphobilinogen synthase; (5) hydroxymethylbilane synthase; (6) uroporphyrinogen III

conversion of the C-3 vinyl group to an acetyl group. The second is the reduction of pyrrole ring B by an enzyme complex similar to the light-independent enzyme that reduces ring D, as discussed in the previous paragraph.

Chlorophyll *b* is made from chlorophyll *a* by oxidation of the methyl group at C-7 to give a formyl group, using a mixed-function oxidase enzyme that depends on O_2 (Tanaka and Tanaka, 2007). Chlorophyll *b* can also be reduced back to chlorophyll *a*. The ratio of the two pigments is tightly regulated and can be adjusted as needed.

4.4 Spectroscopic properties of chlorophylls

The chlorophylls all contain two major absorption bands, one in the blue or near UV region and one in the red or near IR region (Fig. 4.7). The blue absorption band produces a second excited state that very rapidly (picoseconds) loses energy as heat to produce the lowest excited state. The lowest excited state is relatively long-lived (nanoseconds) and is the state that is used for electron transfer and energy storage in photosynthesis.

The lack of a significant absorption in the green region gives the chlorophylls their characteristic green or blue–green color. These absorption bands are $\pi \rightarrow \pi^*$ transitions, involving the electrons in the conjugated π system of the chlorin macrocycle. The absorption and fluorescence spectra of chlorophyll *a* and bacteriochlorophyll *a* are shown in Fig. 4.8. The two lowest-energy transitions are called the Q bands, and the two higher-energy ones are known as the B bands. They are also commonly called **Soret bands**.

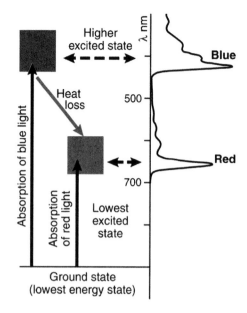

Figure 4.7 Absorption spectrum and simplified energy level diagram for chlorophyll *a*. The blue absorption band populates a second excited state that rapidly converts to the lowest excited state that can also be populated by absorption of a red photon. Energy storage in photosynthesis always occurs from the lowest excited state and the energy difference between the higher excited states and the lowest excited state is lost as heat. *Source:* Blankenship *et al.* (2011) (p. 807)/ American Association for the Advancement of Science.

The spectra can be described theoretically by using a "four orbital" model, originally proposed by Martin Gouterman (1961). The four π molecular orbitals that are principally involved in these transitions are the two highest occupied molecular orbitals (HOMOs) and the two Lowest Unoccupied Molecular Orbitals (LUMOs). The molecular orbital picture and the one-electron transitions

synthase; (7) uroporphyrinogen III decarboxylase; (8) coproporphyrinogen III oxidative decarboxylase; (9) protoporphyrinogen IX oxidase; (10) protoporphyrin IX Mg-chelatase;(11) S-adenosyl-L-methionine:Mg-protoporphyrin IX methyl-transferase; (12)–(14) Mg-protoporphyrin IX monomethyl ester oxidative cyclase; (15) divinyl (proto)chlorophyllide 4-vinyl reductase;(16) light-dependent NADPH:protochlorophyllide oxidoreductase or light-independent protochlorophyllide reductase; (17) chlorophyll synthase. The IUPAC numbering for the tetrapyrrole peripheral substituent positions is shown for protoporphyrin IX. *Source:* Beale (1999) (p. 45)/Springer Nature.

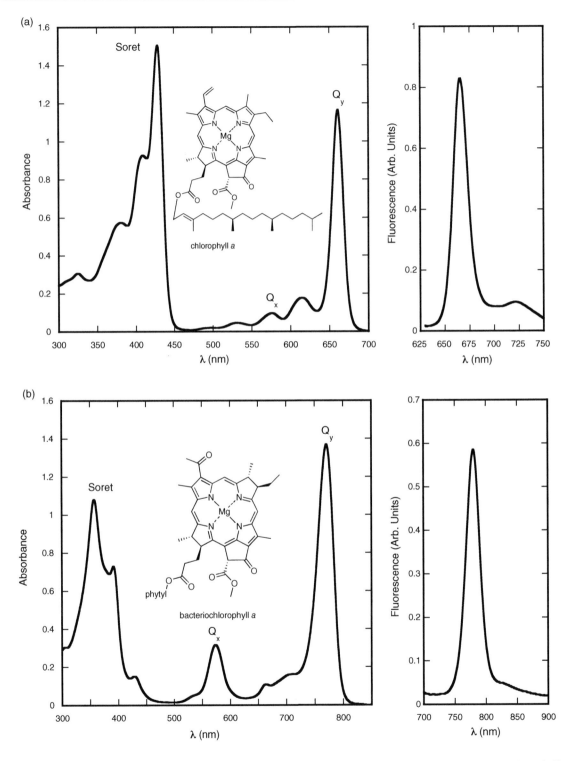

Figure 4.8   Absorption (left) and fluorescence (right) spectra of (a) chlorophyll *a* and (b) bacteriochlorophyll *a* in diethyl ether.

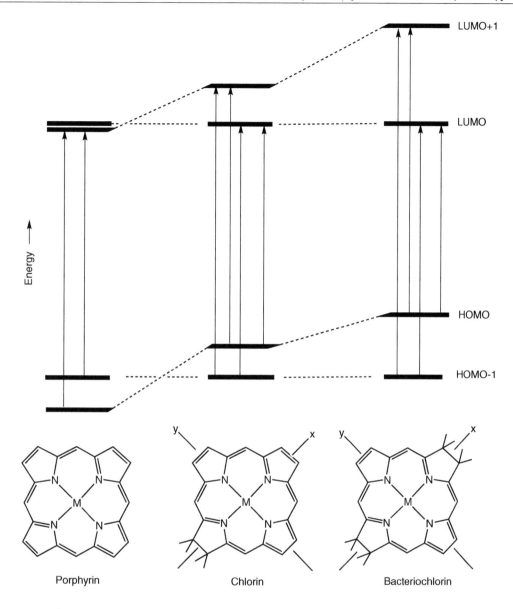

Figure 4.9   Molecular orbital energy level diagram of porphyrin, chlorin, and bacteriochlorin with one-electron transitions that combine via configuration interaction to give the different electronic transitions indicated by arrows.

predicted for a symmetric porphyrin, chlorin, and bacteriochlorin are shown in Fig. 4.9.

The diagram shown in Fig. 4.9 is an oversimplification of what is a complex relationship between electronic states and orbital energies. It is not correct to conclude that electronic transitions reflect a simple promotion of an electron from a HOMO to a LUMO. In reality, several different electronic configurations, including contributions from much higher-energy molecular orbitals can contribute to the electronic transition. This phenomenon is known as configuration interaction. The result is that there is not a simple one-to-one correspondence between orbital occupations and electronic transitions.

However, it is also the case that the majority contribution to the transition can come from a single configuration. This becomes more correct as one goes from the more symmetric porphyrin to the very asymmetric bacteriochlorin, so that the Q_y absorption band in bacteriochlorophyll is described reasonably well by the simple HOMO to LUMO transition, while this is not valid for the more symmetrical porphyrin. The energies and electron densities associated with the four frontier molecular orbitals for a Zn-chlorin are shown in Fig. 4.10. The one-electron transitions that combine via configuration interaction to produce the excited electronic state are shown as arrows, with the major transitions that combine to produce the Q_y transition labeled Y and those that combine to produce the Q_x transition labeled X.

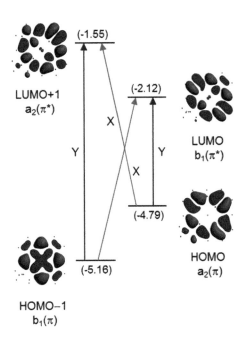

Figure 4.10 Schematic orbital energy diagram for a Znchlorin, illustrated by the calculated orbital energies (in eV) and electron density distributions. The diagram also shows the one-electron promotions that give rise to the electronic states via configuration interaction according to the four-orbital model. *Source: Kee et al.* (2007) (p. 1137)/John Wiley and Sons.

The electronic transitions have transition dipole moments with different strengths and orientations. The longest-wavelength transition is invariably polarized along the y-axis of the molecule and is therefore known as the Q_y transition. The y-axis extends from the N atom in ring A to the N atom in ring C (Fig. 4.2). This means that the absorption will be strongest if the electric vector of plane-polarized exciting light is parallel to the molecular axis of the pigment. The exciting light couples to the π electrons of the molecule and rearranges them somewhat during the transition. The Q_y transition causes a shift in electron density that is directed along the y molecular axis of the molecule. In a similar fashion, the x-axis extends from the N atom in ring B to the N atom in ring D.

For both chlorophyll a and bacteriochlorophyll a, the Q_y transition is strongly polarized along the y molecular axis (see Appendix). The weaker Q_x transition in bacteriochlorophyll is also strongly polarized along the x molecular axis. The Q_x transition in chlorophyll, however, is not as well resolved as in bacteriochlorophyll, and theoretical calculations suggest that it is not polarized directly along the x molecular axis. The Soret bands have a mixed polarization.

In addition to the fundamental Q and B electronic transitions, vibrational overtone transitions can also be observed, especially on the Q_y band. These represent a simultaneous vibrational and electronic transition, with the final state being an excited vibrational state of the excited electronic state. A progression of vibrational states can be observed, with the most intense transition the 0,0 band, with the higher energy satellites termed 0,1; 0,2; and so on. The first number is the vibrational quantum number of the ground electronic state before light absorption, and the second number the vibrational state of the excited electronic state after the transition. Of course, there are many vibrations in a chlorophyll molecule, and only one of these is the one responsible for the vibrational structure in the absorption spectrum.

The fluorescence spectrum of all chlorophylls peaks at slightly longer wavelengths than the absorption maximum. The fluorescence emission is polarized along the y molecular axis, as it is emitted from the Q_y transition. The fluorescence spectrum usually has a characteristic "mirror image" relationship to the absorption. This is because the ground and excited states have similar shapes, so those molecular vibrations that are activated during electronic absorption are also likely to be activated upon fluorescence emission. However, in this case, the initial state is the ground vibrational state of the excited electronic state, and the final state is the excited vibrational state of the ground electronic state. This causes a shift of the emission to the longer-wavelength side of the main transition, in what is known as the **Stokes shift** (see Appendix).

Table 4.2 brings together wavelengths of absorption maxima as well as molar extinction coefficients and fluorescence lifetimes and quantum yields of chlorophylls and bacteriochlorophylls in organic solvents. A comprehensive database of spectra and photophysical properties of chlorophylls and related pigments is available (Taniguchi and Lindsey, 2021). The spectral properties of these pigments are significantly altered when they are incorporated into protein complexes. In every case, the longest-wavelength maximum shifts to longer wavelengths in pigment–proteins; sometimes the shift is more than 100 nm. We will examine the properties of these pigment–protein complexes in more detail in Chapters 5–7.

4.5 Carotenoids

Carotenoids are found in all known native photosynthetic organisms, as well as in many nonphotosynthetic organisms (Britton *et al.*, 1998; Frank *et al.*, 2000; Polívka and Frank, 2010). There are many hundreds of chemically distinct carotenoids, so we will not give a comprehensive list. However, there are some consistent structural features that are common to most photosynthetic carotenoids. They are extended molecules with a delocalized π electron system. Carotenoids from oxygenic organisms usually contain ring structures at each end, and most carotenoids contain oxygen atoms, usually as part of hydroxyl or epoxide groups. Structures of several of the carotenoids found in photosynthetic systems are shown in Fig. 4.11.

Carotenoid biosynthesis consists of the building up of large molecules from a basic building block, the five-carbon branched-chain species isoprene (Britton *et al.*, 1998). It is successively condensed into 10-, 20-, and 40-carbon molecules, ending with the compound phytoene. Phytoene is a hydrocarbon consisting of eight isoprene units attached in a linear fashion. It is colorless, because most of the double bonds are isolated. The second stage in the biosynthesis consists of successive desaturation

Table 4.2   Spectroscopic properties of chlorophylls and bacteriochlorophylls *in vitro*[a]

Pigment	λ_{max} (nm)	ε_{max} (mM^{-1} cm^{-1})[b]	τ_f (ns)	ϕ_f
Chlorophyll *a*	662, 578, 430	90.0	6.3	0.35
Chlorophyll *b*	644, 549, 455	56.2	3.2	0.15
Chlorophyll c_1	640, 593, 462	35.0	6.3	
Chlorophyll *d*	697, 456, 400	63.7	6.2	
Chlorophyll *f*	707, 440, 398	71.1		
Bacteriochlorophyll *a*	773, 577, 358	90.0	2.9	0.2
Bacteriochlorophyll *b*	791, 592, 372	106	2.4	
Bacteriochlorophyll *c*	659, 429	75	6.7	0.29
Bacteriochlorophyll *d*	651, 423	79	6.3	
Bacteriochlorophyll *e*	649, 462	49	2.9	
Bacteriochlorophyll *f*	645, 467		3.4	0.13
Bacteriochlorophyll *g*	762, 566, 365	76	2.7	

Source: Data from Scheer (1991) and Niedzwiedzki and Blankenship (2010).
[a] Most data taken from Scheer (1991) or Niedzwiedzki and Blankenship (2010). Solvents are not the same for all quantities.
[b] Values for ε_{max} are for the longest wavelength absorbing Q_y band.

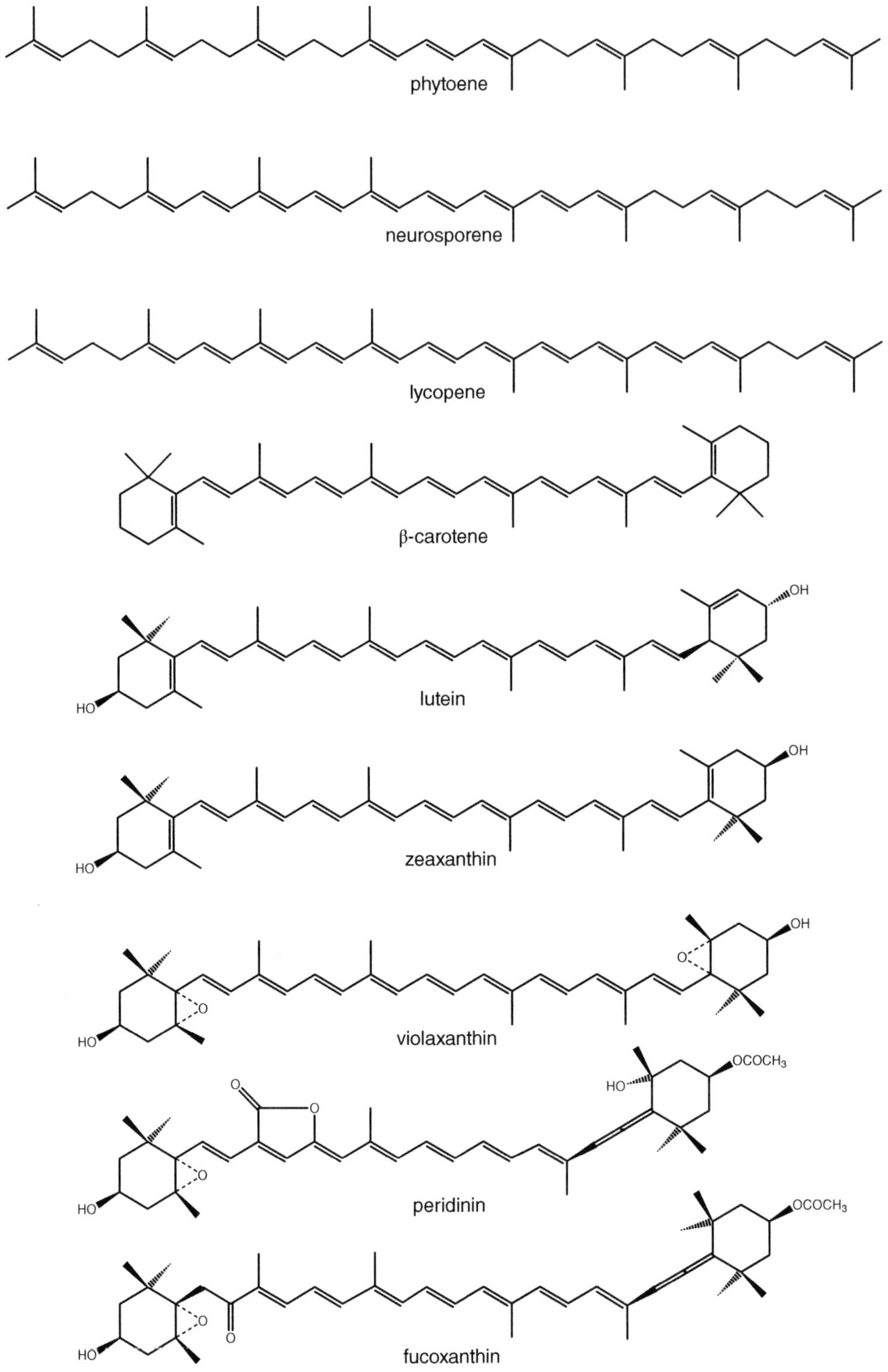

Figure 4.11   Structures of several carotenoids and carotenoid precursors important in photosynthetic systems.

steps, producing a series of intermediates with an increasing number of conjugated double bonds. This has the effect of shifting the absorption into the visible region. The end product of this stage is the compound lycopene, which is responsible for the red color of tomatoes. Some of the intermediates, such as neurosporene, are the end point in carotenoid biosynthesis for some anoxygenic photosynthetic bacteria. In most organisms, there are two additional stages of the biosynthetic pathway: cyclization of the ends of the molecule followed by derivatization by hydroxylation or any of a wide variety of other processes.

Carotenoids have several well-documented essential functions in photosynthetic systems. First, they are accessory pigments in the collection of light, absorbing light and transferring energy to a chlorophyll-type pigment. Most antenna complexes contain carotenoids. Second, carotenoids function in a process called **photoprotection**. Carotenoids rapidly quench triplet excited states of chlorophylls before they can react with oxygen to form the highly reactive and damaging excited singlet state of oxygen. They also quench the singlet oxygen if it is somehow formed. Finally, carotenoids have recently been shown to be involved in the regulation of energy transfer in antennas. These processes, which avoid overexcitation of the photosynthetic system by safely dissipating excess energy, have different mechanisms in different organisms. We will discuss them in more detail in Chapter 5.

Carotenoids have very unusual energetic and spectroscopic properties. They usually exhibit an intense absorption band, typically in the 400–500 nm range, giving them their characteristic orange color. However, this transition is from the ground state (S_0) to the second excited singlet state (S_2), instead of to the first excited singlet state (S_1). The transition from the ground state to the first excited singlet state is forbidden because of the symmetry of the carotenoid molecule. An energy level diagram typical of many carotenoids is shown in Fig. 4.12. The lifetime of S_2 is very short, usually relaxing to S_1 by internal conversion on a subpicosecond time scale. This short lifetime of S_2 means that fluorescence of the S_2 state is highly quenched

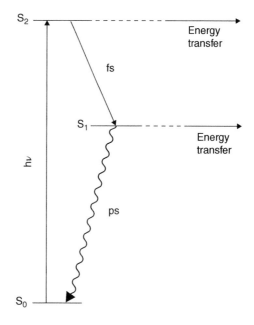

Figure 4.12   Energy Level diagram typical of carotenoids.

and is not observed in most situations. From the S_1 state, the excited carotenoid can relax to the ground state. However, this relaxation is also almost always nonradiative. The fluorescence decay rate constant of an excited state is related to the strength of the absorption that forms the excited state (Eq. A.68). If an absorption transition is extremely weak, such as the S_0 to S_1 carotenoid absorption, then the intrinsic fluorescence decay rate constant between these two states will be very small, and fluorescence will make a negligible contribution to the excited state decay (Eq. A.69). Internal conversion from S_1 to S_0 is typically very efficient, so the S_1 state has a picosecond lifetime. Carotenoids with nine or more double bonds have additional dark states in addition to the three energy levels discussed here (Ostroumov et al., 2013).

The energy of the S_1 state of the carotenoid is very difficult to measure directly, because of the forbidden nature of the S_0 to S_1 transition. One method that can be used to determine this energy is two-photon spectroscopy, in which two photons are absorbed simultaneously, with the sum of their energies equal to the transition energy. The S_0 to S_1 transition is allowed under these conditions (Krueger et al., 1999).

With both the S_2 and S_1 states having exceptionally short excited state lifetimes, it is perhaps surprising that carotenoids are able to carry out energy transfer to chlorophylls before they decay to S_0, releasing heat. Yet, in many cases, carotenoids are efficient antenna pigments, because the energy transfer process is even faster than the deactivation rate. We will explore some of the details of this energy transfer process in complexes where structural information is available in Chapter 5.

4.6   Bilins

Bilins are linear, open-chain tetrapyrrole pigments found in the light-harvesting antenna complexes known as phycobilisomes, which absorb in the spectral region from 550 to 650 nm. Phycobilisomes are well characterized structurally and spectroscopically and are one of the best understood of the various classes of antenna complexes. We will discuss them in more detail in Chapter 5.

The bilins resemble a porphyrin that has been split open and twisted into a linear conformation. Indeed, the bilins are actually formed from heme groups in just this fashion, as described below. The structures of the two most commonly found bilins, phycocyanobilin, and phycoerythrobilin, are shown in Fig. 4.13. The bilins are bound to proteins known as biliproteins. The three main classes of biliprotein antenna complexes are allophycocyanin, phycocyanin, and phycoerythrin. Bilins are the only class of photosynthetic pigments that are covalently attached to proteins. They are linked by thioether bonds to specific cysteine amino acid residues. In most cases, a single thioether linkage on ring A is found, although a dual linkage at both

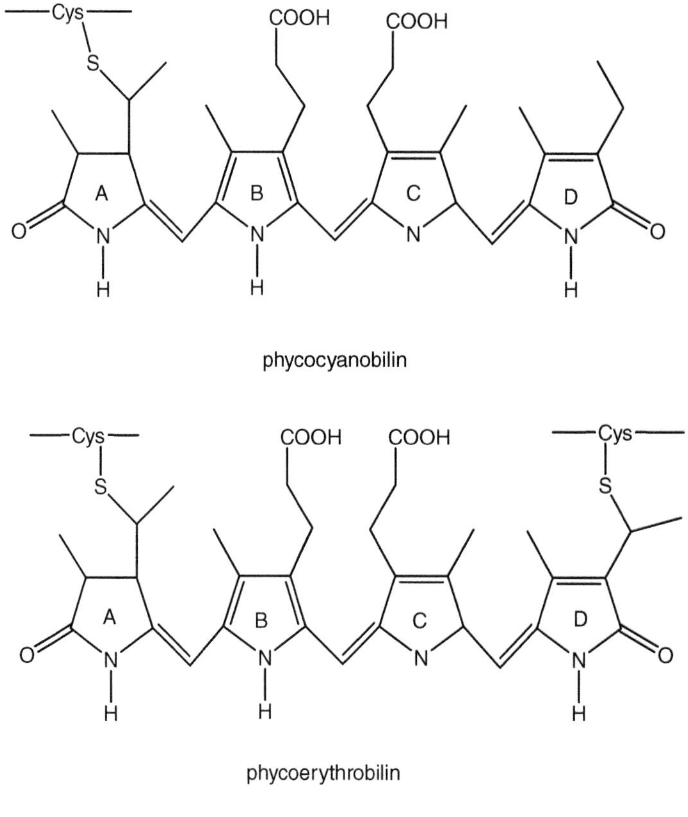

phycocyanobilin

phycoerythrobilin

Figure 4.13   Structures of two of the most common bilins: phycocyanobilin and phycoerythrobilin.

ring A and ring D is found on some pigments in phycoerythrin (MacColl, 1998).

The open-chain tetrapyrrole bilin chromophores are made by a surprisingly complex pathway (Bryant *et al.*, 2020). First, the protoporphyrin IX molecule is synthesized, as described above in the description of chlorophyll biosynthesis. This molecule is converted into a heme by insertion of Fe. The heme is then split open by the action of the enzyme heme oxygenase. Heme oxygenase requires both O_2 and NADPH as substrates, producing the molecule biliverdin, which is subsequently reduced, isomerized, and finally ligated to the apoprotein.

References

Beale, S. I. (1999) Enzymes of chlorophyll biosynthesis. *Photosynthesis Research* 60: 43–73.

Blankenship, R. E., Tiede, D. M., Barber, J., Brudvig, G. W., Fleming, G., Ghirardi, M., Gunner, M. R., Junge, W., Kramer, D. M., Melis, A., Moore, T. A., Moser, C. C., Nocera, D. G., Nozik, A. J., Ort, D. R., Parson, W. W., Prince, R. C., and Sayre, R. T. (2011) Comparing the efficiency of photosynthesis with photovoltaic devices and recognizing opportunities for improvement. *Science* 332: 805–809.

Britton, G., Liaaen-Jensen, S., and Pfander, H., (eds.) (1998) *Carotenoids*, Vol. III. Basel: Birkhäuser-Verlag.

Bryant, D. A., Hunter, C. N., and Warren, M. J. (2020) Biosynthesis of the modified tetrapyrroles—The pigments of life. *Journal of Biological Chemistry* 295: 6888–6925.

Chen, M. (2014) Chlorophyll modifications and their spectral extension in oxygenic photosynthesis. *Annual Review of Biochemistry* 83: 317–340.

Chen, M. and Blankenship, R. E. (2011) Expanding the solar spectrum used by photosynthesis. *Trends in Plant Science* 16: 427–431.

Chen, M., Schliep, M., Willows, R. D., Cai, Z.-L., Brett, A., Neilan, B. A., and Hugo Scheer, H. (2010) A red-shifted chlorophyll. *Science* 329:1318–1319.

Chen, M., Li, Y., Birch, D., and Willows, R. D. (2012) A cyanobacterium that contains chlorophyll *f* – A red-absorbing photopigment. *FEBS Letters* 586: 3249–3254.

Chew, A. G. M. and Bryant, D. A. (2007) Chlorophyll biosynthesis in bacteria: The origins of structural and functional diversity. *Annual Review of Microbiology* 61: 113–129.

Dong, C. S., Zhang, W.-L., Wang, Q., Li, Y.-S., Wang, X., Zhang, M., and Liu, L. (2020) Crystal structures of cyanobacterial light-dependent protochlorophyllide oxidoreductase. *Proceedings of the National Academy of Sciences USA* 117: 8455–8461.

Frank, H., Young, A. J., Britton, G., and Cogdell, R. J., (eds.) (2000) *The Photochemistry of Carotenoids: Applications in Biology*. Dordrecht: Kluwer Academic Publishers.

Gan, F., Zhang, S., Rockwell, N. C., Martin, S. S., Lagarias, J. C., and Bryant, D. A. (2014) Extensive remodeling of a cyanobacterial photosynthetic apparatus in far-red light. *Science* 345: 1312–1317.

Gouterman, M. (1961) Spectra of porphyrins. *Journal of Molecular Spectroscopy* 6: 138–163.

Grimm, B., Porra, R. J., Rüdiger, W., and Scheer, H., (eds.) (2006) *Chlorophylls and Bacteriochlorophylls: Biochemistry, Biophysics, Functions and Applications*. Dordrecht: Springer.

Kee, H. L., Kirmaier, C., Tang, Q., Diers, J. R., Muthiah, C., Taniguchi, M., Laha, J. K., Ptaszek, M., Lindsey, J. S., Bocian, D. F., and Holten, D. (2007) Effects of substituents on synthetic analogs of chlorophylls. Part 2: Redox properties, optical spectra and electronic structure. *Photochemistry and Photobiology* 83: 1125–1143.

Krueger, B. P., Yom, J., Walla, P. J., and Fleming, G. R. (1999) Observation of the S-1 state of spheroidene in LH2 by two-photon fluorescence excitation. *Chemical Physics Letters* 310: 57–64.

Larkum, A. W. D. and Kuhl, M. (2005) Chlorophyll *d*: The puzzle resolved. *Trends in Plant Science* 10: 355–357.

MacColl, R. (1998) Cyanobacterial phycobilisomes. *Journal of Structural Biology* 124: 311–334.

Masuda, T. (2008) Recent overview of the Mg branch of the tetrapyrrole biosynthesis leading to chlorophylls. *Photosynthesis Research* 96: 121–143.

Niedzwiedzki, D. M. and Blankenship, R. E. (2010) Singlet and triplet excited state properties of natural chlorophylls and bacteriochlorophylls. *Photosynthesis Research* 106: 227–238.

Ostroumov, E. E., Mulvaney, R. M., Cogdell, R. J., and Scholes, G. D. (2013) Broadband 2D electronic spectroscopy reveals a carotenoid dark state in purple bacteria. *Science* 340: 52–56.

Ouchane, S., Steunou, A. S., Picaud, M., and Astier, C. (2004) Aerobic and anaerobic Mg-protoporphyrin monomethyl ester cyclases in purple bacteria – A strategy adopted to bypass the repressive oxygen control system. *Journal of Biological Chemistry* 279: 6385–6394.

Polívka, T. and Frank, H. (2010) Molecular factors controlling photosynthetic light harvesting by carotenoids. *Accounts of Chemical Research* 43: 1125–1134.

Raymond, J. and Blankenship, R. E. (2004) Biosynthetic pathways, gene replacement and the antiquity of life. *Geobiology* 2: 199–203.

Scheer, H., (ed.) (1991) *Chlorophylls*. Boca Raton, FL: CRC Press.

Suzuki, J. Y., Bollivar, D. W., and Bauer, C. E. (1997) Genetic analysis of chlorophyll biosynthesis. *Annual Review of Genetics* 31: 61–89.

Tanaka, R. and Tanaka, A. (2007) Tetrapyrrole biosynthesis in higher plants. *Annual Review of Plant Biology* 58: 321–346

Taniguchi, M. and Lindsey, J. S. (2021) Absorption and fluorescence spectral database of chlorophylls and analogs. *Photochemistry and Photobiology*. 97: 136–165.

Tsukatani, Y., Romberger, S. P., Golbeck, J. H., and Bryant, D. A. (2012) Isolation and characterization of homodimeric Type-I reaction center complex from *Candidatus* Chloracidobacterium thermophilum, an aerobic chlorophototroph. *Journal of Biological Chemistry* 387: 5720–5732.

Vogl, K., Tank, M., Orf, G. S., Blankenship, R. E., and Bryant, D. A. (2012) Bacteriochlorophyll *f*: Properties of chlorosomes containing the forbidden chlorophyll. *Frontiers in Microbiology* 3: 1–12.

Wakao, N., Yokoi, N., Isoyama, N., Hiraishi, A., Shimada, K., Kobayashi, M., Kise, H., Iwaki, M., Itoh, S., Takaichi, S., and Sakurai, Y. (1996) Discovery of natural photosynthesis using zinc-containing bacteriochlorophyll in an aerobic bacterium *Acidiphilium rubrum*. *Plant and Cell Physiology* 37: 889–893.

Chapter 5

Antenna complexes and energy transfer processes

5.1 General concepts of antennas and a bit of history

All chlorophyll-based photosynthetic organisms contain light-gathering antenna systems (Green and Parson, 2003; Ruban, 2013; Croce *et al.*, 2018; Bryant and Canniffe, 2018; Croce and van Amerongen, 2020). These systems function to absorb light and transfer the energy in the light to a trap, which quenches or deactivates the excited state. In most cases, the trap is the reaction center itself, and the excited state is quenched by photochemistry with energy storage. In some cases, however, the quenching is by some other process, such as fluorescence or internal conversion.

In contrast to the reaction center complexes, which we will discuss in detail in Chapters 6 and 7, antenna complexes are remarkably diverse, strongly suggesting that they have been invented multiple times during the course of evolution to adapt organisms to widely varied photic environments. In this chapter, we will first explore some general concepts of antenna systems and then examine some of the classes of antennas in more detail, with an emphasis on the structure of antennas and the dynamics of energy transfer and trapping. We will also consider the essential process of regulation of antenna systems.

One of the first hints that pointed the way to the concept of antennas came from the 1932 Emerson and Arnold experiments discussed in Chapter 3. Emerson and Arnold found that one O_2 molecule is produced for only about every 2500 chlorophyll molecules after a short saturating flash of light. However, it was not at all obvious at the time how this result should be interpreted. Several possible explanations were proposed. One was that each chlorophyll molecule indeed carried out photochemistry, but that the product was unstable and was lost if it wasn't quickly processed by a "photoenzyme" that was present in such small amounts and worked so slowly that it could only process the products of one chlorophyll out of 2500 when all were excited with a bright flash. A variant on this idea involved a mobile, but sluggish, photoenzyme, which slowly makes the rounds of chlorophyll molecules, much as a hummingbird sips nectar from hundreds of flowers. The third possibility was proposed by Gaffron and Wohl (1936), who imagined that the energy was transferred from one pigment to another rather than by movement of the products of photochemistry or the processing enzyme.

This **photosynthetic** unit consisted of a collection of many pigments, among which the excitation energy could fluctuate before being trapped or stabilized. However, no physical mechanism was known at the time that could cause this energy to transfer from one chlorophyll to another.

The idea of energy transfer in photosynthetic systems met with severe criticism from James Franck and Edward Teller (of H-bomb fame). In an influential paper (Franck and Teller, 1938), they introduced many important concepts of energy transfer in solids, but ultimately concluded that a photosynthetic unit in which energy is transferred among chlorophylls is impossible. In their analysis, they assumed that the chlorophylls were arranged so that the energy had to diffuse along a linear, or one-dimensional, array of pigments. If this arrangement were correct, then indeed the concept of energy transfer in photosynthesis would not be feasible. One-dimensional diffusion is very inefficient, because many, many transfers are required to move the excitation from one point in the array to another. Of course, we now know that the antenna pigments are arranged in well-defined, three-dimensional structures, so that only a few energy transfer steps are required to connect any two pigments in the array. Figure 5.1 illustrates the difference between these two views. Knox (1996) has discussed the early history of photosynthetic antennas. The theoretical formulation for energy transfer that is now widely applied to photosynthetic systems was provided by Förster in the 1940s and is discussed in detail below.

5.2   Why antennas?

Why isn't every chlorophyll molecule capable of carrying out complete photosynthesis? This would seem to be a simple, efficient arrangement, avoiding much of the complexity that fills this book. In fact, some of the early mechanistic ideas of photosynthesis discussed in Chapter 3 were essentially a direct conversion of CO_2 and H_2O to carbohydrate and O_2, after a photon is absorbed by a chlorophyll.

The answer to the question of "Why antennas?" comes from a consideration of the intensity of sunlight and the economics of cellular processes. We will make a simple calculation to determine how often any individual chlorophyll absorbs a photon. The intensity of full sunlight in the photosynthetically active region (400–700 nm) is approximately $1800 \, \mu E \, m^{-2} \, s^{-1}$ where E is the unit Einstein, which is a mole of photons. This energy is distributed over the solar spectrum according to the curve shown in Fig. 1.1. Although it is an oversimplification, we can visualize the photon flux as a rain of particles on a surface. In order to determine how many photons a molecule absorbs, we need to know only the photon flux and the effective cross-sectional area of the pigment, which is the target size of the molecule being hit by the photons. This target size is not the physical size of the molecule, but rather the effective size, which includes factors such as the wavelength of excitation and how strongly the transition is allowed by quantum mechanics.

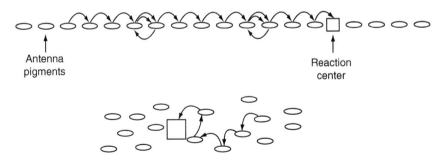

Figure 5.1   One-dimensional and three-dimensional antenna organization models. In the one-dimensional model, excitation must be transferred by many steps before encountering a trap where photochemistry takes place. In the three-dimensional model, the trap is always no more than a few energy transfer steps from any of the pigments in the antenna complex.

The photon flux, or intensity, I, is easy to calculate.

$$I = EN_A \qquad (5.1)$$

where I is the number of photons that strike a surface per unit area per second, E is the number of Einsteins per unit area per second, and N_A is Avogadro's number (6.022×10^{23} items mol^{-1}). Substituting the intensity of full sunlight, we obtain:

$$I = 1.1 \times 10^{21} \text{ photons m}^{-2}\text{s}^{-1} \qquad (5.2)$$

It is easier to visualize the photon flux on a much smaller area, such as an area the size of a typical molecule, with dimensions of Ångstroms. The flux on a surface of 1 Å^2 is:

$$I = 11 \text{ photons Å}^{-2}\text{s}^{-1} \qquad (5.3)$$

So, a chlorophyll molecule, which is roughly square with dimensions of ~10 Å per side, will be illuminated by approximately 1100 photons per second, although not all of these photons will be absorbed.

We next need to calculate the target size that a chlorophyll molecule presents to this incoming rain of photons, in order to determine how many photons will actually be absorbed. The target size is proportional to the strength of the absorption, which we can measure quantitatively by the extinction coefficient. However, the extinction coefficient is a function of wavelength, so we must average over all the absorption bands. For chlorophyll a, the extinction coefficient, ε, at the wavelength of maximum absorption is 90 mM^{-1} cm^{-1} (see Table 4.2). We will make a very rough estimate that the average extinction coefficient over the entire visible region is about 25 mM^{-1} cm^{-1} (see Fig. 4.6). Now, we must convert that ε value into a target size, σ, per molecule.

$$\sigma = 2303\varepsilon / N_A \qquad (5.4)$$

In Eq. (5.4), the factor 2303 arises from the fact that the molar extinction coefficient includes the factor 2.303 for converting from natural logs to base 10 logs (see Eq. A.77), plus a factor of 1000 cm^3 l^{-1}. When we substitute the ε value in Eq. (5.4), we obtain $\sigma = 9.6 \times 10^{-17}$ cm^2, or ~1 Å^2. The number of photon hits per molecule per second is then just $I\sigma$ photons s^{-1}.

$$Hits = I\sigma \qquad (5.5)$$

Substituting the values of I and σ calculated above gives us a final value of roughly 10 photons absorbed per chlorophyll per second. Remember that this is a maximum value at full sunlight and that under most conditions it will be much lower. Viewed in this way, sunlight is a rather dilute energy source.

The target concept is schematically illustrated in Fig. 5.2. This calculation is oversimplified, partly because photons are not so localized as the raindrop analogy suggests. A photon is a strange combination of particle and wave, with a physical size that is roughly similar to the wavelength of the photon, which in the visible region is much larger than a molecule. The photon interacts with a molecule and has

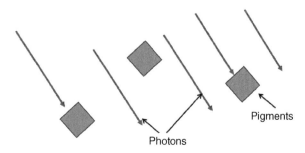

Figure 5.2   Target model of photon absorption. The molecule presents an effective cross section to the incoming photon flux, and the rate of absorption, given by Eq. (5.5), is the product of the two quantities.

its energy deposited in the molecule, whereupon the photon ceases to exist and the molecule is left in an excited state. This process is often called **molecular absorption**. The appendix contains additional information about the process of photon absorption.

We will now consider the cellular economics aspect of the analysis of antennas. We have just calculated that, even with full sunlight, there is approximately a tenth of a second between photons being absorbed by any given chlorophyll molecule, and it can easily be several orders of magnitude longer under most conditions. A tenth of a second is an eternity on the molecular scale. If every chlorophyll had associated with it the entire electron transfer chain and enzymatic complement needed to finish the job of photosynthesis, then these expensive components would sit idle most of the time, only occasionally springing into action when a photon is absorbed. This would obviously be wasteful, and ultimately, such an arrangement would be unworkable. It is as if a factory were to have a number of expensive manufacturing machines sitting idle most of the time while a key raw material is being brought in at a slow pace. It makes more sense to buy only a few expensive machines and somehow to improve the delivery system of raw materials. This is what antennas do for photosynthetic organisms. In addition, in all cases, photochemistry produces unstable initial products that will be lost if a second photochemical event does not take place in a relatively short time. If all pigments were functionally independent, most of these unstable products would be lost. All these factors demand that photons be collected and their energy delivered to and used in a central place. Because of these considerations, every known photosynthetic organism has an antenna of some sort associated with it. However, it appears that in many cases the size of the antenna system is larger than what might be expected to be the most efficient in terms of conversion of solar energy. Many organisms, especially those that live in full sunlight, must dissipate much of the energy that the antenna system collects or risk photodamage. We will return to this issue in later chapters, especially in the context of bioenergy storage.

5.3   Classes of antennas

A remarkable variety of antennas is found in various photosynthetic organisms. In fact, in many cases, the groups of photosynthetic organisms are largely defined by the types of antennas that they contain. There are a number of major classes of antennas, which show no apparent relation to each other in terms of structure or even types of pigments utilized. The different classes of antennas fall into some broad groupings, which are useful for organizing our discussion. We will first set out the general structural motifs that are found in the various classes of antennas and later will consider some representative antenna systems in more detail. Table 5.1 lists the categories of antenna complexes and gives examples of each class.

Almost all antenna complexes are pigment–proteins, in which the chlorophyll or other pigment is specifically associated with proteins in a unique structure. The only known exception to this rule is the chlorosome antenna complex found in the green photosynthetic bacteria, in which pigment–pigment interactions are of primary importance (see below).

Antenna complexes can be broadly divided into **integral membrane antenna complexes** and **extrinsic antenna complexes**. Integral membrane antennas contain proteins that cross the lipid bilayer. The pigments are often deeply buried in the membrane. In extrinsic antennas, the antenna complex is associated with components buried in the membrane, but does not itself span the membrane. It is always attached to one particular side of the photosynthetic membrane. Energy absorbed by the pigments in the extrinsic antenna complex is transferred into the integral membrane antenna complexes and eventually into the reaction center where photochemistry takes place. Examples of extrinsic antenna complexes are the phycobilisome complex found in cyanobacteria and the chlorosome complex found in green sulfur bacteria.

In almost every known case, antenna complexes are attached in some manner to the membrane that contains the reaction center to which the energy is

Table 5.1   Classes of photosynthetic antennas

Extrinsic antennas	Integral membrane antennas		
	Fused	Core	Peripheral
Phycobilisomes of cyanobacteria and red algae	Photosystem I RC	CP43 and CP47 complexes of Photosystem II	LHCII complexes of Photosystem II
Chlorosomes and FMO protein of green bacteria	Green sulfur bacterial RC	LH1 complexes of anoxygenic bacteria	LHCI complexes of Photosystem I
Soluble peridinin-chlorophyll proteins of dinoflagellates and phycobiliproteins of cryptophyte algae	Heliobacterial RC		LH2 complexes of purple bacteria

ultimately delivered. This is because the energy transfer process is intrinsically a short-range process, and if the antenna is physically distant from the reaction center, the efficiency of energy transfer will be very low. This is discussed below in the section on Förster transfer. There are two known cases where antenna complexes are not bound to the membrane. These are the peridinin-chlorophyll protein of dinoflagellate algae and the unusual phycobiliproteins of cryptophyte algae. In both cases, the proteins are soluble and located in the thylakoid lumen. While the pathway of energy transfer in these cases is not known in detail, it may be that these proteins are packed into the lumen at very high concentration so that the distance involved in energy transfer is not too long, despite the fact that there is not a well-defined structure for the energy transfer pathway.

The integral membrane antenna complexes are themselves quite diverse in terms of structure and relative position in the energy transfer sequence. Some of these antenna pigments and their corresponding antenna functions are actually built into the minimal reaction center structure. These pigments cannot be separated biochemically from the electron transfer components, because they are bound to the same polypeptides. We will call these systems **fused antenna/reaction center complexes**. The reasons for this name will become more apparent when we discuss the Photosystem I complex, the best understood example of such a fused system, in detail in Chapter 7.

The class of integral membrane antennas known as **core antenna complexes** are intimately associated with the reaction center, but can usually be separated biochemically from it if care is taken. They generally have a fixed pigment stoichiometry and a well-defined physical arrangement with respect to the reaction center complex itself. Examples are the LH1 complexes found in purple photosynthetic bacteria and the CP43 and CP47 antenna complexes that are part of Photosystem II.

The last group of integral membrane antenna complexes is called peripheral antennas. These complexes are always found in addition to the core, or fused, antennas, not in place of them. They are often present in variable amounts, depending on growth conditions, and may be mobile, in that their physical arrangement with the other antennas and reaction center complexes is not permanent. They are often involved in antenna regulatory processes. The LH2 complexes of purple bacteria and the LHCII complexes of Photosystem II are examples of peripheral antennas.

5.4   Physical principles of antenna function

In this section, we will consider some of the physical measurements and mechanisms that are important in studies of antennas. We will begin with some organizational concepts, then take up experimental aspects of how antennas are investigated and, finally, move on to the more theoretical description of the mechanisms of energy transfer and pigment interactions.

5.4.1 The funnel concept

The analogy of an energy **funnel** is a useful image for visualizing the energy collection in antenna systems. This is illustrated in Fig. 5.3. The more distal parts of the antenna system, often a peripheral antenna complex such as a phycobilisome, maximally absorb photons at shorter wavelengths than do the pigments in the antenna complexes that are proximal to the reaction center. According to the Planck relation $E = hc/\lambda$ (Eq. A.3), the excited states populated by short-wavelength photons are relatively high in energy. Subsequent energy transfer processes are from these high-energy pigments physically distant from the reaction center to lower-energy pigments that are physically closer to the reaction center. With each transfer, a small amount of energy is lost as heat, and the excitation is moved closer to the reaction center. The energy lost in each step provides a degree of irreversibility to the process, so the net result is that the excitation is "funneled" into the reaction center, where some of the energy in it is stored by photochemistry.

It is not absolutely necessary that each individual energy transfer event be downhill energetically. Thermal energy is always available to add to the excitation energy to increase it somewhat, so

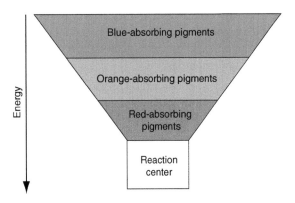

occasional transfers can be from lower-energy excited states to form higher-energy ones. However, the probability of these "uphill" energy transfers becomes exponentially smaller as the energy gap increases, so that, in the end, the excitations will be delivered from the higher-energy pigments to the lower-energy ones.

The funnel concept can be viewed as a mechanism in which a portion of the energy in the excitation is sacrificed as heat during a process in which the energy is delivered in a short time to the proximity of the reaction center trap. A large antenna system that is isoenergetic with the trap might absorb energy well, but much of it might be lost if the excitations had to "wander" around the antenna before eventually finding their way to the trap. For the funnel arrangement to work, there must be both a spatial and an energetic ordering of the antenna pigments, so that the shorter-wavelength absorbing ones are farthest from the trap and the longest-wavelength absorbing ones are nearest the trap. In many cases, this is clearly found to be the case.

In a slightly different form, the funnel concept also applies within an antenna complex. Accessory pigments, most commonly carotenoids, absorb intensely at short wavelengths. These accessory pigments give a broader coverage of the solar spectrum than is possible with just chlorophylls. Energy absorbed by these accessory pigments is rapidly transferred to chlorophylls, usually in the same antenna complex, so some of the energy absorbed by these pigments is available to drive photosynthesis.

The funnel concept seems to be most applicable in cases of extrinsic antenna complexes. For antenna complexes that directly interact with the reaction center, either the integral membrane core antennas or the fused antennas, the funnel model often breaks down. In these situations, the energy of some or all of the antenna pigments is sometimes actually lower than that of the trap, so an uphill energy transfer step is needed before the energy can be trapped. The functional significance of these low-energy antenna pigments is not yet clear.

Figure 5.3 The funnel concept in photosynthetic antennas. Sequential excitation transfers from higher-energy pigments (blue-absorbing) to lower-energy pigments (red-absorbing) deliver excitations to the proximity of the reaction center.

5.4.2   Fluorescence analysis of antenna organization

Measurement and analysis of fluorescence is one of the most powerful ways to probe photosynthetic systems. This fluorescence is from excited states that were lost before photochemistry took place (except delayed fluorescence, which arises from reversal of photochemistry). It usually represents a small fraction of the excited state decay in a functional photosynthetic complex. Nevertheless, the fluorescence is an extremely informative quantity, because it reports on the energy transfer and trapping. Both steady-state and time-resolved fluorescence measurements are widely used methods for probing the organization and functional state of photosynthetic systems. We will encounter many examples of how fluorescence is useful in understanding photosynthesis.

5.4.3   Fluorescence excitation spectra – direct evidence for energy transfer

The basic concept of photosynthetic antennas is that light absorbed by one pigment may subsequently be transferred to other pigments. A convenient way to monitor this energy transfer process is to irradiate a sample with light that is selectively absorbed by one set of pigments and then monitor fluorescence that originates from a different set of pigments. A plot of the intensity of fluorescence emission at a fixed wavelength versus the wavelength of excitation is called a **fluorescence excitation spectrum**. It is an action spectrum for fluorescence emission. If light is absorbed by one set of pigments and emitted by another set, energy transfer must have taken place between the two groups of pigments.

This type of fluorescence excitation experiment can also be used to measure quantitatively the efficiency of energy transfer from one set of pigments to another. For the sake of illustration, we will consider an idealized case in which pigment A

transfers energy to pigment B, which then fluoresces (Fig. 5.4). Pigment B will also fluoresce if it is directly excited. We monitor the fluorescence emission of pigment B at wavelength λ_B. The intensity of emission at λ_B is measured as the excitation wavelength is scanned. The fluorescence excitation spectrum is thereby recorded. The second part of the experiment involves simply measuring the absorption spectrum of the sample. Actually, the proper quantity to use in this comparison is not the absorption spectrum, but a related quantity, the $1 - T$ spectrum, where T is the transmission (Eq. A.72). The $1 - T$ spectrum is the intensity of light that is absorbed as a function of wavelength. For absorbance values of 0.1 or less, the absorption and $1 - T$ spectra have essentially the same shape.

Now, we need to compare the fluorescence excitation and absorption spectra in order to determine the energy transfer efficiency. But how can this be done, as fluorescence and absorption are two very different quantities? It seems as if we are comparing apples and oranges. The solution is to normalize (multiply by a factor that makes the spectra equal at a given wavelength) the absorption and fluorescence excitation spectra at the absorbance maximum of pigment B. For an isolated pigment, the fluorescence excitation and absorption spectra are superimposable, because the only way for the fluorescing excited state to be populated is by absorption of a photon. By normalizing, the efficiency of producing the fluorescence of pigment B is measured relative to the direct excitation of B. Once the spectra are normalized in this manner, their relative amplitudes give the energy transfer efficiency directly, as shown in Fig. 5.4. If pigment A doesn't transfer energy to pigment B, then fluorescence from B is not sensitized by absorption of A. If the transfer is 50% efficient, then the amplitude of the fluorescence excitation spectrum is 50% of the $1 - T$ spectrum. Figure 5.5 shows a measurement of energy transfer efficiency measured by this technique. In this example, the efficiency of transfer from bacteriochlorophyll c absorbing at 740 nm to bacteriochlorophyll a emitting at 900 nm was measured to be 70% (Mimuro et al., 1989).

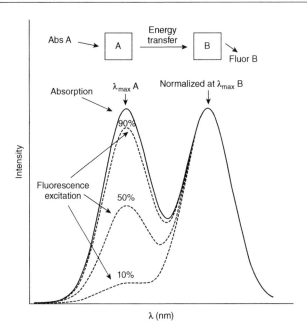

Figure 5.4   Energy transfer efficiency from fluorescence excitation measurements. A schematic picture of absorb-ing (a) and emitting (b) molecules, with energy transfer between them, is shown at the top. Fluorescence excita-tion spectrum compared with absorption spectrum, normalized at the maxima of each spectrum for pigment (b). The curves expected when energy transfer efficiencies from (a) to (b) are 10, 50, and 90% are shown.

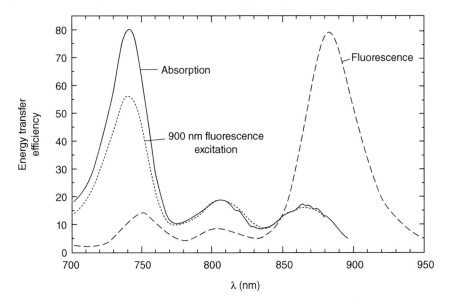

Figure 5.5   Measurement of energy transfer efficiency in whole cells of the green bacterium *Chloroflexus aurantia-cus*. Solid line: absorption spectrum; dotted line: fluorescence excitation spectrum, monitored at 900 nm; dashed line: fluorescence emission spectrum. The energy transfer efficiency of transfer from the bacteriochlorophyll *c* absorbing at 740 nm to the bacteriochlorophyll *a* emitting at 900 nm is 70%, while the transfer efficiency for bac-teriochlorophyll *a* absorbing at 808 nm is 100%. The curves were normalized at 870 nm. *Source:* Mimuro *et al.* (1989) (p. 7504)/American Chemical Society.

5.4.4   Förster theory of energy transfer

So far, we have discussed some general concepts of antennas and some techniques for measuring antenna organization and energy transfer. However, on a deeper level, we have not addressed the details of how energy is transferred. Now, we will discuss physical mechanisms for energy transfer.

The mechanism that is clearly applicable to weakly coupled pigments is the Förster mechanism, which was first proposed by Thomas Förster in the 1940s (Förster, 1965; Sener et al., 2011; Knox, 2012). This **Förster energy transfer** mechanism is a nonradiative resonance transfer process. It can be visualized in a manner similar to the transfer of energy between two tuning forks. Each tuning fork has a characteristic frequency. If one fork is struck, it begins to vibrate. In certain circumstances, energy may be transferred to another tuning fork. For this transfer to take place, the two forks must have some coupling between them. It also depends on their relative orientation and distance.

If two pigments are separated by more than several Ångstroms, and the transitions are allowed, the transfer between energy donor and energy acceptor occurs primarily via a Coulomb (dipole–dipole) mechanism, with rate constant given by Eq. (5.6):

$$k_e = k_f \left(R_0 / R \right)^6 \qquad (5.6)$$

where k_e is the first-order rate constant for energy transfer from the donor to the acceptor, k_f is the rate constant for fluorescence of the energy donor, R is the distance between energy donor and acceptor, and R_0 is the "critical distance" at which energy transfer is 50% efficient. R_0 is given by Eq. (5.7) (in units of Å⁶) (Cantor and Schimmel, 1980):

$$R_0^6 = 8.79 \times 10^{-5} J \kappa^2 n^{-4} \text{Å} \qquad (5.7)$$

In Eq. (5.7), J is an energy overlap factor given by Eq. (5.8), n is the refractive index, and κ^2 is an orientation factor, defined by Eq. (5.9) below.

$$J = \int \varepsilon\left(\lambda\right) F_D\left(\lambda\right) \lambda^4 d\lambda \qquad (5.8)$$

In Eq. (5.8), $\varepsilon(\lambda)$ is the molar extinction coefficient of the energy acceptor on a wavelength scale,

and $F_D(\lambda)$ is the normalized emission spectrum of the energy donor. The overlap parameter J is illustrated schematically in Fig. 5.6. The physical basis underlying the overlap parameter is that the donor and acceptor molecules must have an energy state in common, because when the excitation hops from donor to acceptor, conservation of energy requires that the total energy of the system just before the transfer is the same as the total energy just after the transfer. This can be so only if the two molecules have a common energy state and therefore spectral transitions at the same wavelength.

The requirement for overlap of the fluorescence emission spectrum of the donor and the absorption spectrum of the acceptor sometimes leads to the mistaken impression that the Förster energy transfer process proceeds by emission of a photon by the donor followed by absorption of a photon by the acceptor. This is not the case. The Förster transfer is a nonradiative process, which means that no photon emission or absorption is involved. The energy transfer process becomes one of the many possible decay processes from the excited state (see Appendix). If, as is often the case, it is the dominant decay process, the excited state lifetime of the energy donor is dramatically shortened compared with what it would be in the absence of energy transfer.

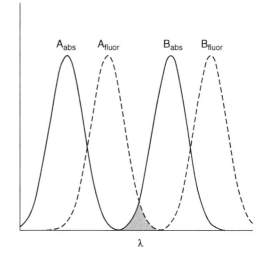

Figure 5.6   Schematic picture of the overlap factor J in the Förster theory. The shaded area indicates the energy levels that the donor and acceptor have in common.

The radiative emission–reabsorption process described above would have no effect on the donor's excited state lifetime.

The orientation factor κ is defined by:

$$\kappa^2 = \left(\cos\alpha - 3\cos\beta_1 \cos\beta_2\right)^2 \tag{5.9}$$

where α is the angle between the two transition dipoles and the βs are the angles between each dipole and the line joining them. The values of κ^2 can range from 0 to 4. For a random orientation of the transition dipoles, the average value of κ^2 is 2/3.

R_0 depends on the properties of both the energy donor and the acceptor, so both must be specified in order to be able to quote a value. Table 5.2 lists typical values of R_0 between chlorophylls and bacteriochlorophylls (assuming $\kappa^2 = 1$) (van Grondelle, 1985). Excitation transfer is intrinsically a longer-range process than is electron transfer (see Chapter 6). This is because of the nature of the dipole–dipole Coulomb coupling effect, which does not require direct overlap of the molecular wave functions as in electron transfer. In fact, the structures of antenna complexes have undoubtedly been fine-tuned by evolution to minimize excited state electron transfer processes. This is accomplished by separating the pigments to a distance that is too large to permit rapid electron transfer, while at the same time keeping them close enough to efficiently transfer energy and ultimately deliver it to the reaction center.

If the pigments are very close to each other, then the simple Förster picture described above breaks down. Electron exchange terms, in which two electrons in different pigments swap places, can also contribute to the energy transfer process. This is thought to be the case for photoprotective quenching of chlorophyll triplet states by carotenoids, where forbidden transitions are involved. In this case, the Coulomb terms are near zero. However, exchange coupling is not usually thought to be important in antenna function.

A number of different theoretical approaches, such as Redfield dynamics, have been developed to describe both the energetics of interacting pigment systems and the dynamics of energy transfer both within antenna complexes and also in the larger systems, including multiple types of antenna and trapping by reaction centers (Cheng and Fleming, 2009; Sener et al., 2011). Experimental data that have been used in the development and testing of these theoretical models include a wide range of structural studies, steady-state and time-resolved fluorescence spectroscopy, transient absorption methods including new types of two-dimensional spectroscopy, and spectroscopic studies on single complexes. These single complex methods avoid the averaging effects that are inherent when a collection of complexes are studied (Cogdell and Köhler, 2009).

5.4.5   Exciton coupling

When pigments are physically very close, which for chlorophylls is often less than 10 Å, the interaction between them is manifested in a qualitatively different manner, known as **exciton coupling** (van Amerongen et al., 2000). Here, we will consider only a dimer of interacting pigments, but the treatment can be extended to larger numbers of interacting pigments. The absorption spectra of the pigments are split and usually a circular dichroism (CD) spectrum is observed. The energy levels of the monomer and the exciton-split dimer are shown in Fig. 5.7a. The magnitude of the splitting and the intensity of the two transitions depend on the distance of the pigments, as well as the relative orientation of the transition dipole moments, as shown for some examples in Fig. 5.7b. The absorption splitting is often too small to be resolved in the

Table 5.2   Förster R_0 values for photosynthetic pigments

Energy donor	Energy acceptor	R_0 (Å)
Chl b	Chl a	100
Chl a	Chl a	80–90
β-Carotene	Chl a	50
B800 in LH2	B850 in LH2	66
Phycoerythrin	Phycocyanin	60
Phycocyanin	Allophycocyanin	64

Source: Data from van Grondelle (1985).

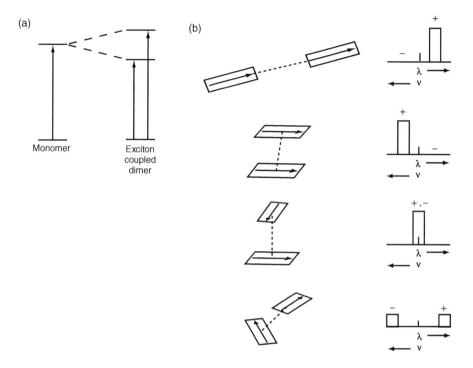

Figure 5.7   (a) Energy level diagram of a monomer and an exciton-split dimer. (b) Absorption spectra of two excitonically coupled pigments at various orientations.

absorption spectrum, but it can be observed in the CD spectrum, where the two exciton bands have opposite signs but equal magnitude. This characteristic derivative-shaped CD spectrum with equal and opposite positive and negative features is called a **conservative spectrum**.

The exciton-coupled dimer (or higher oligomer) is most productively viewed as a supermolecule with delocalized electronic transitions, rather than a collection of individual molecules with localized transitions. The molecular wave functions for this supermolecule are given by Eq. (5.10):

$$\Psi^+ = \left(1/\sqrt{2}\right)\left(\Psi_1 + \Psi_2\right)$$

$$\Psi^- = \left(1/\sqrt{2}\right)\left(\Psi_1 - \Psi_2\right) \qquad (5.10)$$

where Ψ^+ and Ψ^- are the wave functions of the excited dimer, and Ψ_1 and Ψ_2 are the wave functions for the two monomeric pigments.

The energies of the two exciton states are given by Eq. (5.11):

$$E^+ = \left(1/2\right)\left(E_1 + E_2\right) + V_{12}$$

$$E^- = \left(1/2\right)\left(E_1 + E_2\right) - V_{12} \qquad (5.11)$$

where E^+ and E^- are the energies of the exciton states, E_1 and E_2 are the energies of the monomeric pigments, and V_{12} is the coupling energy between them (Fig. 5.7).

5.4.6   Relation of Förster transfer to exciton coupling

We have two pictures of the interactions of pigments: the Förster picture, applicable at long distances and weak interactions, and the exciton picture, applicable at short distances and strong interactions. Can one make a smooth transition between these two seemingly contradictory views of pigment–pigment interactions? A more sophisticated treatment reveals that the two viewpoints are really just two sides of the same coin and that no fundamental difference exists (Knox and Gülen, 1993). However, the coupling manifests itself in somewhat different ways in the two extreme cases. Both Förster and exciton views are essential to understand photosynthetic antenna complexes, as will be illustrated with specific examples below. Real antenna systems are rarely described accurately by either the simple Förster or exciton pictures. As is often the case, the truth is somewhere in between these two extremes.

5.4.7   Quantum coherence effects in antenna systems

There is another level of complexity in the process of energy transfer whose importance is not yet fully understood: the phenomenon of quantum coherence effects (Engel *et al.*, 2007; Chenu and Scholes, 2015; Cao *et al.*, 2020). Energy transfer from one pigment to another is inherently a quantum mechanical process involving wave functions and is therefore not completely described using classical analogies such as the concept of the excitation hopping from one molecule to another as is implied by the Förster theory. A complete description of the process involves not just the probabilities for a system to be in a state but also the amplitude factors, which can interfere either constructively or destructively to either enhance or suppress the energy transfer process. This is analogous to the classical quantum two-slit experiment where electrons or photons are incident on a screen with two slits and the spatial distribution of transmitted electrons or photons is recorded on the other side of the slit. While our common sense tells us that the electron or photon has to go through one or the other slit to get from one side of the screen to the other, the measured distribution tells us that the electron or photon acts as though it goes through both slits simultaneously and that the spatial distribution is described by the interference effects between the two paths. The analogous situation in energy transfer is that the photon may have been absorbed by one pigment and the excitation ended up localized on

another pigment, but in between the path cannot be described as the energy hopping from one pigment to another. In a certain sense, the excitation follows multiple paths simultaneously and the final result depends on quantum interference effects among the various pathways. This result is difficult to reconcile with our everyday experience, but there is nevertheless some evidence that we must consider these effects to gain a deep understanding of photosynthetic energy transfer processes. This topic has been controversial and it now seems likely that some or all of the presumed quantum coherence effects may be due to vibrational effects rather than electronic coherences (Cao *et al.*, 2020).

5.4.8   Trapping models – coupling of antennas to a reaction center

A critical step in the energy storage process is the coupling of the reaction center to the antenna. Several distinct models are possible. Does photochemistry irreversibly quench the excited state each time it is transferred into the reaction center, or do excitations visit the reaction center multiple times, hopping in and out many times before photochemistry finally quenches the excited state? The first of these two extremes is often called a deep trap, while the second is known as a shallow trap. A third possibility is that the trap is so shallow that even after photochemistry has quenched the excited state, reverse electron transfer may take place and recreate the excited state of the reaction center chlorophyll, and the excited state may escape back to the antenna system. Interestingly, each of these three limiting cases appears to apply to particular photosynthetic systems. The understanding of how reaction centers and antennas are coupled has taken a long time to develop and has required input from structural studies, kinetic spectroscopic studies, and theory. In some systems, the processes are relatively well understood, whereas in others, we are still at a very early stage in deciphering how they work.

The excited state lifetimes of most integral membrane photosynthetic antennas connected to reaction centers are on the order of a few tens to hundreds of picoseconds. These are to be compared with lifetimes of up to several nanoseconds for isolated antenna complexes. However, lifetime measurement alone does not reveal the nature of the trapping processes. In the early days of considering this issue, before detailed ultrafast kinetic spectroscopic and structural information became available, it was often assumed that the excitation moved toward the reaction center by means of a series of steps from one pigment to another, with each step taking several picoseconds. The overall lifetime of the excited state was viewed as a **diffusion-limited** process, with the excitation moving progressively nearer the reaction center with each step until it is finally captured. Today, this view is not considered to apply to any systems involving reaction centers and core, or fused, antennas, although it probably does apply to some peripheral antennas, such as phycobilisomes. The current picture is based on a much more rapid excitation movement from one pigment to another within the antenna system, with interpigment transfer times typically in the subpicosecond time range.

If the probability of excitation energy escape from the trap is relatively small, so that once the excitation becomes localized in the core reaction center pigments it almost always leads to stable photochemistry, the system is described as having a deep trap. This happens when the rate constant for photochemistry is much larger than the rate constant for detrapping. The excitation spectrum for fluorescence and the observed excited state lifetime are then strongly dependent on the excitation wavelength. If the excitation is into the antenna system, then the final transfer process into the core reaction center dominates the excited state lifetime. If the excitation is into the core reaction center pigments, then photochemistry rapidly quenches the excitation, and the excited state lifetime largely reflects this process. The purple photosynthetic bacteria are reasonably well described by this model, which is known as **transfer-to-trap-limited** kinetics. A specific example of this behavior is discussed below.

If the trap is shallow, the probability of quenching of the excited state by photochemistry is much lower than the probability of escape of the excitation back into the antenna array. This happens when the rate constant for escape of the excitation back to the antenna is much larger than the intrinsic rate constant for photochemistry. The excitation

will then usually make multiple visits to the reaction center, hopping in and out until finally photochemistry terminates the process. This is known as **trap-limited** kinetics. In this case, the observed excited state lifetime reflects both the size and spectral distribution of the antenna system and the intrinsic rate constant for photochemistry when the excitation is localized in the reaction center. A system that is described by strictly trap-limited kinetics will have an excitation spectrum for photochemistry or fluorescence that is identical to the absorption spectrum and an excited state lifetime that is independent of excitation wavelength. This is because excitation in any pigment, even the pigments that are part of the core electron transfer chain, will result in a high probability of detrapping and repopulating of the antenna system, so it makes little difference which pigment is excited initially. Most current evidence suggests that Photosystem I works in this manner (van Grondelle et al., 1994).

The final case of trapping kinetics occurs when the trap is extremely shallow. Even after photochemistry finally traps the excitation, there is a reasonably large probability of this process being reversed by back electron transfer. This recreates the excited state of the core reaction center pigment, and the excitation is then detrapped back into the antenna system. This is known as **radical-pair-equilibrium**. The radical-pair-equilibrium model is really a special case of the trap-limited kinetics in which the photochemistry is more easily reversible. Photosystem II is thought to follow this behavior (Schatz et al., 1988; van Grondelle et al., 1994).

5.5 Structure and function of selected antenna complexes

We will now discuss the structure, and some functional aspects of selected antenna systems found in a number of different organisms (Mirkovic et al., 2017; Saer and Blankenship, 2017; Bryant and Canniffe, 2018). The systems have been chosen to represent the range of different types of antenna complexes found in various organisms, and the discussion is not intended as a comprehensive treatment.

5.5.1   Purple bacterial LH2 and LH1

The purple photosynthetic bacteria have an antenna system that has been studied extensively. In most organisms, it consists of two types of pigment–proteins known as light-harvesting 1 and 2 complexes, LH1, and LH2 (Cogdell et al., 2006). The LH1 complex is an integral membrane core antenna pigment–protein complex, which is found in fixed proportion to the reaction center and physically surrounding it. The LH2 complex is an integral membrane antenna complex of the type we have classed as a peripheral antenna, in that it is found in some, but not all, organisms. The quantity of LH2 is variable, depending on growth conditions, and it is not directly in contact with the reaction center complex.

The LH2 complex is one of the best-understood photosynthetic antenna complexes. The structure of LH2 has been determined by X-ray diffraction (Fig. 5.8) (McDermott et al., 1995), and it has been imaged in the membrane using atomic force microscopy (Bahatyrova et al., 2004). It has also been studied extensively using a variety of types of spectroscopic measurements and theory (Cogdell et al., 2006). LH2 is built up from minimal units consisting of a heterodimer of two protein subunits, known as the α and β peptides, along with three molecules of bacteriochlorophyll and one molecule of carotenoid. These subunits then aggregate into larger complexes, in which eight or nine subunits assemble into ring-shaped units of diameter ~65 Å.

Some of the bacteriochlorophyll pigments in LH2 are spectrally distinct from others. This is easily seen in the absorption spectrum of the complex, which exhibits two bands centered at 800 and 850 nm (Fig. 5.9). The structure of the complex provides a clear explanation for these two classes of pigments. The pigments that absorb at 800 nm form a ring with the plane of the bacteriochlorophyll molecules parallel to the plane of the membrane in

which the complex is embedded and are known as B800 pigments. The B800 pigments are rather weakly coupled to each other and are at a distance of approximately 21 Å from each other. Their spectral properties are largely consistent with them being isolated molecules.

The pigments that absorb at 850 nm are arranged quite differently. Each subunit complex has two molecules of bacteriochlorophyll arranged as a closely coupled dimer, with the plane of the B850 bacteriochlorophyll molecules approximately perpendicular to the plane of the membrane. These dimers become a band of 16 or 18 pigments when the subunit complexes assemble into the LH2 complex, and the absorbance is shifted to 850 nm by exciton and pigment–protein interactions, so that they are known as B850 pigments. This band of B850 pigments are all strongly exciton-coupled together, so excitations are effectively delocalized over much or all of the entire band, rather than being effectively localized in a single pigment for a short while and then hopping to another pigment, as is the case with the B800 pigments.

The carotenoid molecules have an extended conformation and lie generally perpendicular to the plane of the membrane, with close approach (3.4–3.7 Å) to both the B800 and the B850 pigments. The B850 pigments are located toward the periplasmic side of the membrane, which facilitates energy transfer to LH1, which also has its pigments toward that side of the membrane.

The LH2 complex is thus an excellent example of a case in which some of the pigments are weakly coupled together and can be described using standard Förster theory (the B800 pigments), while others (the B850 pigments) are best described using the exciton formalism.

The LH1 core antenna complex in the purple photosynthetic bacteria is a similar, but clearly distinct, antenna complex to LH2. LH1 surrounds the reaction center complex like a large doughnut, in which the reaction center is stuffed into the hole, forming what is called the RC-LH1 complex. The structure of the LH1 complex subunit is generally similar to that of the LH2 complex, but the aggregation number is larger, with approximately 16 α and

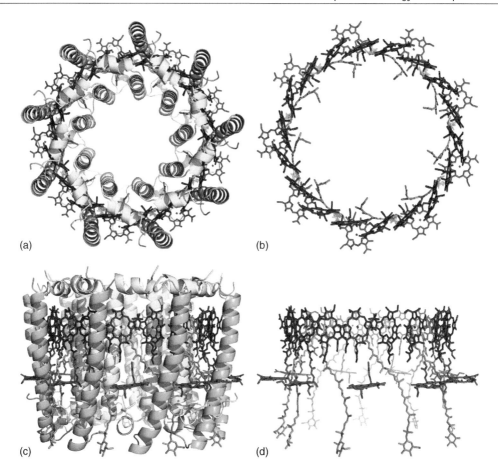

(a)

(b)

(c)

(d)

Figure 5.8   Structure of the LH2 complex in *Rhodopseudomonas acidophila*. (a) and (b), views perpendicular to the membrane plane with and without protein displayed. (c) and (d), views parallel to the membrane plane, with and without protein displayed. The phytyl tails of the bacteriochlorophylls have been omitted for clarity. The α polypeptide is colored yellow, and the β polypeptide is colored green. B800 bacteriochlorophylls are colored red, and B850 bacteriochlorophylls are colored blue. Carotenoids are colored orange. In a and b, the view is from the periplasmic side of the membrane. In (c) and (d), the periplasmic side of the membrane is at the top and the cytoplasmic side is at the bottom. Figure produced from Protein Data Bank file 1NKZ using PyMol.

β subunits combining to form the final LH1 complex. There are no pigments in LH1 that are analogous to the B800 pigments, and the absorption of the band of closely interacting pigments is further shifted to 875 nm; these are known as B875 pigments.

A subunit of LH1 consisting of an α and β peptide pair and two bacteriochlorophyll molecules has been reconstituted from purified components (Parkes-Loach *et al.*, 2004). It has an absorbance maximum at 820 nm, which shifts to 875 nm when

the subunits aggregate together to form the intact LH1 complex.

The structure of the RC-LH1 complex has been determined by X-ray crystallography and also visualized by atomic force microscopy and cryo-EM microscopy (Fig. 5.10) (Roszak *et al.*, 2003; Sturgis *et al.*, 2009; Qian *et al.*, 2013, 2018; Yu *et al.*, 2018). In some cases, the RC-LH1 complex is found as a dimer of two closely interacting RC-LH1 complexes, while in other cases it is monomeric. The LH1 ring around the reaction center is interrupted

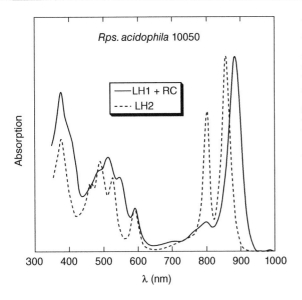

Figure 5.9 Absorption spectra of LH1 plus reaction center and LH2 antenna complexes from *Rhodopseudomonas acidophila* 10050. *Source:* Data courtesy of Richard Cogdell and Niall Fraser.

by a small protein that breaks the symmetry of the ring. In some organisms, the protein that occupies this position has been identified as the PufX gene product, while other organisms lack PufX but have another as yet unidentified protein that occupies the same position. The function of these proteins is thought to be part of a pathway that facilitates the diffusion of quinones in and out of the complex.

Figure 5.11 shows a representation of the purple bacterial photosynthetic unit, including the LH2 accessory antenna, along with the LH1 core antenna surrounding the reaction center of purple photosynthetic bacteria (Sturgis *et al.*, 2009).

5.5.2 Kinetics of energy transfer and trapping in purple bacteria

Figure 5.12 shows a schematic picture of the pigments and energy transfer times in the purple bacterial photosynthetic unit (Fleming and van Grondelle, 1997). After photon excitation in LH2, a series of ultrafast energy transfer processes take place

that deliver energy from LH2 first to LH1 and finally to the reaction center (Sundström *et al.*, 1999; Cogdell *et al.*, 2006). The transfer from the excited B800 to B850 takes approximately 1 ps, which is in excellent agreement with the predictions of Förster theory, given the distances and relative orientations of the pigments from the structure. The B850 pigments are so strongly coupled that the excitation is delocalized over the entire band of 18 B850 pigments in less than 100 fs (van Oijen *et al.*, 1999). The transfer from the B850 ring to the B875 pigments takes approximately 3 ps. The coupling among the pigments in the B875 ring of pigments is also very strong, producing a delocalized excited state. The slow step in the energy transfer sequence is the final transfer from the B875 pigments to the reaction center special pair. This process takes approximately 35–50 ps and is largely irreversible, so once the energy transfer to the core reaction center has taken place, the probability of escape back into the antenna system is relatively low, approximately 10–20%. This can be determined either by steady-state measurements or by direct time-resolved measurements upon selective excitation into the core reaction center. This system thus clearly exhibits transfer-to-trap-limited behavior as described above. The slow step from the B875 pigments to the core of the reaction center is due to the long distance (35–40 Å) between the B875 pigments and the reaction center pigments.

Carotenoids are also components of the bacterial LH1 and LH2 complexes. Excitation of the carotenoid is followed by ultrafast transfer to bacteriochlorophylls. Kinetic measurements in LH2 suggest that in some cases, the energy transfer to bacteriochlorophyll is from the S_2 excited state of the carotenoid, while in others it takes place from the S_1 excited state (Polivka and Frank, 2010).

5.5.3 The LHCII complex from plants and algae

One of the most important antenna complexes in terms of global productivity is the **light-harvesting complex II** (LHCII) found in plants and many algae. Despite the similar name, it is entirely distinct from

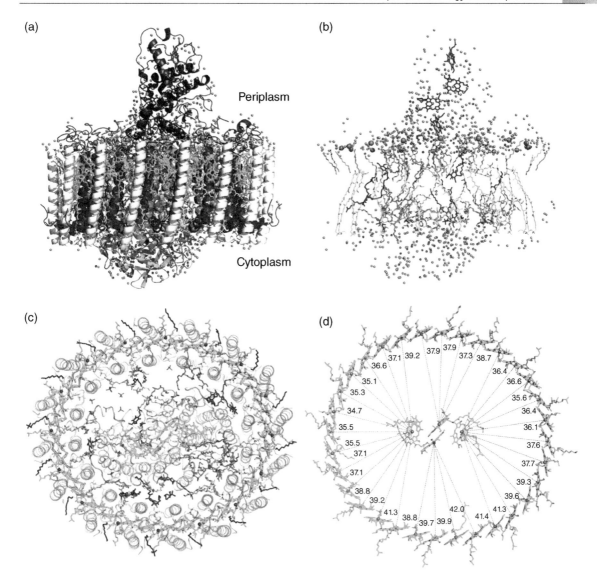

Figure 5.10   Architecture of LH1–RC from *Tch. tepidum* at a resolution of 1.9 Å. (a) View from the direction parallel to the membrane plane. LH1 α-subunit, blue; LH1 β-subunit, light cyan; C subunit, purple. (b) Arrangement of the cofactors and water molecules, with the same view as in (a). (c) Arrangement of the cofactors, viewing perpendicular to the membrane. Protein subunits are depicted in light grey. (d) Distances of the closest BChl pairs between RC and the surrounding LH1 ring. Color codes for cofactors: BChls, green; spirilloxanthin, yellow; Ca²⁺ ions, orange spheres; water molecules, pale pink dots. *Source:* Yu *et al.* (2018)/Springer Nature.

the LH2 complex in purple bacteria and should not be confused with it. The structure of the LHCII complex was first determined by a combination of electron microscopy and electron crystallography (Kühlbrandt *et al.*, 1994). Subsequently, the structure was determined using X-ray crystallography (Liu *et al.*, 2004). It is an integral-membrane antenna of the type we have called peripheral antennas. The structure of the complex is shown in Fig. 5.13. It consists of three transmembrane helices that coordinate

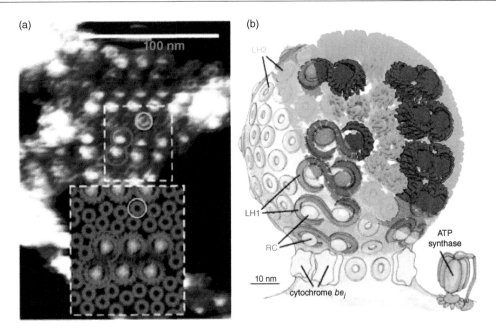

Figure 5.11   Structure of the photosynthetic membrane of *Rhodobacter sphaeroides*. (a) Atomic Force Microscopy structure of the photosynthetic membrane; (b) structural model of the photosynthetic membrane derived from these and other measurements. *Source:* Sturgis *et al.* (2009) (p. 3684). Reproduced with permission of the American Chemical Society.

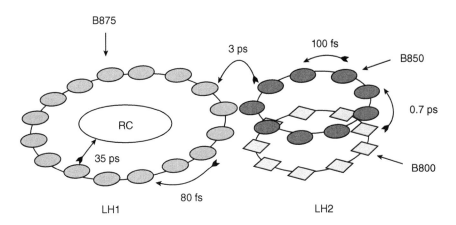

Figure 5.12   Schematic picture of energy transfer kinetics and pathways in the photosynthetic unit of purple bacteria. *Source:* Fleming and van Grondelle (1997) (p. 742)/Elsevier.

seven molecules of chlorophyll *a* and five of chlorophyll *b*. Two lutein carotenoid molecules arranged in an "X" pattern help to hold the complex together. LHCII is associated with Photosystem II and plays a major role in the regulation of the antenna system described below. The spatial arrangement of LHCII with respect to the Photosystem II reaction center is discussed in Chapter 7.

A large number of similar, yet distinct, proteins are found in the membrane of green algae and plants

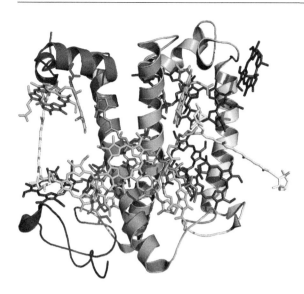

Figure 5.13 Structure of the LHCII antenna complex of pea. The monomeric complex is shown. The protein is color coded with blue at the N-terminal end to red at the C-terminal end. Chlorophyll *a* molecules are shown in green and chlorophyll *b* in blue. Luteins are shown in orange and other carotenoids shown in yellow. Figure produced from Protein Data Bank file 2HBW using PyMol.

(Grossman *et al.*, 1995; Green and Parson, 2003). These proteins are often called LHCI or LHCII if associated with Photosystem I or II, respectively. They were formerly called **cab proteins** (for **chlorophylls *a* and *b***). This nomenclature is not favored, because in some organisms similar proteins contain only chlorophyll *a*, while in others they contain both chlorophyll *a* and chlorophyll *c*. The proteins that make up these complexes are coded for by a large nuclear gene family, known as *lhc* genes. The *lhc* is followed by *a* if the protein is associated with Photosystem I and *b* if it is associated with Photosystem II, followed by a number that identifies the particular gene. For example, *lhca4* is a gene that codes for the Lhca4 protein, which is part of the LHCI complex. Many of the proteins are also often identified by the nomenclature CP (for chlorophyll protein) followed by a number, which is its apparent mass as determined by sodium dodecyl sulfate/polyacrylamide gel electrophoresis (SDS-PAGE); so CP29 is a protein that runs at 29 kDa apparent mass.

This latter nomenclature is complicated by the fact that the same protein may run at different apparent mass on different gel systems, leading to confusion.

5.5.4 Phycobilisomes

Phycobilisomes are large peripheral membrane antenna complexes found in cyanobacteria and red algae (Gantt and Conti, 1966; MacColl, 1998; Watanabe and Ikeuchi, 2013; Adir *et al.*, 2020). Several different types of phycobilisomes are found in various organisms, although the most widely studied type is known as a hemidiscoidal phycobilisome. These complexes consist of two or, more commonly, three types of pigment-proteins known as **biliproteins**, along with a number of additional proteins known as **linkers**. The biliproteins contain covalently linked bilin chromophores, which are attached via thioether linkages to cysteine residues in the proteins (Fig. 4.13). Some of the biliproteins are arranged into six rods, which attach in a fanlike arrangement to a biliprotein core that is attached to the stromal side of the thylakoid membrane, usually in close proximity to Photosystem II. A representation of the overall architecture of the hemidiscoidal phycobilisome found in cyanobacteria, as well as a detailed structural model, are shown in Fig. 5.14. The biliproteins are usually of three major types: phycoerythrin, phycocyanin, and allophycocyanin, which differ in protein identity, chromophore type and attachment, and relative location in the architecture of the phycobilisome complex. The cryo-EM structure of a 14.7 megadalton phycobilisome from the red alga *Porphyridium purpureum* is shown in Fig. 5.15 (Ma *et al.*, 2020). A cyanobacterial energy transfer megacomplex consisting of the phycobilisome, Photosystem I and Photosystem II has been isolated and characterized (Liu *et al.*, 2013).

Phycoerythrin is located on the tips of the rods and is the complex with the shortest wavelength absorption maximum. In the simplest form, it contains phycoerythrobilin chromophores, although in some organisms phycourobilin is also found. Phycocyanin is located in the middle portion of the rods. It contains phycocyanobilin as a chromophore.

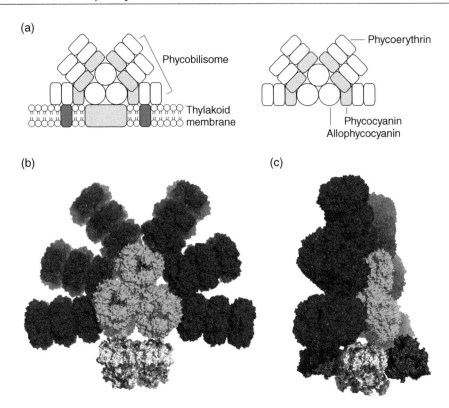

Figure 5.14   (Panel a) Schematic model of phycobilisome structure. *Source:* Reproduced with permission from MacColl (1998). Surface representation of a molecular model of a tricylindrical phycobilisome based on X-ray crystal structures of isolated components and mass spectrometry cross-linking data in combination with structural modeling. Phycocyanin (blue), allophycocyanin (green), and a linker protein (ApcC, orange), linker domain of ApcE (red) with ApcC subunit removed. The model is shown positioned on the stromal side of thylakoid membrane, (Panel b) above a Photosystem II (PDB 3WU2) rendered by surface electrostatic potential (PDB 3WU2), side view of tricylindrical core (cylinder perpendicular to this page), (Panel c) Megacomplex of PBS-PSII-PSI (Photosystem I, purple, PDB 1JB0). *Source:* Panel a: MacColl (1998) (p. 312)/Elsevier. Panels b and c: Courtesy of Dr. Haijun Liu.

Allophycocyanin forms the core of the phycobilisome and also contains phycocyanobilin as a chromophore. While the pigments are the same in phycocyanin and allophycocyanin, the different protein scaffolds impart very different absorption characteristics to the two proteins. All of the biliproteins have sequence and structural homology to each other and form a family of proteins. While the three proteins discussed here are found in a number of organisms that contain phycobilisomes, in many other cases phycobilisomes contain only two major classes of biliproteins: phycocyanin and allophycocyanin. A wide variety of different phycobiliproteins is found in various organisms. In some cases in certain eukaryotic algae, they are found in the thylakoid lumen instead of the stroma.

The structures of several isolated biliproteins are known at high resolution, exemplified by the structure of C-phycocyanin, shown in Fig. 5.16. It consists of a dimer of α and β peptides, which further aggregate to form a trimer of dimers. These trimers associate to form hexamers, which are the basic building blocks of phycobilisomes. These hexamers then associate with particular orientations and stoichiometries to form the complete phycobilisome.

The spectral and kinetic properties of phycobilisome complexes and their constituent phycobiliproteins have been studied extensively, using a wide

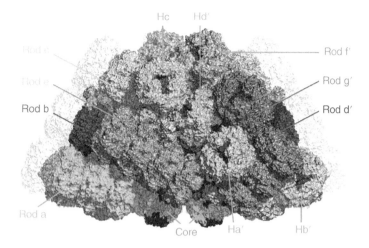

Figure 5.15   Structure of the phycobilisome from the red alga *Porphyridium purpureum*. *Source:* Ma *et al.* (2020) (p. 2)/Springer Nature.

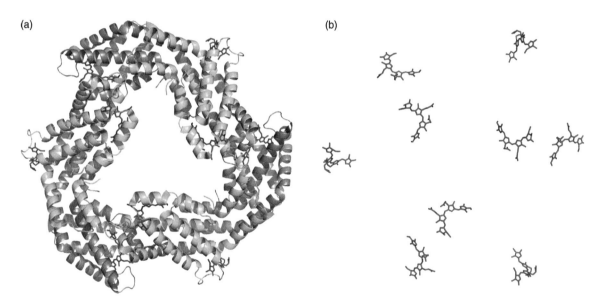

Figure 5.16   Structure of C-phycocyanin from the cyanobacterium *Thermosynechococcus vulcanus*. Panel (a) shows a trimeric complex made up of three α-β subunits. The α subunits are shown in blue and the β subunits are shown in green. The phycocyanobilin pigment cofactors are shown in red. Panel (b) shows the pigments in the same orientation. Figure produced from Protein Data Bank file 3O18 using PyMol.

range of techniques (van Grondelle *et al.*, 1994; MacColl, 1998). The phycobilisome represents a classic example of the funnel concept in photosynthetic antennas, in that the phycoerythrin at the distal ends of the rods absorbs at the shortest wavelength, while the phycocyanin is intermediate, and the allophycocyanin absorbs at the longest wavelength, or lowest energy. Energy transfer processes within the phycobilisome serve to direct the energy toward the membrane, where it is trapped by photochemistry. A second level of control is provided by the linker polypeptides, which tune the absorbance maxima of the biliproteins so as to facilitate the transfer process. Finally, some cyanobacteria can regulate the composition of their phycobilisomes in response to light quality by changing the expression levels of different phycobiliproteins, a phenomenon known as **chromatic acclimation** (Sanfilippo *et al.*, 2019).

Under conditions of Fe limitation, many cyanobacteria express an additional antenna complex, called IsiA (Iron stress induced) (Burnap *et al.*, 1993). The IsiA protein is similar in sequence and structure to the CP43 internal antenna complex from Photosystem II, but primarily is found associated with Photosystem I, where it forms a ring of 18 subunits surrounding a trimeric PSI complex (Bibby *et al.*, 2001; Boekema *et al.*, 2001; Toporik *et al.*, 2019).

5.5.5   Peridinin–chlorophyll protein

Dinoflagellate algae contain a unique peripheral antenna complex known as the peridinin–chlorophyll protein, or PCP. It contains chlorophyll *a* and the unusual carotenoid peridinin (Fig. 4.11).

The structure of the PCP consists of a protein that folds into two domains, each of which surrounds four peridinin molecules and a single chlorophyll (Hofmann *et al.*, 1996). The structure of the PCP is shown in Fig. 5.17. Spectroscopic measurements indicate that energy transfer from the peridinins to the chlorophyll is fast and efficient and that transfer among the chlorophylls is much slower (Bautista *et al.*, 1999). This is nicely explained by the structure, which indicates that the chlorophylls and carotenoids are in van der Waals contact with each other (are touching), while the chlorophyll–chlorophyll distances are 17 Å within the subunit and 40–55 Å between subunits, consistent with their slower energy transfer. The PCP complex transfers energy to an integral membrane chlorophyll *a/c* antenna complex, which, in turn, transfers it to the Photosystem II reaction center. The PCP is very unusual in that it is water-soluble and is thought to reside in the thylakoid lumen, without any specific attachment to the membrane. Usually, antenna complexes are physically associated with the membrane that contains the reaction center that energy is transferred to, because a separation of even a few tens of nanometers will greatly reduce the efficiency of energy transfer.

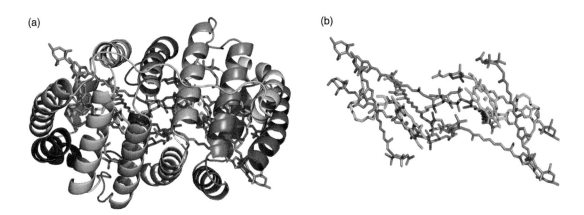

Figure 5.17   Structure of the peridinin–chlorophyll protein from the dinoflagellate *Amphidinium carterae*. (a) Structure of the monomeric PCP complex, including 8 peridinin and 2 chlorophyll *a* pigments. The protein is color-coded with blue at the N-terminal end to red at the C-terminal end. Chlorophylls are colored green and peridinins are colored orange. (b) View of the arrangement of the pigments in the monomeric complex, including two chlorophylls and eight peridinin pigments. Figure produced from Protein Data Bank file 1PPR using PyMol.

5.5.6   Chlorosome and FMO protein

Some photosynthetic bacteria contain a large antenna complex known as a chlorosome (Cohen-Bazire et al., 1964; Olson, 1998; Blankenship and Matsuura, 2003; Oostergetel et al., 2010; Orf and Blankenship, 2013). Three groups of bacteria contain chlorosomes: the green sulfur bacteria, the filamentous anoxygenic phototrophs, and the chloroacidobacteria. Many of these organisms live in conditions of extremely low light levels, and this system is well adapted to energy collection under these conditions.

The chlorosome contains up to 200,000 molecules of bacteriochlorophyll c, d, or e (Fig. 4.4) and is attached to the cytoplasmic side of the inner cell membrane by a baseplate complex that serves as an interface (Montaño et al., 2003). The bacteriochlorophyll c, d, or e pigments in the chlorosome are organized into large oligomers with relatively little protein involvement. Direct pigment–pigment interactions are responsible for the oligomer formation. While these systems are probably not suitable for study by X-ray crystallography due to their disordered nature, the structure of the pigment oligomers has been studied using a variety of other techniques including solid-state NMR and X-ray scattering (Pšenčík et al., 2004; Ganapathy et al., 2009). The pigments in these oligomers are very strongly exciton-coupled, and as a result, their absorption is shifted far to the red. Schematic models of the chlorosome complexes from green sulfur bacteria and FAPs are shown in Fig. 5.18.

In addition to the chlorosome, the green sulfur bacteria and the chloroacidobacteria, but not the FAPs, contain another peripheral antenna complex that has been studied very extensively. It is called the bacteriochlorophyll a protein, or often the FMO protein, after Roger Fenna and Brian Matthews, who first determined its structure, and John Olson, who discovered the protein. The FMO protein is sandwiched between the chlorosome and the reaction center. Its structure, shown in Fig. 5.19, consists of a trimeric complex of subunits, each of which binds seven bacteriochlorophyll molecules for a total of 21 (Matthews and Fenna, 1980; Li et al., 1997). Recently, an eighth BChl molecule has been discovered between subunits, bringing the total to 24 (Tronrud et al., 2009; Wen et al., 2011). These additional pigments are positioned toward the chlorosome and are probably the entry point for excitations from the chlorosome baseplate. The FMO protein was the first chlorophyll-containing protein to have its structure determined. Much of the current understanding of how pigments interact with proteins was obtained from studies of this protein, and it has also been studied thoroughly by spectroscopic and theoretical

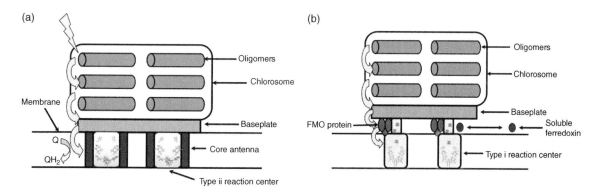

Figure 5.18   Schematic models of chlorosomes from (a) filamentous anoxygenic phototrophs and (b) green sulfur bacteria. The green tubes represent the pigment oligomers. The salmon layer is the baseplate complex. Blue ovals are FMO trimers. Brown ovals are soluble ferredoxins. In panel (a), the core antenna is a circular array of subunits similar to LH1 in purple bacteria. The energy transfer sequence is shown by yellow arrows.

(a)   (b)

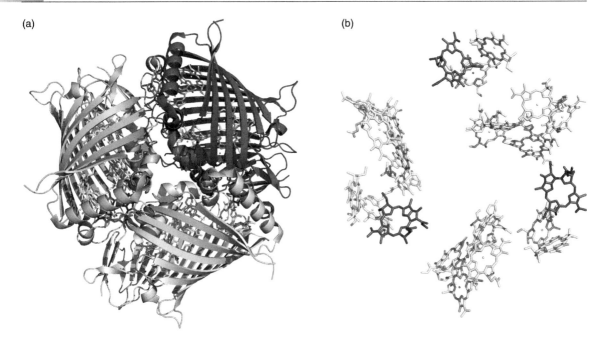

Figure 5.19   Structure of the trimeric Fenna–Matthews–Olson protein from *Prosthecochloris aestuarii*, including protein and bacteriochlorophyll *a* pigments. (a) Trimeric complex with the three identical protein subunits color-coded red, green, and cyan. The seven bacteriochlorophyll molecules in the interior of each subunit are color coded yellow and the three bacteriochlorophyll molecules in the interfaces between subunits are color-coded magenta. (b) The pigments in the same orientation and color-coding as in (a). The phytyl tails of the bacteriochlorophyll molecules have been omitted for clarity. Figure produced from Protein Data Bank file 3EOJ using PyMol.

techniques. The FMO protein was also the first antenna complex found to exhibit quantum coherence effects (Engel *et al.*, 2007).

The FAPs contain chlorosomes but do not contain the FMO protein. In these organisms, the chlorosome transfers energy directly to an integral membrane core antenna complex similar to LH1 and then to the reaction center. The FMO protein is only found in chlorosome-containing organisms that contain a "Type I" reaction center. This reaction center reduces the soluble protein ferredoxin as an electron acceptor. Both the chlorosome and ferredoxin are located on the cytoplasmic side of the membrane. The FMO protein may therefore function as a "spacer" to give ferredoxin diffusional access to the reducing side of the reaction center. The FAPs have a "Type 2" reaction center that uses quinones as electron acceptors. Quinones access the complex from within the membrane bilayer, so

these organisms have no need to have access to the cytoplasmic side of the reaction center and hence do not contain the FMO protein.

5.6   Regulation of antennas

Antenna systems are of immense benefit to the photosynthetic organism, but they carry a significant risk. Under most conditions, light is limiting, and the antenna serves to increase the amount of energy transferred to the reaction center. However, under some conditions, light may be in excess. If the photochemical and electron transfer reactions or the carbon fixation reactions are unable to keep up with the rate of energy collection, then the excess energy must be dissipated, or it may cause massive damage to the organism. An additional constraint for oxygen-evolving organisms is the need to balance energy distribution between the two photosystems, so that one photosystem does not become overexcited, while the other is underexcited. The regulation of energy flow in photosynthesis is one of the most complex, sophisticated aspects of the entire process (van Amerongen and Croce, 2013a, b). It involves processes that range from effects on photophysics of excited states to modulation of gene expression. This regulation is best viewed as a multilayered suite of processes, some of which prevent damage, while others scavenge the products of damaging processes, and others repair damage that results when the first two layers fail (Asada, 1999; Eberhard *et al.*, 2008). In this section, we will discuss only two of these processes that have been identified in the antennas of oxygenic photosynthetic organisms: the phenomena of state transitions and nonphotochemical quenching. Other aspects of the regulation of antenna systems involving gene expression changes and the use of fluorescence techniques to probe physiological aspects of photosynthesis are discussed in Chapters 10 and 11.

5.6.1   State transitions

If the flux of excitations reaching the two photosystems in oxygenic photosynthetic organisms is unequal, an elegant mechanism diverts the excess energy away from the photosystem that is receiving too much and toward the one that is energy-deficient (van Thor *et al.*, 1998; Allen, 2003; Minagawa, 2011). This phenomenon is known as **state transitions** and was first discovered in algae (Bonaventura and Myers, 1969; Murata, 1969). Briefly, state transitions work as follows. If an excess of light energy is being delivered to Photosystem II, then an adjustment takes place so that more is delivered to Photosystem I. The state that results is known as state 2. Conversely, if excess light is delivered to Photosystem I, then an opposite adjustment takes place to shift some of it to Photosystem II, and the state that results is called state 1. These state transitions are easily monitored using fluorescence techniques, because the spectral properties of the two states are slightly different.

When Photosystem II in eukaryotic photosynthetic organisms is receiving excess light, some of the LHCII complexes associated with Photosystem II are phosphorylated by the action of a specific protein kinase. These phosphorylated proteins then move away from the stacked regions of the thylakoids that contain excess Photosystem II and into the stromal membranes that contain an excess of Photosystem I. This transfer of antenna complexes results in more energy being delivered to Photosystem I, and the system enters state 2. The driving force for this movement is thought to be simply charge repulsion between LHCII complexes, which carry an extra negative charge after phosphorylation (Fig. 5.20). The kinase is activated by the reduction of the quinone pool, which tends to become largely reduced when Photosystem II is being driven more than Photosystem I. A phosphatase enzyme, not under redox control, removes the phosphates from the LHCII and permits the opposite transition to state 1 when Photosystem I is being driven more than Photosystem II.

Cyanobacteria also exhibit state transitions, but do not contain LHCII, so the mechanism for how the transition is accomplished must be somewhat different. Although the mechanism has not been fully worked out, it may involve migration or reorientation of phycobilisomes so that excitations are directed to either Photosystem II or Photosystem I. Evidence suggests that it does not involve phosphorylation of one or more proteins, but is triggered by the redox state of the quinone pool (Calzadilla *et al.*, 2019).

5.6.2   Nonphotochemical quenching and the xanthophyll cycle

During the intense light of midday, any organism exposed to full sun will have a significant excess of light energy available. Here, the problem is not one of redistribution but, rather, how to dispose of the excess energy to prevent photodamage. In oxygenic photosynthetic organisms, this photodamage is known as **photoinhibition**. Under excess light conditions, photoinhibition is prevented, and energy is

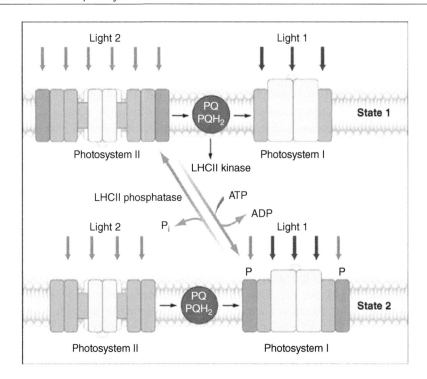

Figure 5.20   Movement of LHCII during state transitions. LHCII phosphorylation leads to migration from the stacked to the unstacked regions of the membrane, as a result of electrostatic repulsion. This migration shifts the balance of energy delivery from Photosystem II toward Photosystem I. *Source:* Allen (2003) (p. 1531)/American Association for the Advancement of Science.

dissipated by a phenomenon known as **nonphotochemical quenching** (NPQ) (Horton *et al.*, 1996; Li *et al.*, 2009; Ruban *et al.*, 2012; Magdaong and Blankenship, 2018; Bennett *et al.*, 2019). NPQ is easily monitored using fluorescence measurements of Photosystem II, which are remarkably informative regarding processes that affect both the antenna and electron transfer activities (Krause and Weiss, 1991). We will discuss these measurements in more detail in Chapter 11. For our present purposes, we will accept that these measurements correctly report on a powerful loss process that is activated in Photosystem II upon excess irradiation. This loss is a quenching process that deactivates the excited states of chlorophylls in the antenna system. It is distinct from the quenching caused by the normal photochemistry, hence the name "nonphotochemical quenching."

A number of changes are observed when the chloroplast makes the transition from low light conditions to being under the influence of NPQ. Most of these changes are localized in the LHCII antenna complexes, as well as some of the minor LHCs, which appear to be the site of action of NPQ. The PsbS antenna protein, which is a minor part of the Photosystem II antenna system, appears to be the locus of much of the NPQ effect, especially in higher plants. The trigger for induction of NPQ seems to be a low pH of the thylakoid lumen induced by electron transport coupled to H^+ translocation.

When NPQ is activated, a significant amount of the epoxide-containing carotenoid violaxanthin is converted to the nonepoxide-containing carotenoid zeaxanthin through the action of a de-epoxidase enzyme associated with the thylakoid membrane.

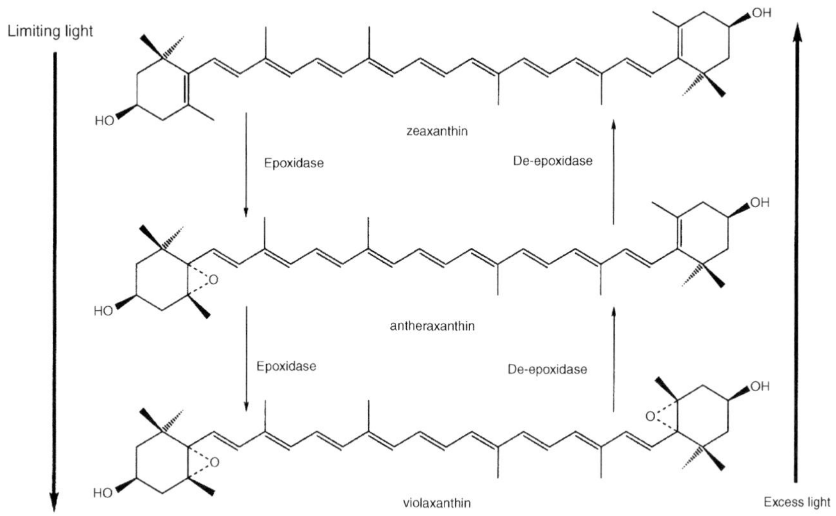

Limiting light

zeaxanthin

Epoxidase   De-epoxidase

antheraxanthin

Epoxidase   De-epoxidase

violaxanthin   Excess light

Figure 5.21   The xanthophyll cycle. Epoxidase and de-epoxidase enzymes interconvert violaxanthin and zeaxanthin in response to changing light conditions. Excess light activates the de-epoxidase enzyme and increases the amount of zeaxanthin, whereas low light favors the epoxidase enzyme and increases the amount of violaxanthin. The monoepoxide compound antheraxanthin is an intermediate in the conversion.

The LHCII complex contains a significant amount of both violaxanthin and zeaxanthin, which are classed as xanthophylls (Fig. 4.10). These carotenoids are located on the periphery of the LHCII complex in the structure shown in Fig. 5.13. An epoxidase enzyme reverses the process, creating a cycle known as the xanthophyll cycle (Fig. 5.21). A large number of studies have correlated the NPQ state with increased amounts of zeaxanthin (Demmig-Adams and Adams III, 1996). Whether this correlation is observed because the zeaxanthin is itself quenching the chlorophyll excited states, or whether it is simply one of several correlated effects but not the cause of NPQ, is a long-standing question. A number of molecular mechanisms have been proposed for how NPQ works, but this has been a controversial issue and is not yet resolved.

Cyanobacteria do not have NPQ that is mediated through the xanthophyll cycle. Instead, many cyanobacteria contain an orange carotenoid protein (OCP) that is involved in photoprotection (Kirilovsky and Kerfeld, 2012; Zhang et al., 2014; Muzzopappa and Kirilovsky, 2020), although the mechanism of action of OCP is not yet understood in detail.

References

Adir, N., Bar-Zvi, S., and Harris, D. (2020) The amazing phycobilisome. *Biochimica et Biophysica Acta* 1861: 14807.

Allen, J. F. (2003) State transitions—A question of balance. *Science* 299: 1530–1532.

van Amerongen, H. and Croce, R. (2013a) Light harvesting in Photosystem I. *Photosynthesis Research* 116: 153–166.

van Amerongen, H. and Croce, R. (2013b) Light harvesting in Photosystem II. *Photosynthesis Research* 116: 251–263.

van Amerongen, H., van Grondelle, R., and Valkunas, L. (2000) *Photosynthetic Excitons*. London: World Scientific.

Asada, K. (1999) The water–water cycle in chloroplasts: Scavenging of active oxygens and dissipation of excess photons. *Annual Review of Plant Physiology and Plant Molecular Biology* 50: 601–639.

Bahatyrova, S., Frese, R. N., Siebert, C. A., Olsen, J. D., van der Werf, K. O., van Grondelle, R., Niederman, R. A., Bullough, P. A., Otto, C., and Hunter, C. N. (2004) The native architecture of a photosynthetic membrane. *Nature* 430: 1058–1062.

Bautista, J. A., Hiller, R. G., Sharples, F. P., Gosztola, D., Wasielewski, M., and Frank, H. A. (1999) Singlet and triplet energy transfer in the peridinin–chlorophyll *a*–protein from *Amphidinium carterae*. *Journal of Physical Chemistry A* 103: 2267–2273.

Bennett, D. I. G., Amarnath, K., Park, S., Steen, C. J., Morris, J. M., and Fleming, G. R. (2019) Models and mechanisms of the rapidly reversible regulation of photosynthetic light harvesting. *Open Biology* 9: 190043.

Bibby, T. S., Nield, J., and Barber, J. (2001) Iron deficiency induces the formation of an antenna ring around trimeric photosystem I in cyanobacteria. *Nature* 412: 743–745.

Blankenship, R. E. and Matsuura, K. (2003) Antenna complexes in green photosynthetic bacteria. In: B. R. Green and W. W. Parson, (eds.) *Light-Harvesting Antennas*. Dordrecht: Kluwer Academic Press, pp. 195–217.

Boekema, E. J., Hifney, A., Yakushevska, A. E., Piotrowski, M., Keegstra, W., Berry, S., Michel, K. P., Pistorius, E. K., and Kruip, J. (2001) A giant chlorophyll-protein complex induced by iron deficiency in cyanobacteria. *Nature* 412: 745–748.

Bonaventura, C. and Myers, J. (1969) Fluorescence and oxygen evolution from *Chlorella pyrenoidosa*. *Biochimica et Biophysica Acta* 189: 366–389.

Bryant, D. A. and Canniffe, D. I. (2018) How nature designs light-harvesting antenna systems: Design principles and functional realization in chlorophototrophic prokaryotes. *Journal of Physics B* 51: 033001.

Burnap, R. L., Troyan, T., and Sherman, L. A. (1993) The highly abundant chlorophyll-protein complex of iron-deficient *Synechococcus* sp. PCC7942 (CP43′) is encoded by the isiA gene. *Plant Physiology* 103: 893–902.

Calzadilla, P. I., Zhan, J., Sétif, P., Lemaire, C., Solymosi, D., Battchikova, N., Wang, Q., and Kirilovsky, D. (2019) The cytochrome b_6f complex is not involved in cyanobacterial state transitions. *The Plant Cell* 31: 911–931.

Cantor, C. R. and Schimmel, P. R. (1980) *Biophysical Chemistry Part II. Techniques for the Study of Biological Structure and Function*. San Francisco: W. H. Freeman.

Cao, J., Cogdell, R. J., Coker, D. F., Duan, H.-G., Hauer, J., Kleinekathöfer, U., Jansen, T. L. C., Mančal, T., Miller, R. J. D., Ogilvie, J. P., Prokhorenko, V. I., Renger, T., Tan, H.-S., Tempelaar, R., Thorwart, M., Thyrhaug, E., Westenhoff, S., and Zigmantas, D. (2020) Quantum biology revisited. *Science Advances* 6: eaaz4888.

Cheng, Y.-C. and Fleming, G. R. (2009) Dynamics of light harvesting in photosynthesis. *Annual Review of Physical Chemistry* 60: 241–262.

Chenu, A. and Scholes, G. D. (2015) Coherence in energy transfer in photosynthesis. *Annual Review of Physical Chemistry* 66: 69–96.

Cogdell, R. J. and Köhler, J. (2009) Use of single-molecule spectroscopy to tackle fundamental problems in biochemistry: Using studies on purple bacterial antenna complexes as an example. *Biochemical Journal* 422: 193–205.

Cogdell, R. J., Gall, A., and Köhler, J. (2006) The architecture and function of the light-harvesting apparatus of purple bacteria: From single molecules to in vivo membranes. *Quarterly Reviews of Biophysics* 39: 227–324.

Cohen-Bazire, G., Pfennig, N., and Kunisawa, R. (1964) The fine structure of green bacteria. *Journal of Cell Biology* 22: 207–225.

Croce, R. and van Amerongen, H. (2020) Light harvesting in oxygenic photosynthesis: Structural biology meets spectroscopy. *Science* 369: eaay2058.

Croce, R., van Grondelle, R., van Amerongen, H., and van Stokkum, I., (eds.) (2018) *Light Harvesting in Photosynthesis*. Boca Raton, FL: CRC Press.

Demmig-Adams, B. and Adams, W. W. III (1996) The role of xanthophyll cycle carotenoids in the protection in photosynthesis. *Trends in Plant Science* 1: 21–26.

Eberhard, S., Finazzi, G., and Wollman, F. A. (2008) The dynamics of photosynthesis. *Annual Review of Genetics* 42: 463–515.

Engel, G. S., Calhoun, T. R., Read, E. L., Ahn, T. K., Mancal, T., Cheng, Y.-C., Blankenship, R. E., and Fleming, G. R. (2007) Evidence for wavelike energy transfer through quantum coherence in photosynthetic systems. *Nature* 446: 782–786.

Fleming, G. R. and van Grondelle, R. (1997) Femtosecond spectroscopy of photosynthetic light-harvesting systems. *Current Opinion in Structural Biology* 7: 738–748.

Förster, Th. (1965) Delocalized excitation and excitation transfer. In: O. Sinanoglu, (ed.) *Modern Quantum Chemistry Istanbul Lectures*, Vol. 3. New York: Academic Press, pp. 93–137.

Franck, J. and Teller, E. (1938) Migration and photochemical action of excitation energy in crystals. *Journal of Chemical Physics* 6: 861–872.

Gaffron, H. and Wohl, K. (1936) Zur Theorie der Assimilation. *Naturwissenschaften* 24: 81–90, 103–107.

Ganapathy, S., Oostergetel, G. T., Wawrzyniak, P. K., Reus, M., Chew, A. G. M., Buda, F., Boekema, E. J., Bryant, D. A., Holzwarth, A. R., and de Groot, H. J. M. (2009) Alternating syn-anti bacteriochlorophylls form concentric helical nanotubes in chlorosomes. *Proceedings of the National Academy of Sciences USA* 106: 8525–8530.

Gantt, E. and Conti, S. F. (1966) Granules associated with chloroplast lamellae of *Porphyridium cruentum*. *Journal of Cell Biology* 29: 423–434.

Green, B. R. and Parson, W. W., (eds.) (2003) *Light-Harvesting Antennas*. Dordrecht: Kluwer Academic Press.

van Grondelle, R. (1985) Excitation energy transfer, trapping and annihilation in photosynthetic systems. *Biochimica et Biophysica Acta* 811: 147–195.

van Grondelle, R., Dekker, J. P., Gillbro, T., and Sundström, V. (1994) Energy transfer and trapping in photosynthesis. *Biochimica et Biophysica Acta* 1187: 1–65.

Grossman, A. R., Bhaya, D., Apt, K. E., and Kehoe, D. M. (1995) Light-harvesting complexes in oxygenic photosynthesis: Diversity, control and evolution. *Annual Review of Genetics* 29: 231–288.

Hofmann, E., Wrench, P. M., Sharples, F. P., Hiller, R. G., Welte, W., and Diedrichs, K. (1996) Structural basis of light harvesting by carotenoids: Peridinin–chlorophyll–protein from *Amphidinium carterae*. *Science* 272: 1788–1791.

Horton, P., Ruban, A. V., and Walters, R. G. (1996) Regulation of light harvesting in green plants. *Annual Review of Plant Physiology and Plant Molecular Biology* 47: 655–684.

Kirilovsky, D. and Kerfeld, C. A. (2012) The orange carotenoid protein in photoprotection of photosystem II in cyanobacteria. *Biochimica et Biophysica Acta* 1817: 158–166.

Knox, R. S. (1996) Electronic excitation transfer in the photosynthetic unit: Reflections on work of William Arnold. *Photosynthesis Research* 48: 35–39.

Knox, R. S. (2012) Förster's resonance excitation transfer theory: Not just a formula. *Journal of Biomedical Optics* 17: 011003.

Knox, R. S. and Gülen, D. (1993) Theory of polarized fluorescence from molecular pairs: Förster transfer at large electronic coupling. *Photochemistry and Photobiology* 57: 40–43.

Krause, G. H. and Weiss, E. (1991) Chlorophyll fluorescence and photosynthesis: The basics. *Annual Review of Plant Physiology and Plant Molecular Biology* 42: 313–349.

Kühlbrandt, W., Wang, D. N., and Fujiyoshi, Y. (1994) Atomic model of plant light harvesting complex by electron crystallography. *Nature* 367: 614–621.

Li, Y. F., Zhou, W. L., Blankenship, R. E., and Allen, J. P. (1997) Crystal structure of the bacteriochlorophyll *a* protein *from Chlorobium tepidum*. *Journal of Molecular Biology* 271: 456–471.

Li, Z., Wakao, S., Fischer, B. B., and Niyogi, K. K. (2009) Sensing and responding to excess light. *Annual Review of Plant Biology* 60: 239–260.

Liu, Z., Yan, H., Wang, K., Kuang, T., Zhang, J., Gui, L., An, X., and Chang, W. (2004) Crystal structure of spinach major light-harvesting complex at 2.72 Å resolution. *Nature* 428: 287–292.

Liu, H., Zhang, H., Niedzwiedzki, D. M., Prado, M., He, G., Gross, M. L., and Blankenship, R. E. (2013) Phycobilisomes feed both photosystems in one megacomplex in *Synechocystis* sp. PCC 6803. *Science* 342: 1104–1107.

Ma, J., You, X., Sun, S., Wang, X., Qin, S., and Sui, S.-F. (2020) Structural basis of energy transfer in *Porphyridium purpureum* phycobilisome. *Nature* 579: 146–151.

MacColl, R. (1998) Cyanobacterial phycobilisomes. *Journal of Structural Biology* 124: 311–334.

Magdaong, N. C. M. and Blankenship, R. E. (2018) Photoprotective, excited-state quenching mechanisms in diverse photosynthetic organisms. *Journal of Biological Chemistry* 293: 5018–5025.

Matthews, B. W. and Fenna, R. E. (1980) Structure of a green bacteriochlorophyll protein. *Accounts of Chemical Research* 13: 309–317.

McDermott, G., Prince, S. M., Freer, A. A., Hawthornthwaite-Lawless, A. M., Papiz, M. Z., Cogdell, R. J., and Isaacs, N. W. (1995) Crystal structure of an integral membrane light-harvesting complex from photosynthetic bacteria. *Nature* 374: 517–521.

Mimuro, M., Nozawa, T., Tamai, N., Shimada, K., Yamazaki, I., Lin, S., Knox, R. S., Wittmershaus, B. P., Brune, D. C., and Blankenship, R. E. (1989) Excitation energy flow in chlorosome antennas of green photosynthetic bacteria. *Journal of Physical Chemistry* 93: 7503–7509.

Minagawa, J. (2011) State transitions—The molecular remodeling of photosynthetic supercomplexes that controls energy flow in the chloroplast. *Biochimica et Biophysica Acta* 1807: 897–905.

Mirkovic, T., Ostroumov, E. E., Anna, J. M., van Grondelle, R., Govindjee, and Scholes, G. D. (2017) Light absorption and energy transfer in the antenna complexes of photosynthetic organisms. *Chemical Reviews* 117: 249–293.

Montaño, G. A., Bowen, B. P., LaBelle, J. T., Woodbury, N. W., Pizziconi, V. B., and Blankenship, R. E. (2003) Characterization of *Chlorobium tepidum* chlorosomes- A calculation of bacteriochlorophyll *c* per chlorosome and oligomer modeling. *Biophysical Journal* 85: 2560–2565.

Murata, N. (1969) Control of excitation transfer in photosynthesis. I. Light-induced change in chlorophyll *a* fluorescence in *Porphyridium cruentum*. *Biochimica et Biophysica Acta* 172: 242–251.

Muzzopappa, F. and Kirilovsky, D. (2020) Changing color for photoprotection: The orange carotenoid protein. *Trends in Plant Science* 25: 92–104.

van Oijen, A. M., Ketelaars, M., Kohler, J., Aartsma, T. J., and Schmidt, J. (1999) Unraveling the electronic structure of individual photosynthetic pigment–protein complexes. *Science* 285: 400–402.

Olson, J. M. (1998) Chlorophyll organization and function in green photosynthetic bacteria. *Photochemistry and Photobiology* 67: 61–75.

Oostergetel, G. T., van Amerongen, H., and Boekema, E. J. (2010) The chlorosome: A prototype for efficient light harvesting in photosynthesis. *Photosynthesis Research* 104: 245–255.

Orf, G. S. and Blankenship, R. E. (2013) Chlorosome antenna complexes from green photosynthetic bacteria. *Photosynthesis Research* 116: 315–331.

Parkes-Loach, P. S., Majeed, A. P., Law, C. J., and Loach, P. A. (2004) Interactions stabilizing the structure of the core light-harvesting complex (LH1) of photosynthetic bacteria and its subunit (B820). *Biochemistry* 43: 7003–7016.

Polivka, T. and Frank, H. A. (2010) Molecular factors controlling photosynthetic light harvesting by carotenoids. *Accounts of Chemical Research* 43: 1125–1134.

Pšencík, J., Ikonen, T. P., Laurinmäki, P., Merckel, M. C., Butcher, S. J., Serimaa, R. E., and Tuma, R. (2004). Lamellar organization of pigments in chlorosomes, the light harvesting complexes of green photosynthetic bacteria. *Biophysical Journal* 87: 1165–1172.

Qian, P., Papiz, M. Z., Jackson, P. J., Brindley, A. A., Ng, I. W., Olsen, J. D., Dickman, M. J., Bullough, P. A., and Hunter, C. N. (2013). Three-dimensional structure of the *Rhodobacter sphaeroides* RC-LH1-PufX complex: Dimerization and quinine channels promoted by PufX. *Biochemistry* 52: 7575–7585.

Qian, P., Siebert, A., Wang, P., Caniffe, D. P., and Hunter, C. N. (2018) Cryo-EM structure of the *Blastochloris viridis* LH1-RC complex at 2.9 Å. *Nature* 556: 203–208.

Roszak, A. W., Howard, T. D., Southall, J., Gardiner, A. T., Law, C. J., Isaacs, N. W., and Cogdell, R. J. (2003) Crystal structure of the RC-LH1 core complex from *Rhodopseudomonas palustris*. *Science* 302: 1969–1972.

Ruban, A. (2013) *The Photosynthetic Membrane*. Chichester, UK: Wiley.

Ruban, A. V., Johnson, M. P., and Duffy, C. D. P. (2012) The photoprotective molecular switch in the photosystem II antenna. *Biochimica et Biophysica Acta* 1817: 167–181.

Saer, R. and Blankenship, R. E. (2017) Light-harvesting in phototrophic bacteria: Structure and function. *Biochemical Journal* 474: 2107–2131.

Sanfilippo, J. E., Garczarek, L., Partensky, F., and Kehoe, D. M. (2019) Chromatic acclimation in cyanobacteria: A diverse and widespread process for optimizing photosynthesis. *Annual Review of Microbiology* 73: 407–433.

Schatz, G. H., Brock, H., and Holzwarth, A. R. (1988) A kinetic and energetic model for the primary processes in photosystem II. *Biophysical Journal* 54: 397–405.

Sener, M., Strümpfer, J., Hsin, J., Chandler, D., Scheuring, S., Hunter, C. N., and Schulten, K. (2011) Förster energy transfer theory as reflected in the structures of photosynthetic light-harvesting systems. *ChemPhysChem* 12: 518–531.

Sturgis, J. N., Tucker, J. D., Olsen, J. D., Hunter, C. N., and Niederman, R. A. (2009) Atomic force microscopy studies of native photosynthetic membranes. *Biochemistry* 48: 3679–3698.

Sundström, V., Pullerits, T., and van Grondelle, R. (1999) Photosynthetic light-harvesting: Reconciling dynamics and structure of purple bacterial LH2 reveals function of photosynthetic unit. *Journal of Physical Chemistry B* 103: 2327–2346.

van Thor, J. J., Mullineaux, C. W., Matthijs, H. C. P., and Hellingwerf, K. J. (1998) Light harvesting and state transitions in cyanobacteria. *Botanica Acta: Journal of the German Botanical Society* 111: 430–443.

Toporik, H., Li, J., Williams, D., Chiu, P.-L., and Mazor, Y. (2019) The structure of the stress-induced photosystem I–IsiA antenna supercomplex. *Nature Structural and Molecular Biology* 26: 443–449.

Tronrud, D. E., Wen, J., Gay, L., and Blankenship, R. E. (2009) The structural basis for the difference in absorbance spectra for the FMO antenna protein from various green sulfur bacteria. *Photosynthesis Research* 100: 79–87.

Watanabe, M. and Ikeuchi, M. (2013) Phycobilisome: Architecture of a light-harvesting supercomplex. *Photosynthesis Research* 116: 256–276.

Wen, J., Zhang, H., Gross, M. L., and Blankenship, R. E. (2011) Native electrospray mass spectrometry reveals the nature and stoichiometry of pigments in the FMO antenna protein. *Biochemistry* 50: 3502–3511.

Yu, I. J., Suga, M., Wang-Otomo, Z.-Y., and Shen, J. R. (2018) Structure of photosynthetic LH1–RC supercomplex at 1.9 Å resolution. *Nature* 556: 209–213.

Zhang, H., Niedzwiedzki, D. M., Liu, H., Prado, M., Jiang, J., Gross, M. L., and Blankenship, R. E. (2014) The molecular mechanism of orange carotenoid proteinmediated photoprotection in cyanobacteria. *Biochemistry* 53: 13–19.

Chapter 6

Reaction centers and electron transport pathways in anoxygenic phototrophs

In this chapter, we will explore the electron transfer processes that take place in anoxygenic phototrophic organisms. We will first discuss the reaction center, where electron transfer takes place after energy transfer from the antenna system delivers energy to the complex. The initial electron transfer process creates a charge-separated or ion-pair state that is the first point where electromagnetic energy has been converted into chemical energy. This primary photochemistry is then followed by a series of secondary electron and proton transfer processes that stabilize some of the energy trapped in the initial process.

The reaction center is an elegant molecular energy-transduction device. It converts the energy in sunlight to a form that an organism can use to power its life processes. In order to understand how this process works at a deep level, we will need to understand both the big picture and the details. First, we will discuss some general principles that apply equally well to all reaction centers. This discussion builds on the brief introduction in Chapter 1 and the physical principles presented in the Appendix. Then, we will turn to the details of the structure of the complexes in anoxygenic phototrophic organisms, the pathways and kinetics of the electron transfer processes, and how proton and electron transfers are often coupled together. We also need some theoretical framework to understand the distance, energy, and temperature dependence of the various processes. The reaction centers and electron and proton transport mechanisms of oxygenic organisms are discussed in Chapter 7.

We will begin our study of photosynthetic reaction center complexes with a detailed analysis of what is surely the best understood of all reaction centers, the complexes found in the anoxygenic purple bacteria. These complexes will serve as a case study, in that by studying this one complex in significant detail, we hope to establish some general principles that will apply to all photosynthetic reaction centers. Our discussions of the other classes of reaction centers can then be somewhat briefer and focus on the unique aspects of each group.

Extensive reviews of reaction center structure and function can be found in specialized books and articles by Deisenhofer and Norris (1993), Blankenship *et al.* (1995), Fromme (2008), Hunter *et al.* (2009), and Jones (2021).

6.1 Basic principles of reaction center structure and function

Reaction center complexes are integral membrane pigment–proteins that span the membrane in a vectorial fashion, so that the complex is always oriented the same way with respect to the sidedness of the membrane. All reaction centers carry out light-driven electron transfer reactions, resulting in charge separation across the membrane. In addition, some reaction centers also pump protons, which are coupled to the electron transfer reactions involving quinones.

The chemical properties of an excited molecule may be very different from those of the same molecule in the ground, or lowest energy state. In particular, the oxidation-reduction potential for electrons either being added to the molecule or given up by it is very different in the excited state compared with the ground state (Blankenship and Prince, 1985). The basis of this effect is discussed in more detail in the Appendix. The net result is that the excited molecule is an extremely strong reducing agent, which readily gives up an electron to a nearby electron acceptor molecule. This excited-state electron transfer process is the primary photochemical process in photosynthesis and is the moment at which light energy is converted into chemical energy. The product of this electron transfer event is an oxidized dimer of chlorophylls and a reduced electron acceptor molecule. Both these species are ions, with a charge, and also free radicals, with an unpaired electron. The primary ion-pair state consists of the oxidized, and now positively charged, donor and the reduced, and now negatively charged, acceptor. It is, however, a highly unstable system, and the energy is easily lost if it is not stabilized by subsequent secondary processes, discussed below.

The reaction center thus takes light energy and uses it to drive electron transfer reactions. In the instant just after the primary electron transfer process, the system is poised at a critical juncture. The oxidized primary electron donor is positioned next to the reduced acceptor. The most likely outcome according to the laws of thermodynamics is for the electron to simply transfer back to the donor. This process, called **recombination**, results in the conversion of all the energy of the photon to heat, without the opportunity for any to be stored. To avoid this fate, an ultrafast series of secondary electron reactions separate the oxidized and reduced species. The result is that the positive and negative charges are spatially separated from each other, and the chances of recombination are greatly reduced. If the secondary stabilization reactions are much faster than the recombination process, then most of the complexes will avoid recombination and produce stable products (See Fig. 1.4). Energetic factors may also be important, as discussed below in the section on the theory of electron transfer processes.

6.2 Development of the reaction center concept

Now it seems almost self-evident that the reaction center is a well-defined complex with an identity that is distinct and separable from the rest of the photosynthetic apparatus, with a fixed stoichiometry of proteins, pigments, and other cofactors. However, this view emerged only slowly, and not until after many years of hard work by a large number of dedicated researchers. It is remarkable how difficult it is for "self-evident" concepts to take form and come to fruition. The idea of a reaction center grew out of the Emerson and Arnold experiments, as discussed in Chapter 3. These experiments clearly established that not all chlorophylls carry out complete photosynthesis. Important early absorbance change measurements on purple bacteria by Louis Duysens in the 1950s further laid the groundwork for the idea that some of the pigments were present in a special environment that was somehow different from that of the majority.

The purple bacteria were also the first to have their reaction centers biochemically isolated and characterized. The idea that the reaction center is an isolatable entity was first formulated with studies on these organisms. Roderick Clayton, working at Oak Ridge National Laboratory in the early 1960s, observed that in old cultures of a carotenoid-less mutant of the purple bacterium *Rhodobacter sphaeroides* (then called *Rhodopseudomonas sphaeroides*), most of the bacteriochlorophyll had been converted to bacteriopheophytin. However, a small amount of bacteriochlorophyll was resistant to this treatment and showed large photoactivity. This suggested that the antenna and the reaction center were distinct biochemical entities. Clayton then began a quest to isolate and purify the reaction center complex and finally succeeded in the late 1960s. Much of the early characterization of these purified complexes was carried out by Clayton and, independently, by George Feher, who have both written interesting accounts of their work (Clayton, 1988; Feher, 1998).

6.3   Purple bacterial reaction centers

6.3.1   Structural features of purple bacterial membranes and reaction centers

The purple bacterial reaction centers are located in a specially modified portion of the inner cell membrane of the organism called the **intracytoplasmic membrane**. The membrane invaginates and sometimes forms tubes, vesicles, or flat lamellar membranes (Drews and Golecki, 1995). The reaction centers are oriented uniquely in the membrane, so that the primary electron donor bacteriochlorophylls face the periplasm of the cell. The electron-donating cytochrome (see below) is then in the periplasmic space and topologically outside the cell.

When cells are disrupted for study, the broken membranes often reseal into small vesicles called **chromatophores**, in the process trapping cytochromes and other periplasmic components in the interior of the vesicle (Fig. 6.1). Chromatophores are thus opposite in orientation compared with intact cells. Considerable evidence also indicates that some chromatophore vesicles that are not attached to the cytoplasmic membrane are also present in *Rhodobacter sphaeroides* (Tucker *et al.*, 2010).

The reaction centers are purified by dissolving the membranes with gentle detergents and then purifying the pigment–proteins using a variety of biochemical techniques, such as column chromatography. The purified reaction centers in purple bacteria are composed of one copy each of three or four protein subunits, depending on the species. The constant subunits are known as L (light), M (medium), and H (heavy). In many species, a fourth subunit, known as C (cytochrome) is present. The designations L, M, and H date from the time when the true molecular masses of the subunits were not known, the labels being assigned on the basis of the mobility of the peptides as determined by SDS-PAGE (Feher, 1998). Ironically, the true masses are significantly different from the apparent masses, with H being the peptide with the lowest mass, L the next, and M the heaviest of the three. However, the L, M, and H designations are so well established that it seems best to retain this nomenclature. It also serves as a reminder that SDS-PAGE molecular masses are not very accurate, and that the technique often significantly underestimates masses for very hydrophobic integral membrane proteins.

In addition to the protein subunits, reaction centers contain a number of noncovalently associated cofactors, including pigments, quinones, and metal ions. Purple bacterial reaction centers contain four molecules of bacteriochlorophyll (BChl), two molecules of bacteriopheophytin (BPh), one metal ion (in most cases Fe^{2+}), two quinones, and, in most cases, one carotenoid.

High-resolution X-ray structures of reaction centers from three different species of purple bacteria have been obtained. The first was the

(a)   (b)

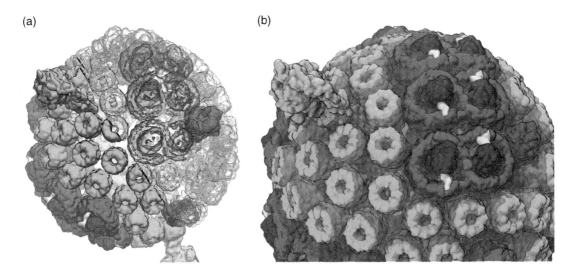

Figure 6.1   Schematic model of a purple bacterial chromatophore from *Rhodobacter sphaeroides*, based on AFM, EM, crystallography, mass spectroscopy, proteomics, and optical spectroscopy data. Color coding: LH2 complexes (green), RC–LH1–PufX complexes (LH1:red; RC:blue; PufX:lime), cyt bc_1 (magenta), and ATP synthase (orange). (a) Proteins and BChls of the chromatophore. (b) Close-up of chromatophore showing its lipid membrane (transparent). *Source:* Adapted from Sener *et al.* (2016).

four-subunit complex (LMHC) from *Blastochloris viridis* (Fig. 6.2). This constituted a landmark in biochemistry, as it was the first time that a high-resolution structure of an integral membrane protein had been obtained. It earned Hartmut Michel, Hans Deisenhofer, and Robert Huber the 1988 Nobel prize for chemistry (Deisenhofer and Michel, 1989). Subsequently, structures for the three-subunit reaction center (LMH) from *Rhodobacter sphaeroides* and the LMHC reaction center from *Thermochromatium tepidum* were determined by several groups. The structures for the LMH parts of all the complexes are extremely similar.

The L and M subunits of the reaction center are threaded through the membrane five times each, and the structure of the complex shows that the portion of the protein that crosses the membrane is almost purely α-helical in secondary structure. The transmembrane helical sections of these proteins are composed of mostly hydrophobic amino acids. This ensures that the protein is firmly positioned in the membrane in a unique position and with a well-defined orientation. The H subunit is significantly less hydrophobic and has only one transmembrane segment. This anchors it to the membrane, with the bulk of its mass on the cytoplasmic side. There are thus eleven transmembrane helices, with five each from L and M, and one from the H subunit. In some species, the C subunit has a single transmembrane helix.

The L and M proteins have a pseudo-twofold axis of symmetry, which is perpendicular to the plane of the membrane. A pseudo-twofold symmetry results when the two halves of the complex are similar but not identical. When the complex is rotated 180 degrees around the symmetry axis, the resulting complex has a structure that is generally similar to that before the rotation, but not identical. The symmetry is broken by the H subunit, which has no symmetry-related counterpart, and also by the fact that the L and M subunits have only about 25–30% sequence identity.

The reaction center protein forms a scaffold upon which the cofactors are arranged. The redox cofactors are also arranged with a pseudo-twofold symmetry to form two branches, labeled A and B in Fig. 6.2. A variety of evidence indicates that electron transfer

(a)

(b)

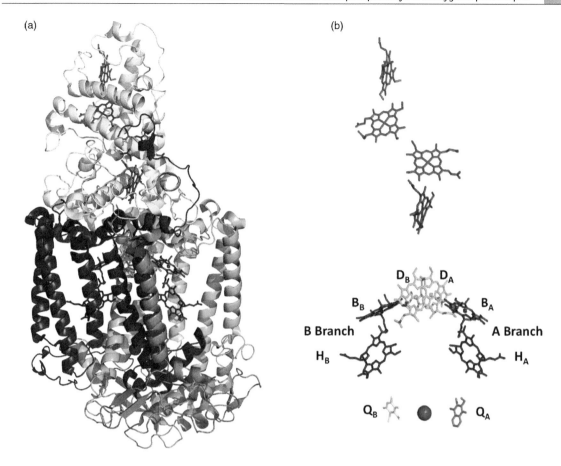

Figure 6.2 Structure of the LMHC reaction center complex from *Blastochloris viridis*. (a) The entire complex, including L, M, H, and C protein subunits and cofactors. The protein subunits are colored L-green, M-blue, H-orange, C-yellow. (b) Cofactors only. Heme-red, special pair BChl-green, accessory BChl-purple, menaquinone-orange, ubiquinone-yellow. The polyisoprenoid tails of the pigments and quinones have been removed for clarity. *Source:* Produced from Protein Data Bank file 3T6D using Pymol.

proceeds only down the pigments on the A branch and then crosses over from the A to the B branch at the quinones. The function of the pigments in the B branch is not clear and may be primarily structural.

The two pigments labeled D_A and D_B in Fig. 6.2 are bacteriochlorophyll molecules that are very closely interacting. They are called P###, after the wavelength maximum of their Q_y absorbance band. This is 870 nm (hence P870) for those purple bacteria that contain bacteriochlorophyll *a* and 960 nm (P960) for those that contain bacteriochlorophyll *b*. This "special pair" of pigments together form the photoactive pigment whose excited state loses an electron to form the primary ion-pair state. The EPR (electron paramagnetic resonance) and ENDOR (electron-nuclear double resonance) spectra of the oxidized special pair have been studied extensively (Hoff and Deisenhofer, 1997). These techniques are sensitive to the unpaired electron that remains on the special pair after electron transfer forms the cation radical. The unpaired electron is approximately equally shared between the two pigments, resulting in a reduction of the electron-nuclear hyperfine coupling constants by a factor of two, compared with the cation radical of the monomeric pigment in solution.

The C subunit is found in many purple bacterial reaction centers, but is absent in others. In most species, it has no transmembrane segments. Instead, a covalently attached fatty acid molecule serves to anchor the C subunit to the membrane. Nearly all of the mass of the C subunit is found on the periplasmic side of the membrane. However, it is tightly associated with the rest of the complex. The most distinctive feature of the C subunit is the four heme groups, which are covalently attached to cysteine amino acid side chains. The hemes alternate with high and low redox potentials, with the highest-potential heme near the LM complex and the lowest-potential heme at the end of the chain. Cytochrome c_2 donates an electron to the low-potential heme at the far end of the chain, and the electron is subsequently transferred through the other hemes until it reduces the oxidized special pair (Ortega $et\ al.$, 1999). The kinetics and energetics of this process are remarkable. The midpoint redox potential of the electron donor cytochrome c_2 is +285 mV, while the first heme group has a potential of −60 mV. The electron transfer is thus energetically uphill by +345 mV, yet still takes place in 60 μs when the cytochrome c_2 is bound to the tetraheme cytochrome, because of a favorable positioning of the two heme groups. The overall electron transfer process from cytochrome c_2 to P960 is thermodynamically downhill, but has these surprising uphill intermediate steps. In those species of purple bacteria that lack a bound cytochrome subunit, the soluble cytochrome c_2 donates electrons directly to the oxidized special pair, after first docking to the surface of the reaction center (Axelrod $et\ al.$, 2009).

6.3.2 Kinetics and pathways of electron transfer within purple bacterial reaction centers

A wide variety of techniques have been used to investigate the bacterial reaction center system, including almost every imaginable kind of spectroscopy (Amesz and Hoff, 1996; Hoff and Deisenhofer, 1997; Artsma and Matysik, 2008), as well as a range of biochemical and genetic manipulations. Here it is only possible to give a brief summary of some of the results. The focus will be on the first few reactions following excitation of the special pair, either directly by photon absorption or indirectly by excitation transfer from antenna or other reaction center pigments. The technique of picosecond-absorbance transient-difference spectroscopy, as described in the Appendix, has been especially informative with respect to elucidating the pathway of electron flow in these complexes and will therefore be emphasized. Figure 6.3 summarizes the kinetics and energetics of photochemical and early secondary reactions that take place in isolated reaction centers of purple bacteria. A variety of evidence indicates that the electron transfer pathway and kinetics in isolated reaction centers are not significantly altered from their behavior $in\ vivo$.

Following excitation of the special pair, the excited state of P870 has a lifetime of about 3 ps at room temperature, decreasing to about 1 ps at cryogenic temperatures. This excited state (P870*) is conveniently monitored by measuring stimulated emission in the 900 nm region. The P870* excited state is a very strong reductant, with an estimated excited state redox potential of −940 mV $vs.$ NHE

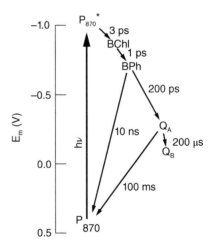

Figure 6.3 Energy-kinetic diagram describing the energetics and reaction times for electron transfer processes in reaction centers from $Rhodobacter\ sphaeroides$.

(Table A.3). It decays primarily by losing an electron to become the cation radical $P870^+$. Whenever a pigment is excited or either gains or loses an electron, its ground state absorption bands will bleach. New absorption bands that are characteristic of the excited, oxidized, or reduced species will appear at the same time as the ground state bands bleach. Analysis of the kinetics and spectral characteristics of these spectral changes helps elucidate the electron transfer pathway.

As $P870^*$ decays to form $P870^+$, absorbance bands bleach at 535 and 760 nm, which are assigned to Q_x and Q_y transitions of the bacteriopheophytin on the A branch. At the same time, a broad new band appears in the 650 nm region, which is assigned to the anion radical BPh_A^-, and a band appears at 1250 nm, which is assigned to $P870^+$. The new state formed is the radical ion-pair state $P870^+ BPh_A^-$, in which an electron has been transferred from the excited special pair to the bacteriopheophytin on the A branch. The A branch bacteriopheophytin, instead of the one on the B branch, is identified as the acceptor largely by the difference spectra in the Q_x spectral region. The two bacteriopheophytin molecules have slightly different spectral signatures, due to differences in the amino acid polarities in their immediate environments. If it were not for this fortuitous spectral difference between the two BPh molecules, it would be very difficult to determine that one electron transfer branch is active while the other is inactive.

The $P870^+ BPh_A^-$ ion-pair state decays in 200–300 ps to form the $P870^+ Q_A^-$ state. Finally, the electron is transferred to Q_B in 100–200 µs to form the state $P870^+ Q_B^-$. This state is characterized by bleaching of the absorption bands due to P870 and Q_B, plus the appearance of new bands due to the cation and anion, respectively. The $Q_A^- \rightarrow Q_B^-$ reaction is thermally activated, while all the others are nearly independent of temperature, actually speeding up slightly at cryogenic temperatures. A variety of inhibitors, including herbicides such as atrazine, block this electron transfer process by displacing Q_B from the binding site. Proton uptake accompanies electron transfer in the quinone acceptor complex. This is discussed in more detail below.

In parallel with the electron transfer from Q_A^- to Q_B^-, electron transfer from a cytochrome rereduces the oxidized special pair. This is either the bound cytochrome subunit in those organisms that contain it, or a small soluble cytochrome, in most cases cytochrome c_2. The bound cytochrome donates an electron in a reaction that takes only a few hundred nanoseconds and also continues to take place at cryogenic temperatures. At low temperatures, this temperature-independent electron donation is clear evidence for quantum-mechanical tunneling processes and was the first such evidence in biological systems (DeVault and Chance, 1966).

The arrangement of electron transfer cofactors shown in Fig. 6.2 suggests that an earlier, charge-separated state may precede the formation of the $P870^+ BPh_A^-$ state. This earlier state is $P870^+ BChl_A^-$, where $BChl_A$ is the accessory bacteriochlorophyll labeled B_A, and is likely because of the spatial position of $BChl_A$ between P870 and BPh_A. This state almost certainly participates in the primary electron transfer process, although the nature of its involvement has been controversial. The data indicate that the rate constant for its decay is faster than that of its formation, thereby precluding the buildup of substantial amounts of the state (see the discussion of the kinetics of sequential reactions in the Appendix). The energy of the $P870^+ BChl_A^-$ state is probably above that of the $P870^*$ state (Parson and Warshel, 2009). In that case, it contributes to the electron transfer by what is known as a **superexchange process**, in which the $BChl_A$ anion can be viewed as a virtual state. The distances and calculated electronic couplings between the special pair and BPh_A are sufficiently long and weak, respectively, that it is extremely unlikely that the electron transfer process takes place directly from $P870^*$ to BPh_A (see below).

Perhaps the most striking aspect of the comparison of the structural and kinetic data on reaction centers is the apparent twofold symmetry of the structure, contrasted with the clear evidence for asymmetry in the electron transfer pathway. Estimates of the ratio for electron transfer probability down the A branch compared with the B branch are about 100:1. A considerable effort, including analysis of many mutants, has gone into trying to understand the factors that direct electron transfer down the A branch instead of the B branch (Wakeham and Jones, 2005; Faries *et al.*, 2012; Laible *et al.*, 2020). The reasons are most probably due to the energetics of the charge-separated states on the A and B branches, due to the environmental effects of the protein environment. The possible functional role of the unidirectional electron transfer is also not clear, but almost certainly has to do with the one-electron to two-electron conversion involved in quinone reduction.

Another interesting feature of the kinetic constants, shown in Fig. 6.3, is that the rates of recombination processes, in which the electron returns from one of the acceptors directly to the oxidized special pair, are invariably a factor of 50 or more slower than are the rate constants for the forward reaction. It is because of this kinetic control of the rates of wasteful processes that the quantum yield for photochemistry is so high. This kinetic "steering" of the system towards the productive sequence of charge-separation reactions and away from the wasteful recombination reactions is the critical factor that makes the primary processes of photosynthesis so efficient. Again, the molecular reasons for this remarkable behavior have proved difficult to elucidate. One of the major factors thought to be involved in the slow recombination rates is the large energy gap involved. This is best understood by considering some theoretical aspects of biological electron transfer reactions.

6.4 Theoretical analysis of biological electron transfer reactions

Biological electron transfer reactions have been the subject of a great deal of both experimental and theoretical analysis in a wide range of systems (Marcus and Sutin, 1985; Moser *et al.*, 1992; Bendall, 1996; Page *et al.*, 1999; Gray and Winkler, 2003). Of these systems, the bacterial photosynthetic reaction center is attractive, because it exhibits in a single complex a number of unusual features, including the ability to follow the reactions from liquid helium temperatures to above room temperature, and the ability to modify the free energy of the process by chemical, genetic, and electrical methods. The structure of the complex is known at high resolution, diffusion processes are not involved, and a wealth of kinetic and other spectroscopic data is available. No other biological system possesses all these features.

The theoretical description of biological electron transfer has grown out of the pioneering work of Rudolph Marcus on inorganic reactions in solution, for which he received the 1992 Nobel prize for chemistry. The electron transfer process can be viewed as a nonradiative relaxation process from an initial state with the electron donor (D) reduced and the acceptor (A) oxidized to a final state with the donor oxidized and the acceptor reduced, according to Eq. (6.1).

$$D_{red} + A_{ox} \rightarrow D_{ox} + A_{red} \qquad (6.1)$$

The Fermi golden rule describes the first-order rate constant for the electron transfer process, according to Eq. (6.2). This equation is in many ways similar to that for the Förster theory of energy transfer discussed in Chapter 5. An initial state in which an electron (or an excitation) is localized makes a nonradiative transition to a final state in which the electron (or excitation) has moved to another molecule.

$$k_{et} = \frac{2\pi}{\hbar} \left| \left\langle \Psi_i \,|\, \widetilde{V} \,|\, \Psi_f \right\rangle \right|^2 \delta \left(E_i - E_f \right) \qquad (6.2)$$

In Eq. (6.2), the transition is between two particular microscopic states of the system, which includes details such as the vibrational state and the electronic properties. Note that the wave functions for the initial and final states (Ψ_i and Ψ_f) include both donor and acceptor species. The parameter $\widetilde{V}$ is the electronic coupling between the two states and has units of energy, usually given as wavenumbers (cm^{-1}). The delta function ensures conservation of energy, in that it has a value of 1 if the energies of the two states are the same, and 0 if they are different. This means that only initial and final states of the same energy contribute to the observed rate. However, there are many possible initial states and many possible final states of the system. Many of these will differ only by the vibrational state of the molecules. In order to recover the overall rate of electron transfer, it is necessary to sum over all possible initial states and also to weight their contributions according to how likely they are to contribute. This is given in Eq. (6.3).

$$k_{et} = \frac{2\pi}{\hbar} \sum_i P_i \left| \left\langle \Psi_i \mid \tilde{V} \mid \Psi_f \right\rangle \right|^2 \delta\left(E_i - E_f\right) \quad (6.3)$$

In Eq. (6.3) the summation is over all the vibrational substates of the initial state i, weighted according to their probability P_i, times the square of the electron transfer matrix element in brackets.

Making the Born–Oppenheimer approximation for the separation of nuclear and electron wave functions simplifies this equation, as described in the Appendix. This results in Eq. (6.4), in which $\tilde{V}_e$ is the electronic coupling matrix element between the electronic states of the reactants and those of the product, and FC is the Franck–Condon factor (implicitly including the summation over all states).

$$k_{et} = \frac{2\pi}{\hbar} \tilde{V}_e^2 \, FC \quad (6.4)$$

The electronic coupling matrix element experimentally depends primarily on the distance and orientation of the reacting species. A variety of evidence indicates that this parameter depends exponentially on the distance between the reacting groups, as described by Eq. (6.5).

$$\tilde{V}_e^2 = \tilde{V}_e^{0^2} e^{-\beta d} \quad (6.5)$$

In Eq. (6.5), $\tilde{V}_e^{0^2}$ is the maximum possible electronic coupling when the molecules are in contact, β is a parameter describing the distance dependence of the coupling, and d is the distance. There has been a great deal of discussion about the proper value of β, including whether it is even appropriate

to have a single parameter that describes the distance dependence. A consensus seems to have been reached that a β value of $1.4 \, \text{Å}^{-1}$ is reasonable for protein-mediated electron transfer processes. Note the difference between the exponential distance dependence of electron transfer and the inverse sixth power distance dependence of the Förster energy transfer process. This difference arises because the Förster process is mediated by a dipole–dipole coupling interaction, whereas the electron transfer is mediated by coupling that reflects the overlap of the wave functions of the reacting species, which characteristically fall off in an exponential manner.

The Franck–Condon factor depends on the overlap of the nuclear wave functions of the initial and final states, suitably weighted by the Boltzmann factor P_i. Again, this term is almost identical to the Franck–Condon factor discussed in the Appendix in the context of electronic transitions. This term includes the effects of temperature and the free-energy change of the reaction. Unfortunately, this term is difficult to calculate, although simplifications are available. A useful way to describe the Franck–Condon factor is by means of the reorganization energy, λ, which can be considered to be the amount of energy required to distort the geometry of the reactants into that of the products without the electron transfer actually taking place. A simple equation describing how the Franck–Condon factor depends on the energy change in the reaction was proposed by Marcus and is given in Eq. (6.6).

$$FC = \left(4\pi\lambda kT\right)^{-1/2} exp - \left[\left(-\Delta G^\circ - \lambda\right)^2 / 4\lambda kT\right] \quad (6.6)$$

where ΔG° is the standard state free-energy change in the reaction. Three general regimes can be considered for the free-energy dependence of the electron transfer rate constant, shown schematically in Fig. 6.4.

If $-\Delta G^\circ < \lambda$, the process exhibits normal thermally activated behavior, illustrated by Fig. 6.4a. This is the case for endergonic to slightly exergonic reactions. If the reaction is significantly exergonic, so that $-\Delta G^\circ = \lambda$, then the potential energy curves for the reactants and products intersect near the bottom of the reactant potential well, and the reaction is in the "activationless" regime (Fig. 6.4b). The Franck–Condon factor is at a maximum, and the reaction is extremely fast, with a weak temperature dependence. Remarkably, the earliest reactions in photosynthetic reaction centers are all in this regime. Finally, for extremely exergonic reactions with $-\Delta G° > \lambda$, the system is in the "inverted" regime, and the Franck–Condon factor, and therefore the rate, decreases again (Fig. 6.4c).

The size of the Marcus λ parameter depends on the nature of the reacting groups. If both the donor and the acceptor are large molecules such as chlorophylls, which do not change the structure or vibrational frequencies much upon oxidation or reduction, then the value for λ will be relatively small, typically a few tenths of an electron volt. If, however, the molecules are very different, such as one being a quinone and the other a chlorophyll, then the value for λ will be large, as much as 1.5 eV.

The Marcus inverted region has been invoked as a major reason why the recombination reactions from the early charge-separated states in photosynthetic reaction centers are so slow compared with the forward reaction at the same point (Fig. 6.3). This may indeed be the case, although many other factors are involved that have not been considered in this brief discussion.

In many cases, a simplified picture of the electron transfer process is useful, one in which the rate of the electron transfer process is represented as primarily dependent on the distance between the donor and the acceptor (Moser et al., 1992). This formulation assumes that the free-energy change of the reaction and the Marcus λ parameter are equal, so that the process is optimal, and observed rate differences reflect only the distance between the two species that react. This is often found to be the case, or corrections can be made to account for the differences. This view supposes that most proteins are similar enough in

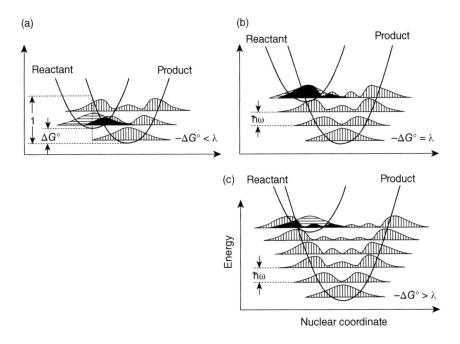

Figure 6.4   Potential energy diagrams for electron transfer reactions according to the Marcus theory. (a) Normal region, with $-\Delta G° < \lambda$, (b) Activationless region, with $-\Delta G° = \lambda$. (c) Inverted region, with $-\Delta G° > \lambda$. *Source:* Moser *et al.* (1992) (p. 797)/Springer Nature.

dielectric properties to behave in the same way, and the details of the protein structure are averaged away, so that the rate is a function only of the distance between the electron donor and acceptor. An alternative view emphasizes the detailed structure of the electron transfer proteins and analyzes potential pathways that the electron might follow from donor to acceptor (Gray and Winkler, 2003). Both approaches have advantages and disadvantages, so which is used depends on the precise questions being asked and the system under consideration.

6.5 Quinone reductions, the role of the Fe and pathways of proton uptake

The quinone molecules in the reaction center function as a two-electron gate. Figure 6.5 shows the structures of the quinones found in reaction centers, along with structures of the oxidized quinone form, the singly reduced semiquinone form, and

Figure 6.5   (a) Structures of quinones found in various photosynthetic reaction centers. Ubiquinone is found in all purple bacteria; menaquinone is found in some purple bacteria, and in all green bacteria; phylloquinone is found in Photosystem I; plastoquinone is found in Photosystem II. (b) Structure of oxidized quinone, partially reduced semiquinone, and fully reduced quinol of a generic quinone.

the fully reduced quinol form of a generic quinone. Each of these species is formed during the reaction sequence. The semiquinone form can be present as the anionic semiquinone, or a proton can bind to form the neutral semiquinone. The pKa of this proton binding is in the physiological range and is affected by the environment of the quinone. The oxidized and fully reduced forms do not exhibit acid–base behavior in the physiological range.

The photochemistry of the reaction center takes place one electron at a time. However, one of the products of the electron transfer process is a reduced ubiquinone, also known as a quinol, which has taken up two electrons as well as two protons. To form this species, the reaction center must turn over twice, with two electrons entering the complex by donation of cytochrome to the oxidized special pair and two protons are taken up from the cytoplasm. This is shown schematically in Fig. 6.6. The first electron transfer produces the semiquinone species Q_A^-. In the same time regime as the oxidation of Q_A^-, the donor D (P870$^+$) is reduced by electron donation from the cytochrome. The electron is rapidly transferred to Q_B, reoxidizing Q_A to the quinone form and making it capable of going through the reaction cycle again. The Q_B^- state is relatively stable, lasting until a second photon activates the reaction center complex.

A pathway of protonatable amino acid residues connects the Q_B binding site in the interior of the protein with the aqueous cytoplasmic compartment. The particular residues involved in the proton transport pathway have been identified in large part using site-directed mutants (Okamura et al., 2000). Figure 6.7 shows the proposed pathway of H$^+$ transfer in Rhodobacter sphaeroides. After a second turnover of the reaction center, a fully reduced neutral ubiquinol is formed in the Q_B site. This quinol is released from the complex into the hydrocarbon portion of the membrane and is

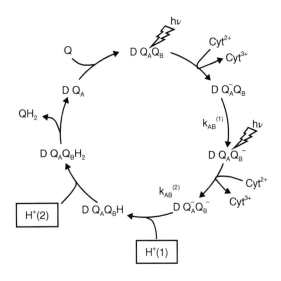

Figure 6.6 The quinone reaction cycle in bacterial reaction centers. Q_B is reduced in two one-electron reactions, $k_{AB}^{(1)}$ and $k_{AB}^{(2)}$, and takes up two protons to form the quinol. The oxidized donor is rereduced by cytochrome after each photoactivation. The reduced quinol leaves the reaction center and is replaced by a new molecule of oxidized quinone, and the system is ready for another cycle. Source: Okamura et al. (2000) (p. 151)/Elsevier.

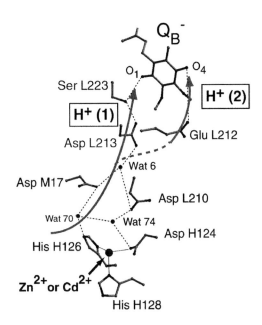

Figure 6.7 Pathway of H$^+$ uptake in Rhodobacter sphaeroides. The solid line is the pathway for the first H$^+$ to bind to Q_B; the dashed line is the pathway to protonate the other oxygen by the second H$^+$. Binding of metal ions at the site indicated blocks proton transfer. Source: Okamura et al. (2000) (p. 159)/Elsevier.

replaced by an oxidized quinone. The overall process transfers two electrons from the cytochrome to the quinone and takes up two protons from the cytoplasm. When the quinol is reoxidized by the cytochrome bc_1 complex, the protons are expelled into the periplasm. This reaction is described in detail below. The reaction center and the cytochrome bc_1 complexes, working together, use the energy in light to pump protons from the cytoplasm into the periplasm, usually against a concentration gradient. The free energy stored in this proton gradient plus the electrical potential across the membrane is converted into phosphate bond energy in ATP by the ATP synthase enzyme. This process is considered in more detail in Chapter 8.

The non-heme Fe in the reaction center is centrally positioned between the two quinone molecules. It has a distorted octahedral geometry, with four histidine nitrogen ligands and two oxygen ligands from a glutamic acid. The Fe is low-spin Fe^{2+} and exhibits no EPR spectrum unless one of the quinones is in the semiquinone state. Under these conditions, the Fe^{2+} and Q^- spins couple to give rise to a very broad EPR signal centered at $g = 1.8$. The position of the Fe^{2+} between the two quinones prompted suggestions that it may be directly involved in the electron transfer between the two quinones. However, a variety of evidence suggests that this is not the case. It may be that the Fe^{2+} serves mainly a structural function, to hold the two halves of the complex together, as it is one of the few groups to be coordinated to both the L and M subunits.

6.6   Organization of electron transfer pathways

Now we will put the pieces together to complete the electron transport chains of anoxygenic phototrophic organisms. Again, we will use the purple bacteria as a case study, although many of the principles apply equally well to all photosynthetic organisms. The reaction center carries out photochemistry and some early stabilization reactions, but additional processes must be carried out before long-term energy storage can take place. This is the job of the electron transport chain.

The reaction center creates and stabilizes oxidized and reduced species. However, if the oxidized reaction center pigment and the reduced acceptor are not regenerated, then electron transfer cannot take place a second time. It is absolutely essential for there to be a series of processes that restore the system to the state prior to the photochemical electron transfer. Electrons must be donated to the donor and extracted from the acceptor. There are two ways to accomplish this. One is to have a **cyclic electron transfer** process in which the reduced species ultimately rereduces the oxidized reaction center, so that there is no net oxidation or reduction of the substrate, as shown in Fig. 1.4. This will work as a successful energy storage strategy only if some part of the energy of the photon can be captured during the cyclic electron transfer process. The purple phototrophic bacteria operate largely in this mode, as will be described in detail below.

The other way is to have a **noncyclic electron transfer** pattern, in which a substrate feeds electrons into the system, becoming oxidized in the process, and a second substrate becomes reduced. This is shown in Fig. 1.5.

As discussed earlier, a soluble c-type cytochrome rereduces the oxidized special pair in those reaction centers that do not contain a tightly bound tetraheme cytochrome. Cytochromes are heme-containing electron transfer proteins that are alternately oxidized and reduced during their function (Moore and Pettigrew, 1990; Scott and Mauk, 1996; Cramer and Kallas, 2016). The c-type cytochrome heme group is covalently attached to the protein through two cysteine thioether linkages, as shown in Fig. 6.8. In other types of cytochromes, the hemes are not covalently attached. A simple mnemonic that helps in remembering this pattern is that "c stands for covalent."

In all cytochromes, the redox changes are localized on the iron in the center of the porphyrin ring structure. This is in contrast to chlorophyll-like pigments, in which the electron is added to or removed from the ring itself. Figure 6.8 shows the

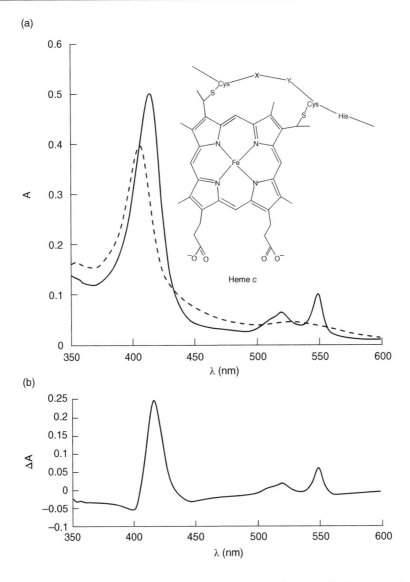

(a)

(b)

Figure 6.8   (a) Absorption spectrum of oxidized (dashed line) and reduced (solid line) cytochrome c, with the structure of the heme cofactor for c-type cytochromes. The heme group is covalently attached to the protein via two thioether linkages to cysteine residues. The lower panel (b) shows the reduced-minus-oxidized difference spectrum.

absorption spectrum of oxidized and reduced cytochrome c. Reduced cytochromes contain three prominent absorption bands, which change significantly upon oxidation. The longest-wavelength absorption band is the sharp α band, which is located in the 550–555 nm wavelength range for c-type cytochromes, and a few nanometers longer for b-type cytochromes, discussed below. The β band is broader and weaker and is located at about 525 nm. The third and strongest absorption band is the γ band, usually called the Soret band. Upon oxidation of the cytochrome, α and β bands bleach, and the Soret band shifts to shorter wavelengths. These absorbance changes make it convenient for following the electron transfer processes in which the cytochrome is oxidized and reduced.

The structure of cytochrome c_2 from *Rhodobacter sphaeroides* is shown in Fig. 6.9. The heme group is located in a crevice in the protein, with most of its surface area located in the interior of the protein, but with one edge exposed to the outside. Surrounding this exposed heme edge is a group of lysine amino acid residues, which are positively charged at physiological pH. These positive charges interact with complementary negative charges on the reaction partner to create a tightly bound complex. Cytochrome c_2 thereby binds to a specific site on the surface of the reaction center, with the heme group positioned optimally for electron transfer with respect to the special pair (Axelrod *et al.*, 2009).

The flash-induced oxidation kinetics of cytochrome c_2 *in vivo* in *Rhodobacter sphaeroides* are observed to be multiphasic (Mathis, 1994). A fast phase, with the time constant of ~1 μs, reflects the oxidation of cytochrome c_2 that is already bound to the reaction center complex. Slower phases reflect complexes in which the reduced cytochrome is not bound to the reaction center prior to electron transfer. The slower kinetics, which are second-order, result from the diffusion process that must take place before a new reduced cytochrome is in place and ready to be oxidized.

In those purple bacterial reaction centers that contain tightly bound tetraheme cytochromes, such as in *Blastochloris viridis*, the cytochrome is always properly positioned for electron transfer. The heme group closest to the special pair is oxidized in a few hundred nanoseconds after flash excitation. Subsequent interheme transfers reduce the heme in microseconds. Finally, a cytochrome c_2 reduces the tetraheme cytochrome. In some organisms, alternate electron donor proteins to cytochrome c_2 are found, such as cytochrome c_y, cytochrome c_8, or a high-potential iron–sulfur protein (HiPIP) (Menin *et al.*, 1998; Myllykallio *et al.*, 1998).

The overall process carried out by the reaction center is given by Eq. (6.7):

$$2\text{cyt}\,c_{2\text{red}} + \text{UQ} + 2\text{H}^+ + 2h\nu \rightarrow 2\text{cyt}c_{2\text{ox}} + \text{UQH}_2 \tag{6.7}$$

In this reaction, protons are taken up from the cytoplasmic side of the membrane and the cytochrome c_2 is oxidized on the periplasmic side. The quinones are found in the nonpolar region of the membrane. The reaction center can be considered to be a light-driven proton-pumping cytochrome c_2–ubiquinone oxidoreductase. The reactions that complete the cyclic electron transport chain are mediated by the cytochrome bc_1 complex, described next.

6.7 Completing the cycle – the cytochrome bc_1 complex

The cytochrome bc_1 complex is a large, multisubunit, integral membrane protein complex, which is located in the intracytoplasmic membrane of the purple bacteria (Xia *et al.*, 1997; Zhang *et al.*, 1998; Berry *et al.*, 2000; Crofts, 2004; Kramer *et al.*, 2009; Sarewicz *et al.*, 2021). A similar complex is found in many types of nonphototrophic bacteria and also in mitochondria of eukaryotic cells. The cytochrome

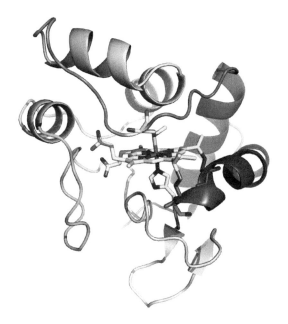

Figure 6.9 Structure of cytochrome c_2 from *Rhodobacter sphaeroides*. The axial methionine and histidine ligands to the Fe and the heme group are shown. The backbone is color-coded with the N terminus blue and the C terminus red. Figure produced from Protein Data Bank file 1CXC using Pymol.

b_6f complex found in oxygenic photosynthetic organisms is also generally similar in both structure and function and will be described in Chapter 7. The cyt bc_1 complex consists of a minimum of three protein subunits, known as cytochrome b, cytochrome c_1, and the "Rieske" iron–sulfur protein, (hereafter called ISP), named after the scientist who first described it. In some organisms, additional subunits are found. The structure of the cyt bc_1 complex from mitochondria and other sources has been determined by X-ray diffraction. The structure of the cyt bc_1 complex from *Rhodobacter sphaeroides* is shown in Fig. 6.10. Several lines of evidence indicate that the mitochondrial and bacterial complexes are very similar in terms of the core structure that contains the electron transfer cofactors. The mitochondrial cyt bc_1 complex contains several additional subunits whose function is not known with certainty but may be involved in assembly or regulation.

Cytochrome b is an integral membrane protein of 40–45 kDa mass, containing eight transmembrane helical segments that firmly anchor it in the membrane. Two noncovalently associated protoheme cofactors (heme b) are buried in the membrane-spanning portion of the complex and are coordinated by imidazole groups from four histidine residues, giving each heme bis-His ligation. One heme is closer to the periplasmic side, and the other is closer to the cytoplasmic side, with both heme planes roughly perpendicular to the membrane plane, as shown schematically in Fig. 6.11, with the pathways of electron transfer superimposed. The redox properties of the two heme groups are significantly different. The heme nearer the periplasmic side has a low midpoint potential of $-100\,\text{mV}$ and is known as cytochrome b_L, with the L subscript indicating a low redox potential. The heme nearer the cytoplasmic side has a significantly higher redox potential of $+50\,\text{mV}$ and is known as cytochrome b_H. Cytochrome b also contains two quinone-binding sites. One of these sites, known as the Q_o site the quinone-oxidizing (also outside) site is located near the periplasmic side of the cytochrome b. The other site is near the cytoplasmic side of the membrane and is where the quinone is reduced. It is usually called the Q_i site (*i* for inside).

(a)   (b)

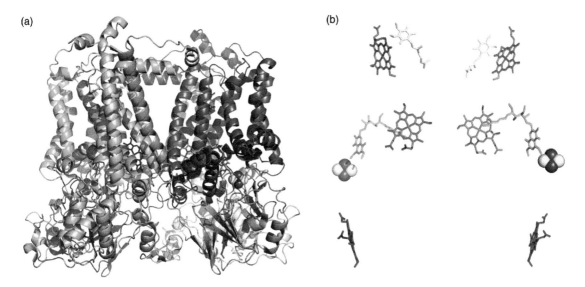

Figure 6.10   Structure of the cytochrome bc_1 complex from *Rhodobacter sphaeroides*. (a) Structure of the entire dimeric complex. The cytochrome b subunits are shown in green and purple, the cytochrome c_1 subunits in orange and cyan, the ISP protein subunits in red and olive. (b) Cofactor arrangement. The b-type hemes are shown in blue, the c-type hemes in red, and the FeS cluster is shown in a space-filling representation with Fe in red and sulfur is yellow. Ubiquinone is shown as yellow and the inhibitor stigmatellin is shown in green. Figure produced from Protein Data Bank file 2QJY using Pymol.

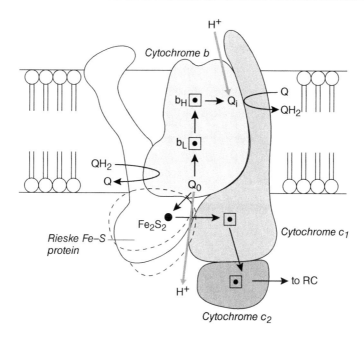

Figure 6.11   Schematic structure of the cytochrome bc_1 complex from purple bacteria. The pathway of electron transfer is overlaid on the structure. The dashed lines indicate movement of the Rieske protein.

The ISP is largely located on the periplasmic side of the membrane, with a single N-terminal transmembrane segment anchoring it to the membrane. The total mass is about 20 kDa. The ISP has an unusual 2Fe–2S Fe–S cluster, in which the ligands for the Fe atoms are two Cys and two His residues (typical Fe–S centers have only Cys ligands), as shown in Fig. 6.12. This difference in ligation causes the Fe–S center to have a much higher redox potential than that of other Fe–S centers, with a nominal midpoint potential of +280 mV, although the potential may change as the complex moves. The ISP undergoes a large-amplitude motion during the catalytic cycle. This motion has been inferred from X-ray studies, which reveal large differences in the location of the ISP under different conditions, such as inhibitor addition.

The cytochrome c_1 subunit also has most of its mass on the periplasmic side of the membrane, anchored by a C-terminal membrane-spanning helix. The structure of the cytochrome c_1 subunit is generally similar to those of the soluble c-type cytochromes such as cytochrome c_2, with the exception of the anchor. The redox potential of cytochrome c_1 is +290 mV, so it can be reduced by the ISP and can in turn reduce cytochrome c_2, which has a midpoint redox potential of +285 mV versus NHE.

Figure 6.10 shows the structure of the cytochrome bc_1 complex, as determined by X-ray crystallography. The complex is structurally dimeric. Each half of the dimer can operate independently, although communication between the two-electron transport chains has been demonstrated (Swierczek *et al.*, 2010).

6.7.1   The mechanism of electron and proton transfer in the cytochrome bc_1 complex

The detailed mechanism of electron and proton flow through the cytochrome bc_1 complex is not yet fully understood, but a mechanism known as the

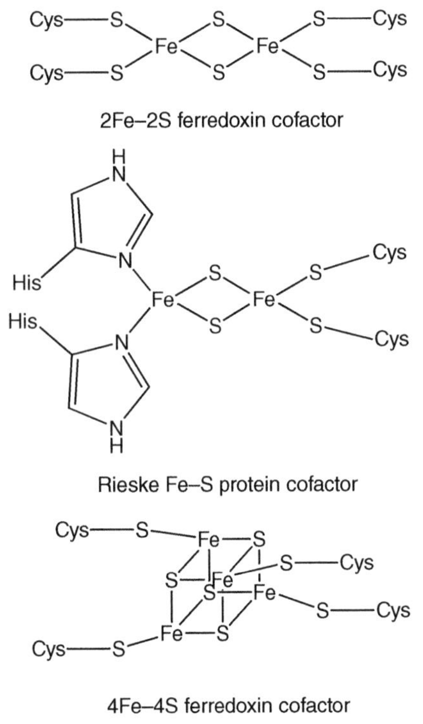

Figure 6.12   Structure of the Fe–S cofactors in various Fe–S proteins. Top: 2Fe–2S center from a soluble plant-type ferredoxin. Middle: structure of the Rieske Fe–S protein cofactor of the ISP. Bottom: 4Fe–4S center in the bacterial-type ferredoxins and the bound Fe–S centers of Photosystem I.

modified Q cycle accounts for most of the observations (Crofts, 2004; Sarewicz et al., 2021). In this mechanism, two molecules of ubiquinol are oxidized, and one is reduced. The protons resulting from quinone oxidation are delivered to the periplasmic side of the membrane, and those taken up during reduction come from the cytoplasmic side. In addition, electrons are transferred to cytochrome c_2 and eventually are used to rereduce the oxidized special pair of the reaction center. The overall reaction for two turnovers of the cyt bc_1 complex is given by Eq. (6.8):

$$2\,UQH_2 + 2\,cyt\,c_{2ox} + UQ + 2\,H^+ \rightarrow 2\,UQ \\ + 4\,H^+ + 2\,cyt\,c_{2red} + UQH_2 \tag{6.8}$$

UQ and UQH_2 appear on both sides of this equation. The quinone species shown in regular

type are involved in the quinone oxidation phase of the mechanism, while those in bold type are involved in the quinone reduction phase. The stoichiometry of protons pumped per electron transferred through the chain is $2H^+/e^-$ (this is most easily calculated by focusing on the two electrons that exit the complex, compared with the four protons released on the periplasmic side).

The reaction sequence is as follows and is shown schematically in Fig. 6.11:

1. A ubiquinol molecule binds to the quinine-oxidizing (Q_o) site, located near the periplasmic side of cytochrome b.
2. An electron is transferred to the ISP Fe–S center, leaving a ubisemiquinone in the Q_o site. One proton is released to the periplasm.
3. The ubisemiquinone transfers the second electron to heme b_1 and is then transferred on to heme b_H. The second proton is released to the periplasm.
4. The ISP moves away from the cytochrome b and towards the cytochrome c_1. The motion is large, so the Fe–S center is displaced by up to 20 Å.
5. A ubiquinone molecule binds to the Q_i site and is reduced to the semiquinone by the electron on heme b_H.
6. The oxidized ubiquinone dissociates from the Q_o site.
7. The ISP Fe–S center reduces cytochrome c_1, which goes on to reduce cytochrome c_2. The ISP moves back to its original position.

At this point, one quinol molecule has been oxidized, and one of the electrons has traversed the electron transport chain, while the other has gone to reduce an oxidized ubiquinone to the semiquinone state. The system is now poised for a second turnover.

8. A new ubiquinol molecule binds to the Q_o site and is oxidized, with the electrons transferred to the ISP, cytochrome c_1 and cytochrome c_2 on one branch and heme b_L and b_H from cytochrome b on the other branch, as in steps 2–7.
9. The electron on heme b_H reduces the ubisemiquinone to the quinol form, taking up two protons from the cytoplasmic side of the membrane. The reduced quinol dissociates from the complex.

One of the most remarkable aspects of the proposed mechanism is the large-amplitude movement of the ISP, swinging back and forth from the cytochrome b to the cytochrome c_1. While this may seem unnecessarily complex, it actually neatly solves a serious problem that would otherwise significantly reduce the proton-pumping efficiency of the complex. After the first electron is transferred to the ISP, what is to prevent it from rapidly reacting with the cytochrome c_1 and then also taking the second electron from the ubisemiquinone? This would short-circuit the cycle through the b hemes and would reduce the proton-to-electron ratio from two to one. This would cut the energy storage in half. The reduced ISP cannot accept a second electron from the quinone, so the electron must go to heme b_L. The ISP then moves and delivers a single electron to cytochrome c_1 with essentially no chance of a double transfer. This "swinging door," or "ratchet," mechanism may be essential in preventing this unwanted process, which otherwise would be very likely to significantly reduce the efficiency of energy storage in bacterial photosynthesis and in other electron transport chains utilizing similar complexes. Whether this explanation for the possible functional role of the motion of the ISP is correct is not yet clear, but numerous studies have clearly shown that the motion is necessary for the complex to function (Sarewicz *et al.*, 2021).

Several inhibitors have been important in understanding the cytochrome bc_1 complex. Three of the most important are antimycin A, stigmatellin, and myxothiazol. Antimycin A blocks electron transfer from the heme b_H, thereby preventing the reduction of the quinone at the Q_i site, and is observed to bind to that region of the cytochrome b that makes up the Q_i site. The other two inhibitors bind in the region of the Q_o site, in similar, but not identical, sites. They both interfere with quinol oxidation. Additional insight has come from the analysis of numerous mutants of the various components of the complex.

The net result of two turnovers of the cytochrome bc_1 complex is that two electrons are transferred to cytochrome c_2, two ubiquinols are oxidized to the quinone form, and one oxidized ubiquinone is reduced to the hydroquinone quinol form. In addition, four protons are transferred from the cytoplasmic to the periplasmic side of the membrane.

In this way, electron flow connecting the acceptor side of the reaction center to the donor side gives rise to a proton motive force across the membrane, due to H^+ concentration differences on the two sides of the membrane. This proton motive force is used to power the synthesis of ATP, as will be discussed in more detail in Chapter 8.

6.8   Membrane organization in purple bacteria

Considerable evidence now suggests that the LH1 antenna complex that surrounds the reaction center is in many cases interrupted by a small protein called PufX (Qian *et al.*, 2008). This protein is thought to provide a channel by which quinone molecules can enter and leave the reaction center. In *Rhodobacter sphaeroides*, the PufX protein facilitates the dimerization of the RC–LH1 complex into dimers, while in other cases PufX is not present and the RC–LH1 complex is monomeric.

6.8.1   Other electron transport pathways in purple bacteria

The light-driven cyclic electron transport pathway described in detail in the previous section is the principal source of cellular energy when cells are grown under photosynthetic conditions. Under these conditions of little or no oxygen and the presence of light, the only product of light-driven electron flow is an electrochemical potential. There is no net oxidation or reduction of substrate compounds. This potential can be transformed into ATP, and this serves to power a large number of cellular processes. However, ATP is not sufficient as the sole source of cellular free energy. A large number of cellular reactions involve net oxidation or reduction of various compounds. For these to take place, it is necessary to have a source and a sink for electrons. The most important example of this is in the reduction of CO_2 to form sugars, which takes place via the Calvin–Benson cycle. This will be described in detail in Chapter 9. Here it is sufficient to point out that the substrates for this set of reactions are CO_2, ATP, and NADH (NADPH in oxygen-evolving organisms). The CO_2 is taken up from the environment, and the ATP is synthesized as a result of light-driven cyclic electron flow. What is the source of the reductant NADH? It is necessary to provide a continuing source of this strong reducing agent, as it is consumed during CO_2 assimilation. These organisms are extremely versatile metabolically and can utilize a variety of electron donors as sources of electrons, including H_2S, organic compounds such as succinate, and even H_2, as shown in Fig. 6.13. Their oxidation feeds electrons into the ubiquinone pool, reducing it to the quinol form. However, the reduced quinone is not a sufficiently strong reducing agent to reduce NAD^+ (+110 vs. −320 mV). To accomplish this, an energy-dependent electron transfer takes place, in which reduced quinone is the electron donor and NAD^+ is the acceptor. This process is almost certainly mediated by the NADH dehydrogenase (Complex I) running backward, although this has not been extensively studied. The energy for this uphill electron transfer is supplied by a transmembrane chemiosmotic potential. This potential was, of course, built up by the light-driven cyclic electron transport system. The reduction of NAD^+ is thereby coupled to photosynthetic electron flow, although the coupling is indirect, via the chemiosmotic potential. The key experiment that reveals this indirect coupling is that uncouplers, compounds that dissipate the chemiosmotic potential, block light-dependent reduction of NAD^+ (Knaff, 1978).

6.9   Electron transport in other anoxygenic phototrophic bacteria

The filamentous anoxygenic phototrophic bacteria have a cyclic pattern of electron transport similar to that of purple bacteria, although there are a number of differences. The quinone is menaquinone instead of ubiquinone, and this has a significantly lower redox potential. In addition, these organisms lack soluble cytochromes similar to cytochrome c_2. A small blue copper protein known as auracyanin, similar to the plastocyanin described below in oxygenic photosynthetic organisms, probably rereduces a tetraheme cytochrome that in turn reduces P870 in the manner of cytochrome c_2. There is also no cytochrome bc_1 complex in this group of organisms. Instead, they contain a structurally unrelated complex known as alternative complex III. Figure 6.14 illustrates the

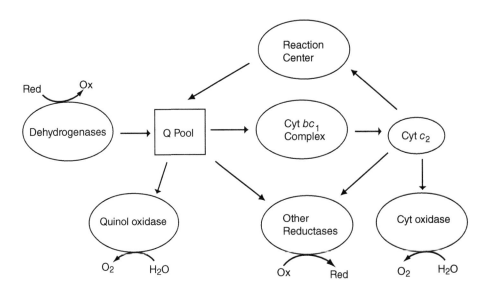

Figure 6.13   Alternate pathways for electron flow in purple photosynthetic bacteria.

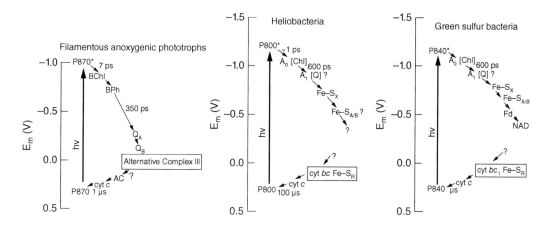

Figure 6.14 Pathway of electron transfer in the filamentous anoxygenic bacteria, heliobacteria, and green sulfur bacteria. The involvement of quinones in the electron transfer pathways in heliobacteria and green sulfur bacteria is uncertain.

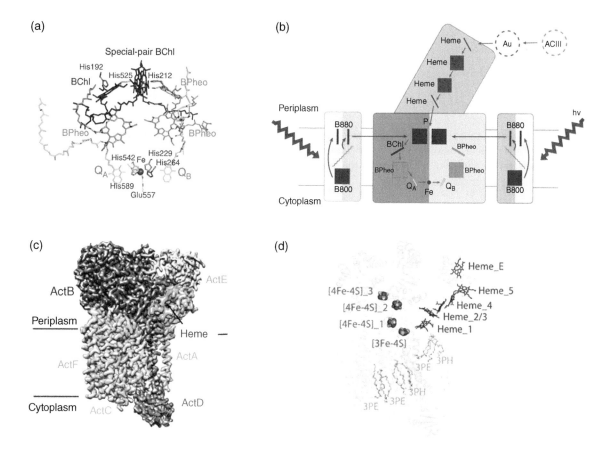

Figure 6.15 Structure of the reaction center and alternative complex III from the Filamentous Anoxygenic Phototroph *Roseiflexus castenholzii*. (a) Reaction center electron transfer cofactors (bound cytochrome and LH1 antenna complexes not shown). (b) Cartoon of electron transfer pathway. (c) Structure of alternative complex III. (d) Overlay of electron transfer cofactors in alternative complex III. *Source:* Adapted from Xin et al. (2018) and Shi et al. (2020).

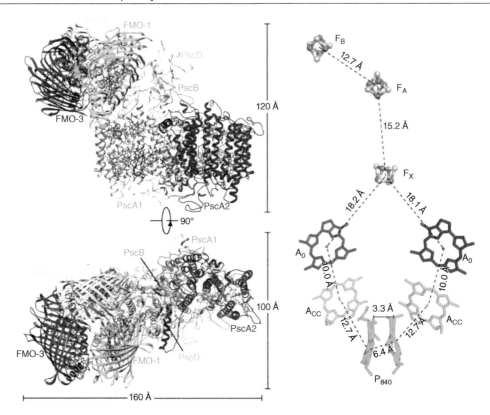

Figure 6.16   Cryo-EM structure of the homodimeric green sulfur bacterial reaction center. Left: Structure of the protein part of the complex viewed from the membrane (top) and from the cytoplasmic side of the membrane (bottom). The cofactor arrangement is shown in the right panel. *Source:* Chen *et al.* (2020) (p. 2–3)/American Association for the Advancement of Science.

electron transport pathway in *Chloroflexus aurantiacus*, a representative member of this group of organisms (Blankenship, 1994). Figure 6.15 shows the cryo-EM structure of the reaction center and the alternative complex III from the FAP *Roseiflexus castenholzii*, a closely related organism (Xin *et al.*, 2018; Shi *et al.*, 2020).

Green sulfur bacteria have a single photosystem similar to Photosystem I, such that it can directly reduce ferredoxin and then NAD$^+$ without the need for energy-dependent reverse electron flow. This reaction center is a homodimer, consisting of two identical copies of a core subunit (Büttner *et al.*, 1992; He *et al.*, 2014). A structural model of the green sulfur bacterial reaction center is shown in Fig. 6.16 (Chen *et al.*, 2020). Reduced compounds such as H_2S or thiosulfate serve as electron donors, with CO_2 as the electron acceptor for noncyclic electron flow. The pathway of CO_2 fixation in the green sulfur bacteria is the reverse tricarboxylic acid cycle. This cycle is very different from the Calvin–Benson cycle, which is the pathway utilized by most other photosynthetic organisms. This is discussed in more detail in Chapter 9. These organisms can probably also carry out cyclic electron flow, but this is not very well documented. The pattern of electron transport in *Chlorobaculum tepidum*, a representative of this group, is shown in Fig. 6.14 (Sakurai *et al.*, 1996).

Heliobacteria, which have a homodimeric reaction center that is structurally similar to that of green sulfur bacteria, have a significantly different

(a)   (b)

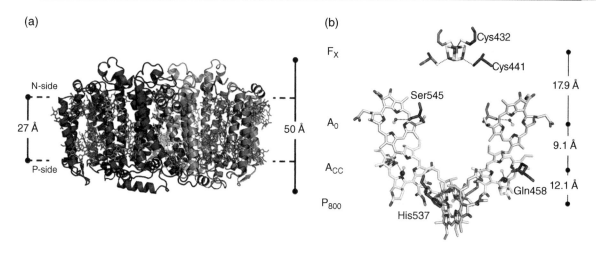

Figure 6.17 Structure of the homodimeric photosystem from *Heliobacterium modesticaldum*. (a) Overall structure of the photosystem. (b) Electron transfer cofactors. *Source:* Adapted from Gisriel *et al.* (2017).

pathway of electron transfer. They carry out cyclic electron transfer using only membrane-bound cytochromes (Fig. 6.14). It is not yet clear if they are able to reduce NAD(P)$^+$ using noncyclic electron flow (Sattley and Blankenship, 2010). The structure of the reaction center from *Heliobacterium modesticaldum* is shown in Fig. 6.17 (Gisriel *et al.*, 2017). In addition to the electron transfer cofactors, it contains about 50 antenna bacteriochlorophylls.

References

Amesz, J. and Hoff, A. J., (eds.) (1996) *Biophysical Techniques in Photosynthesis*. Dordrecht: Kluwer Academic Publishers.

Artsma, T. and Matysik, J., (eds.) (2008) *Biophysical Techniques in Photosynthesis*, Vol. II. Dordrecht: Springer.

Axelrod, H., Miyashita, O., and Okamura, M. (2009) Structure and function of the cytochrome c_2: Reaction center complex from *Rhodobacter sphaeroides*. In: C. N. Hunter, F. Daldal, M. C. Thurnauer, and J. T. Beatty, (eds.) *The Purple Photosynthetic Bacteria*. Dordrecht: Springer, pp. 323–336.

Bendall, D. S., (ed.) (1996) *Protein Electron Transfer*. Oxford: Bios Scientific Publishers.

Berry, E. A., Guergova-Kuras, M., Huang, L.-S., and Crofts, A. R. (2000) Structure and function of cytochrome *bc* complexes. *Annual Review of Biochemistry* 69: 1005–1075.

Blankenship, R. E. (1994) Protein structure, electron transfer and evolution of prokaryotic photosynthetic reaction centers. *Antonie van Leeuwenhoek Journal of Microbiology* 65: 311–329.

Blankenship, R. E. and Prince, R. C. (1985) Excited state redox potentials and the Z scheme of photosynthesis. *Trends in Biochemical Sciences* 10: 382–383.

Blankenship, R. E., Madigan, M. T., and Bauer, C. E., (eds.) (1995) *Anoxygenic Photosynthetic Bacteria*. Dordrecht: Kluwer Academic Publishers.

Büttner, M., Xie, D., Nelson, H., Pinther, W., Hauska, G., and Nelson, N. (1992) Photosynthetic reaction center genes in green sulfur bacteria and in photosystem-1 are related. *Proceedings of the National Academy Sciences USA* 89: 8135–8139.

Chen, J.-H., Wu, H., Xu, C., Liu, X.-C., Huang, Z., Chang, S., Wang, W., Han, G., Kuang, T., Shen, J.-R., and Zhang, X. (2020) Architecture of the photosynthetic complex from a green sulfur bacterium. *Science* 370: 931.

Clayton, R. K. (1988) Memories of many lives. *Photosynthesis Research* 19: 207–224.

Cramer, W. A. and Kallas, T., (eds.) (2016) *Cytochrome Complexes: Evolution, Structures, Energy Transduction and Signaling*. Dordrecht: Springer.

Crofts, A. R. (2004) The cytochrome bc_1 complex: Function in the context of structure. *Annual Review of Physiology* 66: 689–733

Deisenhofer, J. and Michel, H. (1989) The photosynthetic reaction center from the purple bacterium *Rhodopseudomonas viridis*. *Science* 245: 1463–1473.

Deisenhofer, J. and Norris, J. R. (1993) *The Photosynthetic Reaction Center*. San Diego: Academic Press.

DeVault, D. and Chance, B. (1966) Studies of photosynthesis using a pulsed laser. I. Temperature dependence of cytochrome oxidation rate in *Chromatium*. Evidence for tunneling. *Biophysical Journal* 6: 825–847.

Drews, G. and Golecki, J. R. (1995) Structure, molecular organization and biosynthesis of membranes of purple bacteria. In: R. E. Blankenship, M. T. Madigan, and C. E. Bauer, (eds.) *Anoxygenic Photosynthetic Bacteria*. Dordrecht: Kluwer Academic Publishers, pp. 231–257.

Faries, K. M., Kressel, L. L., Wander, M. J., Holten, D., Laible, P. D., Kirmaier, C., and Hanson, D. K. (2012) High throughput engineering to revitalize a vestigial electron transfer pathway in bacterial photosynthetic reaction centers. *The Journal of Biological Chemistry* 287: 8507–8514.

Feher, G. (1998) Three decades of research in bacterial photosynthesis and the road leading to it: A personal account. *Photosynthesis Research* 55: 1–40.

Fromme, P., (ed.) (2008) *Photosynthetic Protein Complexes*. Weinheim: Wiley-Blackwell.

Gisriel, C., Sarrou, I., Ferlez, B., Golbeck, J. H., Redding, K. E., and Fromme, R. (2017) Structure of a symmetric photosynthetic reaction center-photosystem. *Science* 357: 1021–1025.

Gray, H. B. and Winkler, J. R. (2003) Electron tunneling through proteins. *Quarterly Reviews of Biophysics*, 36: 341–372.

He, G., Zhang, H., King, J. D., and Blankenship, R. E. (2014) Structural analysis of the homodimeric reaction center complex from the photosynthetic green sulfur bacterium *Chlorobaculum tepidum*. *Biochemistry* 53: 4924–4930.

Hoff, A. J. and Deisenhofer, J. (1997) Photophysics of photosynthesis. Structure and spectroscopy of reaction centers of purple bacteria. *Physics Reports* 287: 1–247.

Hunter, C. N., Daldal, F., Thurnauer, M. C., and Beatty, J. T., (eds.) (2009) *The Purple Photosynthetic Bacteria*. Dordrecht: Springer.

Jones, M. (2021) Purple bacteria: Photosynthetic reaction centers. In: J. Jez, (ed.) *Encyclopedia of Biological Chemistry*, 3rd Edn. Amsterdam: Elsevier.

Knaff, D. B. (1978) Reducing potentials and the pathway of NAD^+ reduction. In: R. K. Clayton and W. R. Sistrom, (eds.) *The Photosynthetic Bacteria*. New York: Plenum, pp. 629–640.

Kramer, D. M., Nitschke, W., and Cooley, J. W. (2009) The cytochrome bc_1 and related bc complexes: The rieske/cytochrome b complex as the functional core of a central electron/proton transfer complex. In: C. N. Hunter, F. Daldal, M. C. Thurnauer, and J. T. Beatty, (eds.) *The Purple Photosynthetic Bacteria*. Dordrecht: Springer, pp. 451–473.

Laible, P. D., Hanson, D. K., Buhrmaster, J. C., Tira, G. A., Faries, K. M., Holten, D., and Kirmaier, C. (2020) Switching sides—Reengineered primary charge separation in the bacterial photosynthetic reaction center. *Proceedings of the National Academy of Sciences USA* 117: 865–871.

Marcus, R. A. and Sutin, N. (1985) Electron transfers in chemistry and biology. *Biochimica et Biophysica Acta* 811: 265–322.

Mathis, P. (1994) Electron-transfer between cytochrome c_2 and the isolated reaction-center of the purple bacterium *Rhodobacter sphaeroides*. *Biochimica et Biophysica Acta* 1187: 177–180.

Menin, L., Gaillard, J., Parot, P., Schoepp, B., Nitschke, W., and Vermeglio, A. (1998) Role of HiPIP as electron donor to the RC-bound cytochrome in photosynthetic purple bacteria. *Photosynthesis Research* 55: 343–348.

Moore, G. R. and Pettigrew, G. W. (1990). *Cytochromes c: Evolutionary, Structural, and Physicochemical Aspects*. Berlin: Springer.

Moser, C. C., Keske, J. M., Warncke, K., Farid, R. S., and Dutton, P. L. (1992) Nature of biological electron transfer. *Nature* 355: 796–802.

Myllykallio, H., Drepper, F., Mathis, P., and Daldal, F. (1998) Membrane-anchored cytochrome c_y mediated microsecond time range electron transfer from the cytochrome bc_1 complex to the reaction center in *Rhodobacter capsulatus*. *Biochemistry* 37: 5501–5510.

Okamura, M. Y., Paddock, M. L., Graige, M. S., and Feher, G. (2000) Proton and electron transfer in bacterial reaction centers. *Biochimica et Biophysica Acta* 1458: 148–163.

Ortega, J. M., Drepper, F., and Mathis, P. (1999) Electron transfer between cytochrome c_2 and the tetraheme cytochrome c in *Rhodopseudomonas viridis*. *Photosynthesis Research* 59: 147–157.

Page, C. C., Moser, C. C., Chen, X. X., and Dutton, P. L. (1999) Natural engineering principles of electron tunneling in biological oxidation-reduction. *Nature* 402: 47–52.

Parson, W. W. and Warshel, A. (2009) Mechanism of charge separation in purple bacterial reaction centers.

In: C. N. Hunter, F. Daldal, M. C. Thurnauer, and J. T. Beatty, (eds.) *The Purple Photosynthetic Bacteria.* Dordrecht: Springer, pp. 355–377.

Qian, P., Bullough, P. A., and Hunter, C. N. (2008) Three-dimensional reconstruction of a membrane-bending complex – The RC-LH1-PufX core dimer of *Rhodobacter sphaeroides. Journal of Biological Chemistry* 283: 14002–14011.

Sakurai, H., Kusumoto, N., and Inoue, K. (1996) Function of the reaction center of green sulfur bacteria. *Photochemistry and Photobiology* 64: 5–13.

Sarewicz, M., Pintscher, S., Pietras, R., Borek, A., Bujnowicz, Ł., Hanke, G., Cramer, W., Finazzi, G., and Osyczka, A. (2021) Catalytic reactions and energy conservation in the cytochrome bc_1 and b_6f complexes of energy-transducing membranes. *Chemical Reviews* 121: 2020-2108.

Sattley, W. M. and Blankenship, R. E. (2010) Insights into heliobacterial photosynthesis and physiology from the genome of *Heliobacterium modesticaldum. Photosynthesis Research* 104: 113–122.

Scott, R. A. and Mauk, A. G., (eds.) (1996) *Cytochrome c: A Multidisciplinary Approach.* Sausalito, CA: University Science Books.

Sener, M., Strumpfer, J., Singharoy, A., Hunter, C. N., and Schulten, K. (2016) Overall energy conversion efficiency of a photosynthetic vesicle. *eLife* 5: e09541.

Shi, Y., Xin, Y., Wang, C., Blankenship, R. E., Sun, F., and Xu, X. (2020) Cryo-EM structures of the air-oxidized and dithionite-reduced photosynthetic alternative complex III from *Roseiflexus castenholzii. Science Advances* 6: eaba2739.

Swierczek, M., Cieluch, E., Sarewicz, M., Borek, A., Moser, C. C., Dutton, P. L., and Osyczka, A. (2010) An electronic bus bar lies in the core of cytochrome bc_1. *Science* 329: 451–454.

Tucker, J. D., Siebert, C. A., Escalante, M., Adams, P. G., Olsen, J. D., Otto, C., Stokes, D. L., and Hunter, C. N. (2010) Membrane invagination in *Rhodobacter sphaeroides* is initiated at curved regions of the cytoplasmic membrane, then forms both budded and fully detached spherical vesicles. *Molecular Microbiology* 76: 833–847.

Wakeham, M. C. and Jones, M. R. (2005) Rewiring photosynthesis: Engineering wrong-way electron transfer in the purple bacterial reaction centre. *Biochemical Society Transactions* 33: 851–857.

Xia, D., Yu, C. A., Kim, H., Xian, J. Z., Kachurin, A. M., Zhang, L., Yu, L., and Deisenhofer, J. (1997) Crystal structure of the cytochrome bc_1 complex from bovine heart mitochondria. *Science* 277: 60–66.

Xin, Y., Shi, Y., Niu, T., Wang, Q., Niu, W., Huang, X., Ding, W., Yang, L., Blankenship, R. E., Xu, X., and Sun, F. (2018) Cryo-EM structure of the RC-LH core complex from an early branching photosynthetic prokaryote. *Nature Communications* 9: 1568.

Zhang, Z., Huang, L., Shulmeister, V. M., Chi, Y. I., Kim, K. K., Hung, L. W., Crofts, A. R., Berry, E. A., and Kim, S. H. (1998) Electron transfer by domain movement in cytochrome bc_1. *Nature* 392: 677–684.

Chapter 7

Reaction centers and electron transfer pathways in oxygenic photosynthetic organisms

In this chapter, we discuss the structure and mechanism of the reaction centers and electron transport complexes found in oxygenic photosynthetic organisms. We will begin with some aspects of membrane organization and then examine the structural and mechanistic aspects of the Photosystem II reaction center, including the important process of H_2O oxidation to form O_2. We will then follow electrons from Photosystem II, through the cytochrome $b_6 f$ complex, to Photosystem I all the way to reduction of $NADP^+$ to produce NADPH, which is used as the reductant in the process of CO_2 reduction. Assembly, regulation, and repair of photosynthetic complexes will be discussed in Chapter 10.

7.1 Spatial distribution of electron transport components in thylakoids of oxygenic photosynthetic organisms

The thylakoid membranes of oxygenic photosynthetic organisms contain both Photosystem I and Photosystem II with their associated antennas, as well as the cytochrome $b_6 f$ complex (Ort and Yocum, 1996; Wydrzynski and Satoh, 2005; Golbeck, 2006; Fromme, 2008; Cramer and Kallas, 2016). As discussed in Chapter 2, the thylakoid membranes in higher plant chloroplasts form large, interconnected stacks of membranes called **grana lamellae**, connected by nonstacked membranes called **stroma lamellae**.

A variety of lines of evidence has established that when these highly stacked membranes are present, the distribution of membrane-bound complexes is highly nonuniform (Allen and Forsberg, 2001; Anderson, 2002; Nevo *et al.*, 2012; Nagy *et al.*, 2014; Kirchhoff *et al.*, 2017). Perhaps the most direct evidence comes from immunocytochemistry coupled with electron microscopy studies. In these experiments, antibodies that recognize a particular protein complex are allowed to react with thin sections of membranes. The antibodies are then decorated with a secondary antibody that is conjugated to an electron-dense material such as gold, and the sample is visualized using electron microscopy. The distribution of the label is then compared with the membrane architecture, to determine the lateral distribution of complexes in the membrane. The results show a remarkable nonuniform distribution

Molecular Mechanisms of Photosynthesis, Third Edition. Robert E. Blankenship.
© 2021 Robert E. Blankenship 2021 by John Wiley & Sons Ltd.
Companion website: https://www.wiley.com/go/blankenship/molecularphotosynthesis3e

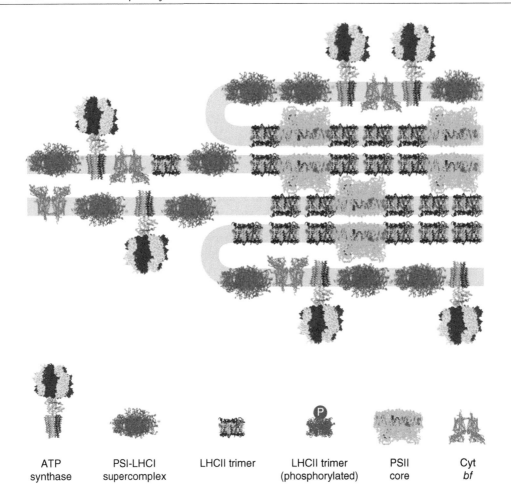

| ATP synthase | PSI-LHCI supercomplex | LHCII trimer | LHCII trimer (phosphorylated) | PSII core | Cyt *bf* |

Figure 7.1   Lateral heterogeneity of protein complexes in stacked and unstacked thylakoid membranes. *Source:* Adapted from Nagy *et al.* (2014). Reproduced with permission of US National Academy of Sciences.

of complexes. Photosystem II is found almost exclusively in the stacked regions of the membranes, whereas Photosystem I is found primarily in the stroma lamellae, as well as on the ends of the grana stacks. The cytochrome $b_6 f$ complex is nearly equally distributed between the two types of membranes, and the ATP synthase enzyme complex is entirely localized to the stroma lamellae. Figure 7.1 illustrates the lateral distribution of components in a typical higher plant chloroplast.

This spatial distribution of membrane components imposes severe constraints on the pathways of electron and proton transfer in the thylakoids. In normal noncyclic electron flow, water oxidation in Photosystem II generates electrons that reduce plastoquinone. Because of the equal distribution of the cytochrome $b_6 f$ complex, the oxidation of plastoquinone by this complex and the reduction of plastocyanin can take place either in the stacked grana membranes or in the unstacked stroma membranes. The oxidation of plastocyanin and the reduction of ferredoxin by Photosystem I take place in the unstacked stroma membranes. If the oxidation of plastoquinone takes place in the grana, then the reduced plastocyanin must diffuse in the thylakoid lumen to the stroma, which can be tens or hundreds of nanometers distant. The thylakoid lumen is a very crowded space where diffusion is restricted (Kirchhoff *et al.*, 2011). If the oxidation of plastoquinone takes place in the cytochrome b_6f complex that is located in the stroma, then the plastoquinone must diffuse within the lipid bilayer from the grana membranes to the stromal membranes. In either case, a long-distance diffusion process is necessary to complete the traversal of the electron transport chain. Mutant studies suggest that plastocyanin diffusion in the lumen is the major pathway for electrons to be shuttled between the photosystems during noncyclic electron flow (Höhner *et al.*, 2020).

In addition to the long-distance electron transport described above, a similar problem exists for the protons produced during water oxidation and the cytochrome b_6f function. The ATP synthase enzyme complex is exclusively located in the unstacked stromal membranes, yet many of the protons are generated in the grana and must thus diffuse within the aqueous luminal space tens to hundreds of nanometers to reach the ATP synthase.

The possible functions of this remarkable lateral distribution of membrane-bound complexes and their mechanistic consequences are still not very well understood. Cyanobacteria, algae, and some mutants of higher plants do not exhibit the extremes of grana stacking and lateral distribution of photosynthetic complexes observed in higher plants, yet all these organisms seem to function efficiently. Most indications are that the function of the lateral segregation is related to the regulation of antenna function (Kana, 2013). The lateral movement of the LHCII antenna discussed in Chapter 5 is an example of this regulation.

7.2   Noncyclic electron flow in oxygenic organisms

The major mode of electron transport in oxygenic photosynthetic organisms is noncyclic electron transfer, in which water is oxidized to molecular oxygen and NADP$^+$ is reduced to NADPH, driven by two sequential photoreactions in Photosystems II and I. A modern version of the Z scheme for electron flow in oxygenic photosynthetic organisms is shown in Fig. 1.5, along with a cartoon of the complexes that are found in the thylakoid membrane. The overall reaction for this four-electron process is given by Eq. (7.1):

$$2H_2O + 2NADP^+ + 2H^+$$
$$\rightarrow O_2 + 2NADPH + 4H^+ \tag{7.1}$$

The species shown in bold type are either located near or taken up from the stromal side, whereas those in normal type are from the luminal side. Equation (7.1) shows only the net electron transfer reaction; it does not indicate the transmembrane proton motive force that is also generated. The proton pumping across the membrane carried out by the cytochrome b_6f complex will be discussed below.

The minimum expected quantum requirement for this reaction is eight, with one photon absorbed by each photosystem for each electron transported through the chain. Measured values for the quantum requirement of oxygen evolution in intact leaves under optimum conditions are typically nine to ten, which also includes the subsequent carbon fixation reactions, where losses can occur (Evans, 1987; Skillman, 2008). This is remarkably close to the theoretical value, considering all the competing reactions that can go on.

7.3   Photosystem II overall electron transfer pathway

Photosystem II from oxygenic organisms is unique among photosynthetic reaction centers, in that it has the capability to oxidize H_2O to O_2 (Wydrzynski and Satoh, 2005). The oxygen produced by Photosystem II is the source of the oxygen in the atmosphere and has fundamentally changed the development of life on Earth. Photosystem II is the only biological system known that is capable of oxidizing water to molecular oxygen, so it has been intensely studied by scientists trying to understand how the natural system works, but also in the context of artificial systems for energy conversion.

The overall reaction carried out by Photosystem II is given in Eq. (7.2).

$$2H_2O + 2PQ + 4H^+ \rightarrow O_2 + 4H^+ + 2PQH_2 \qquad (7.2)$$

where PQ refers to the oxidized plastoquinone, and PQH_2 is the fully reduced plastoquinol. The species in bold type are part of the water oxidation process, which takes place in the thylakoid lumen. The species in normal type are the part of the quinone reduction part of the process. The quinones are localized entirely within the hydrophobic part of the membrane, while the protons are taken up from the stromal side of the membrane. The overall architecture of the core pigments and proteins, and especially the quinone reduction part of the Photosystem II reaction center, is remarkably similar to that discussed in Chapter 6 for the purple bacterial reaction center.

The unique part of the Photosystem II reaction center is the oxygen-evolving complex (OEC) (see below). The OEC contains a tetranuclear Mn cluster, which accumulates four oxidizing equivalents from four photochemical turnovers of the photoactive pigments. The structure of the OEC and the proposed mechanism of the water oxidation process are considered in detail below.

7.4   Photosystem II forms a dimeric supercomplex in the thylakoid membrane

Structures of various Photosystem II supercomplexes have been obtained using electron microscopy techniques (Rhee *et al.*, 1998; Kouřil *et al.*, 2012; van Bezouwen *et al.*, 2017; Sheng *et al.*, 2019). These studies have given important insights into the structure of the complex. Photosystem II *in vivo* is dimeric, in that two entire photosystems are closely associated, forming a

single large structural unit. Interestingly, there is little evidence that Photosystem II is functionally dimeric, so each of the two reaction centers appears to function independently of the other. Figure 7.2 shows a structural model of the dimeric Photosystem II LHCII–PSII supercore complex from a green alga obtained using cryo-electron microscopy, in which a number of proteins have been identified (Sheng *et al.*, 2019). The entire complex includes the D1 and D2 proteins, CP43 and CP47, several additional light-harvesting complexes, including the LHCII complex discussed in Chapter 5, some other related LHC proteins, and many additional smaller proteins. The LHC proteins are not present in cyanobacterial Photosystem II complexes. Instead, the phycobilisome antenna complex discussed in Chapter 5 is attached to the cytoplasmic side of the complex.

Photosystem II is considerably more complicated in protein composition than the reaction centers found in purple bacteria and includes more than 25 distinct polypeptides (Nelson and Yocum, 2006; Shi *et al.*, 2012; Pagliano *et al.*, 2013; Vinyard *et al.*, 2013; Nelson and Junge, 2015; Shen, 2015). The polypeptide complement of cyanobacterial Photosystem II is generally similar to that found in eukaryotic organisms, although there are some important differences. Table 7.1 lists the proteins associated with Photosystem II, along with their known or suspected functions. The D1 and D2 proteins comprise the core of the Photosystem II reaction center. The functional role of some of the other proteins associated with Photosystem II is not yet clear, as mutants in which many of these proteins have been deleted appear to grow nearly as well as wild-type organisms. It is possible to remove most of the proteins in Photosystem II by biochemical methods and still retain the primary photochemical activity.

Several well-characterized preparations enriched in Photosystem II have been described. A highly purified Photosystem II reaction center complex, consisting of two chlorophyll-binding proteins known as D1 and D2, along with the α and β subunits of a membrane-bound cytochrome $b559$, was reported by Nanba and Satoh (1987) and has been extensively studied. It is capable of primary

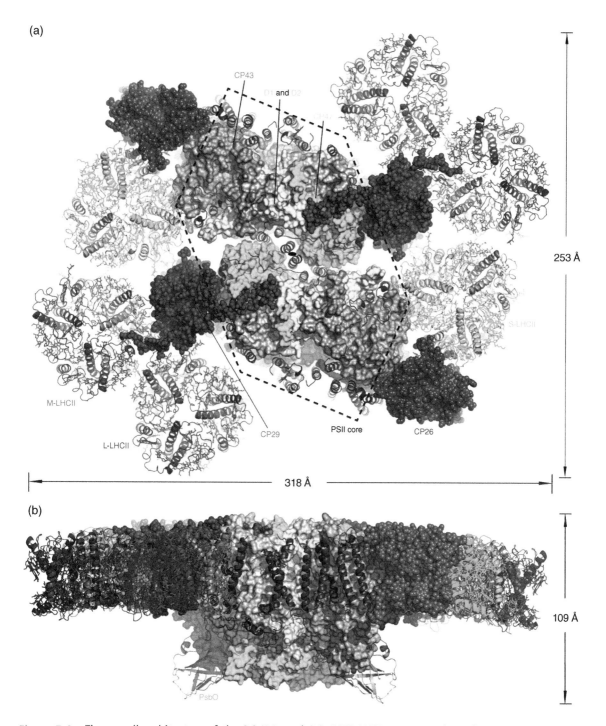

Figure 7.2 The overall architecture of the $C_2S_2M_2L_2$ and C_2S_2 PSII–LHCII supercomplexes from *C. reinhardtii*. (a) Top view of the $C_2S_2m_2L_2$ supercomplex from the stromal side. The central elliptical symbol indicates the C_2 axis running perpendicular to the membrane plane. (b) Side view along the membrane plane. The four large subunits (D1, D2, CP43 and CP47) of the PSII core are shown as surface models. CP26 and CP29 are shown as sphere models, whereas LHCII trimers (S-, m- and L-types) are shown as cartoon and stick models. The small intrinsic subunits and PsbO are shown as cartoon models in brown and gold, respectively. *Source:* Adapted from Sheng *et al.* (2019). Reproduced with permission of *Nature Plants*.

Table 7.1   Subunit structure of Photosystem II

Subunit name	Gene	Gene location[a]	Mass (kDa)[b]	Cofactors	Function
PSII-A (D1)	psbA	C	39	Chlorophyll, pheophytin, quinone, β-carotene, Fe	Core reaction center of Photosystem II
PSII-B (CP47)	psbB	C	56	Chlorophyll, β-carotene	Core antenna
PSII-C (CP43)	psbC	C	51	Chlorophyll, β-carotene	Core antenna
PSII-D (D2)	psbD	C	39	Chlorophyll, pheophytin, quinone, β-carotene, Fe	Core reaction center of Photosystem II
PSII-E (cyt b-559α)	psbE	C	9	Heme	Core reaction center of Photosystem II
PSII-F (cyt b-559β)	psbF	C	4	Heme	Core reaction center of Photosystem II
PSII-H	psbH	C	8	Phosphate	Photoprotection, Q_A to Q_B regulation
PSII-I	psbI	C	4		Core reaction center of Photosystem II
PSII-J	psbJ	C	4		Assembly of Photosystem II
PSII-K	psbK	C	4		Role in PSII assembly
PSII-L	psbL	C	4		Role in Q_A binding
PSII-M	psbM	C	4		Role in PSII stability
PSII-N	psbN	C	5		Role in PSII stability
PSII-O (OE33)	PsbO	N	27		Stabilizes Mn cluster, Ca^{2+} and Cl^- binding
PSII-P (OE23)[c]	PsbP	N	20		Ca^{2+} and Cl^- binding
PSII-Q (OE16)[c]	PsbQ	N	17		Ca^{2+} and Cl^- binding
PSII-R[c]	PsbR	N	10		?
PSII-S (CP22)	PsbS	N	22	Chlorophyll, carotenoids	Antenna regulation by xanthophyll cycle
PSII-T (ycf8)	psbT	C	3		Role in PSII stability
PSII-U[d]	psbU		14		Role in O_2 evolution
PSII-V (cyt c-550)[d]	psbV		15	Heme	Role in O_2 evolution
PSII-W[c]	PsbW	N	6		Role in PSII stability
PSII-X	psbX	C	4		Role in Q_A function
PSII-Y (ycf32)	PsbY	N	4		Mn binding?
PSII-Z (ycf9)	psbZ	C	9		Antenna-reaction center interaction
Psb27	slr1645		12		Role in PSII assembly
Psb28	sll1398		13		Role in PSII assembly
Psb29	slll414		27		Role in PSII repair
Psb30 (ycf12)	sll0047		3		Stabilizes PSII
Psb31[e]	psb31		12		Role in O_2 evolution
Psb32	sll1390		22		Role in PSII repair
Psb33	psb33	N	23		Role in association of LHCII to PS II core
Psb34	ts10063	N	6		Role in PSII assembly
LHCII-outer[c]	Lhcbl	N	30	Chlorophyll, carotenoids	Antenna function
LHCII-outer[c]	Lhcb2	N	31	Chlorophyll, carotenoids	Antenna function
LHCIIa-outer[c]	Lhcb3	N	25	Chlorophyll, carotenoids	Antenna function
LHCII-inner (CP29)[c]	Lhcb4	N	35	Chlorophyll, carotenoids	Antenna function
LHCII-inner (CP26)[c]	Lhcb5	N	36	Chlorophyll, carotenoids	Antenna function
LHCII-inner (CP24)[c]	Lhcb6	N	18	Chlorophyll, carotenoids	Antenna function

[a] Gene location applies only to eukaryotic organisms. C, chloroplast; N, nucleus.
[b] Mass is actual mass based on gene sequence.
[c] Found only in eukaryotic organisms.
[d] Found only in cyanobacteria.
[e] Found only in marine diatom.

photochemistry but not O_2 evolution or secondary electron transfer involving the quinones. The possible function of the cytochrome b_{559} subunit has not yet been definitively established, but it appears to function primarily to protect against photodamage (Shinopoulos and Brudvig, 2012).

The Photosystem II reaction center core complex has a slightly different pigment composition compared with the purple bacteria, with six chlorophyll *a* molecules, two pheophytin *a* molecules, and two β-carotenes. Two of the chlorophylls are rather easily removed from the complex and are bound near the periphery. These pigments are called Chl_Z and are also important in photoprotection.

A second preparation that has been widely studied is called the BBY complex, named after Berthold, Babcock, and Yocum, who first reported it (Berthold *et al.*, 1981). It is an enriched membrane preparation capable of oxygen evolution at high rates, and it contains all the components of the reaction center described above, plus two core antenna complexes known as CP43 and CP47 (for chlorophyll protein with apparent mass of 43 and 47 kDa, respectively), as well as LHCII antenna complexes. In addition, this complex contains some of the peripheral proteins that are associated with the OEC.

A series of X-ray structures of increasing resolution of the Photosystem II reaction center from the thermophilic cyanobacteria *Thermosynechococcus elongatus* and *T. vulcanus* have been determined (Zouni *et al.*, 2001; Ferreira *et al.*, 2004; Umena *et al.*, 2011; Suga *et al.*, 2019). These structures have provided an increasingly detailed understanding of the mechanism of Photosystem II. Figure 7.3 shows the structure of the complex at 1.9 Å resolution. The structure of the cofactors of the Photosystem II reaction center is shown in Fig. 7.4.

The D1 and D2 proteins are functionally analogous to the L and M proteins in the purple bacterial reaction center and exhibit weak but definite sequence homology to L and M (Michel and Deisenhofer, 1988). The similarity between the two types of reaction centers is strongest on the reducing, or acceptor, side, with clear evidence for pheophytin as an early electron acceptor in Photosystem II and two quinones operating as a two-electron gate, analogous to the function of Q_A and Q_B in the

bacterial complex (Müh *et al.*, 2012). There is also a nonheme Fe in the Photosystem II complex, which perturbs the EPR spectra of Q_A^- and Q_B^- in a manner very similar to that observed in the purple bacterial system. Many of the same herbicide inhibitors that block the Q_A^- to Q_B reaction in bacteria by displacing Q_B also inhibit the same reaction in Photosystem II. The most widely used of these inhibitors is 3-(3′,4′-dichlorophenyl)-1,1-dimethyl urea (DCMU) (which is not effective in anoxygenic bacteria). One significant difference between the bacterial photosystem and Photosystem II is the presence of a tightly bound bicarbonate ion in Photosystem II, which is bound to the Fe atom (Shevela *et al.*, 2012). The functional asymmetry in the directionality of electron transfer down only one of two apparent electron transfer chains is also found in Photosystem II, similar to what was described in Chapter 6 for the purple bacterial reaction center. The overall similarity between Photosystem II and the purple bacterial reaction center breaks down on the oxidizing, or donor, side, as the bacterial complex does not evolve oxygen.

Photosystem II produces and utilizes perhaps the most strongly oxidizing species known in biology, $P680^+$. It is therefore not surprising that unwanted and damaging side reactions can take place there. The D1 protein of Photosystem II is rapidly damaged and degraded during the normal functioning of the photosystem. All oxygenic photosynthetic organisms have a sophisticated mechanism for replacing the D1 protein and restoring Photosystem II to a functional form. This damage-recovery process is discussed in more detail in Chapter 10.

7.5   The oxygen-evolving complex and the mechanism of water oxidation by Photosystem II

Without question, the most fascinating and complicated aspect of Photosystem II is the oxygen-evolving activity (Cox *et al.*, 2013; Linke and

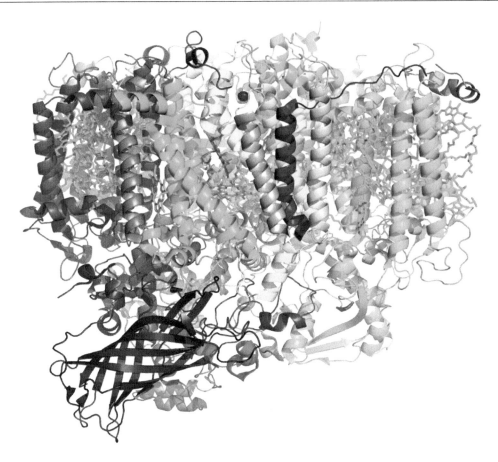

Figure 7.3 Structure of Photosystem II from the thermophilic cyanobacterium *Thermosynechococcus vulcanus*. The structure of one monomer of the dimeric complex is shown from the side, with the luminal side of the membrane at the bottom and the stromal side at the top of the figure. Chlorophylls and pheophytins are shown as green and blue stick structures, respectively. The nonheme Fe is shown as a blue sphere. Subunits are color-coded: D1 protein, green; D2 protein, yellow; CP43 pink; CP47 cyan; PsbO protein, purple; PsbU protein, blue; PsbI protein, orange; PsbL protein, red; PsbT protein, gray; PsbM protein, dark yellow; PsbH protein, blue-gray. Some other subunits are not visible. *Source:* Courtesy of Jian-Ren Shen.

Ho, 2013; Yano and Yachandra, 2014; Shen, 2015; Vinyard and Brudvig, 2017; Cox *et al.*, 2020). In this section, we will discuss the current picture of the structure of the OEC and the mechanism of water oxidation.

The chemical reaction carried out by the OEC is the oxidation of water to molecular oxygen, as shown in Eq. (7.3).

$$2H_2O \rightarrow O_2 + 4H^+ + 4e^- \qquad (7.3)$$

The thermodynamics of this redox process are formidable. Water is an extremely poor electron donor, which is as expected, because oxygen is such a powerful electron acceptor. The redox potential for the half-reaction shown in Eq. (7.3) is +0.82 V at pH 7 and is somewhat higher in the acidic environment in which water oxidation takes place. In order to oxidize water, it is essential that an even stronger oxidant be available. This is provided by the oxidized reaction center chlorophyll P680+,

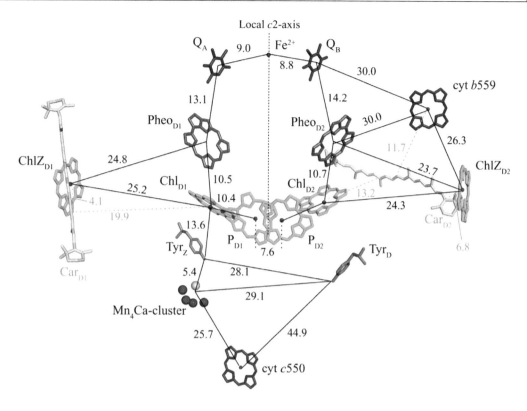

Figure 7.4 Structure of the cofactors in the Photosystem II reaction center complex. Center-to-center distances in Å are indicated. The different types of cofactors are color-coded and labeled. *Source:* Courtesy of Jian-Ren Shen.

which has a redox potential estimated to be as high as +1.2 V The electrons that re-reduce P680+ ultimately come from water, and the protons that result from water oxidation are expelled into the thylakoid lumen. They are released into the lumen because of the vectorial nature of the membrane and the fact that the OEC is localized on the interior surface of the thylakoid. These protons are eventually released from the lumen to the stroma through the process of ATP synthesis. In this way, the electrochemical potential formed by the release of protons during water oxidation contributes to ATP formation.

The chemistry of water oxidation is inherently a four-electron process, as four electrons must be extracted to make a single molecule of oxygen. The photochemistry of the reaction center takes place one electron at a time, so there is a fundamental mismatch between the intrinsic chemistries of

these two processes. One could imagine several possible solutions to this problem, including at one extreme the cooperation of four reaction centers that all contribute to oxidize a single OEC, or at the other extreme, the storage of four oxidizing equivalents by a single reaction center/OEC working alone, which then oxidizes water to oxygen in a concerted step.

The latter possibility was given dramatic support in a classic series of experiments performed in the 1960s, first by Pierre Joliot and slightly later by Bessel Kok. Their experiments were a variation on the flashing light oxygen measurements made in the Emerson and Arnold experiment described in Chapter 3. However, these workers devised extremely sensitive electrochemical methods capable of detecting oxygen produced in a single flash. When this technique was applied to a sample that had been allowed to dark-adapt for several minutes, a remarkable pattern of oxygen production was observed, as shown in Fig. 7.5 (Kok *et al.*, 1970). Little or no oxygen is produced on the first two flashes; a large amount is produced on the third, with lesser maxima with a period of four, until the oxygen production finally damps to a constant value by about the twentieth flash.

Kok and coworkers proposed a schematic model (Fig. 7.5) explaining these observations. The model consists of a series of five states, known as S_0 to S_4, which represent successively more oxidized forms of the OEC. Light flashes advance the system from one S state to the next, until state S_4 is reached. State S_4 produces O_2 without further light input and returns the system to S_0. Occasionally, a center does not advance to the next S state upon flash excitation, and, less frequently, a center is activated twice by a single flash. These "misses" and "double hits" cause the synchrony achieved by dark adaptation to be lost, and the oxygen yield eventually damps to a constant value. After this steady state has been reached, a complex has the same probability of being in any of the states S_0 to S_3 (S_4 is unstable and occurs only transiently) and the yield of O_2 becomes constant. States S_2 and S_3 decay in the dark, but only as far back as S_1, which is stable in the dark. Therefore, after dark adaptation, approximately three-fourths of the oxygen-evolving complexes appear to be in state S_1 and one-fourth in state S_0. This distribution of states explains why the maximum yield of O_2 is observed after the third of a series of flashes given to dark-adapted chloroplasts.

This **S state mechanism** formally explains the observed pattern of O_2 release, but not the chemical nature of the S states or the actual chemical mechanism of the process. It has been known for many years that Mn is an essential cofactor in the water-oxidizing process, and for many years it was suspected that the S states represent successively oxidized states of an Mn-containing enzyme (Debus, 1992). This hypothesis has been confirmed by a variety of experiments, most notably X-ray absorption and EPR studies, both of which detect the Mn directly (McEvoy and Brudvig, 2006; Yano *et al.*, 2006). Analytical experiments indicate that four Mn atoms are associated with each oxygen-evolving complex. Other experiments have shown that Cl^- and Ca^{2+} ions are essential for O_2 evolution, although their precise mechanistic roles are not yet certain. A structural model for the Mn cluster is shown in Fig. 7.6. Many of the amino acid residues that serve as ligands for the Mn cluster come from the D1 protein, as well as the C terminal carboxyl terminus of the D1 protein. Consequently, the OEC

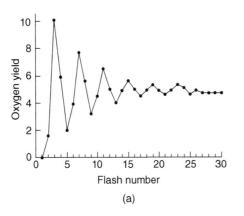

(a)

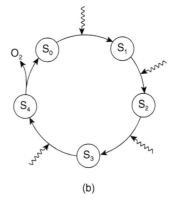

(b)

Figure 7.5   (a) Pattern of oxygen evolution in flashing light. (b) Kok S state model for oxygen production by Photosystem II. The states are successively oxidized by photon absorption, beginning with S_0. When the S_4 state is reached, O_2 is produced, and the system resets to S_0. States S_2 and S_3 are metastable and decay back to S_1 in the dark, producing a 3:1 ratio of S_1 to S_0 in the dark. The synchrony is gradually lost by misses and double hits, which either fail to advance the system or advance it by two S states on a single flash.

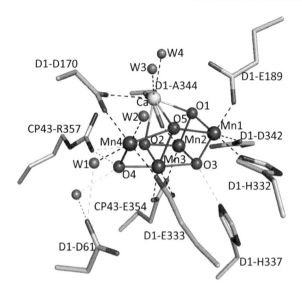

Figure 7.6 Structure of the oxygen-evolving complex (OEC), as determined by X-ray crystallography. The complex is shown including Mn (purple), Ca (yellow), bridging O (red), and nearby waters (orange). Amino acids that are either ligands to the metals or otherwise thought to be important in the mechanism are shown, with the protein identification, residue number, and one-letter amino acid code given. *Source:* Courtesy of Dr. Jian-Ren Shen.

is located off center towards the D1 side of the core reaction center complex (Fig. 7.3).

An additional protein that is important in the oxygen evolution process is a 33 kDa peripheral membrane protein (PsbO protein). This protein is found in all oxygenic photosynthetic organisms. Removal of this protein does not abolish oxygen evolution activity entirely, but it destabilizes the Mn cluster and perturbs the system. Two additional peripheral proteins, with masses of 23 and 17 kDa (PsbP and PsbQ), are part of the complex that stabilizes the OEC in eukaryotic organisms, but are not found in cyanobacteria. Instead, two different proteins, PsbU and PsbV, are found in this position. All these extrinsic proteins, as they are often called, help to stabilize the OEC and protect it against damage, but they do not directly bind the Mn cluster (Bricker *et al.*, 2012).

There is one intermediate electron carrier between P680 and the OEC. After P680 is oxidized

in the primary photochemistry, it is re-reduced in tens of nanoseconds by electron transfer from a tyrosine amino acid side chain (Y161) from the D1 protein, thus forming a tyrosine radical (Styring *et al.*, 2012). This tyrosine residue is known as Y_Z, and it serves to connect the OEC to P680 (Fig. 7.5). The Y_Z radical can be observed by EPR (Blankenship *et al.*, 1975). A second tyrosine in the symmetry-related position in the D2 protein is known as Y_D. It appears not to be involved in mainstream electron flow, but, surprisingly, it is usually found in the oxidized radical form.

Current ideas about the detailed chemical mechanism of oxygen evolution incorporate information from the structure of the OEC, biochemical data, theoretical calculations, and principles of inorganic chemistry (Siegbahn, 2013; Yano and Yachandra, 2014; Vinyard and Brudvig, 2017; Cox *et al.*, 2020). Many mechanisms have been put forward to explain how the OEC accomplishes the difficult chemistry of oxygen evolution. Most of these mechanisms involve the essential role of the tyrosine in a coupled proton and electron (hydrogen atom) transfer process (Hoganson and Babcock, 1997; Hammes-Schiffer, 2018). This proton-coupled electron transfer (PCET) mechanism stresses the importance of maintaining charge neutrality in the OEC. Otherwise, the energetics of the removal of additional electrons from a highly positively charged center found in the higher S states would be prohibitively difficult. Some evidence suggests that bicarbonate also functions as a proton acceptor on the donor side of Photosystem II, in addition to its role on the acceptor side (Shevela, *et al.*, 2020).

The technique of serial femtosecond crystallography has in recent years transformed our understanding of the mechanism of oxygen evolution (Kupitz *et al.*, 2014; Kern *et al.*, 2018; Suga *et al.*, 2019; Ibrahim *et al.*, 2020). In this technique, a given number of preflashes are given to microcrystals of Photosystem II, producing a well-defined S state of the system. The crystals are then subjected to X-ray irradiation to produce a structural model of the system in different intermediate stages of the water oxidation process. Perhaps not surprisingly, the structure of the Mn complex changes significantly as the system advances through the S states. While the molecular details of the mechanism of O_2 evolution are still not agreed on by all researchers, the basic outlines of the mechanism are starting to emerge. Figure 7.7 shows two possible mechanisms that have been proposed (Cox *et al.*, 2020).

7.6   The structure and function of the cytochrome $b_6 f$ complex

The cytochrome $b_6 f$ complex is an essential player in noncyclic and cyclic electron flow (Baniulis *et al.*, 2008; Cramer *et al.*, 2011; Hasan *et al.*, 2013; Sarewicz *et al.*, 2021). The cytochrome $b_6 f$ complex

is similar in most ways to the cytochrome bc_1 complex discussed in Chapter 6. However, there are some important differences. The structure of the cytochrome $b_6 f$ complex is shown in Fig. 7.8. Table 7.2 gives the identity and masses of the proteins of the cytochrome $b_6 f$ complex. In addition to the proteins, the complex contains chlorophyll and carotenoid molecules of unknown function.

Cytochrome f is a c-type cytochrome that serves a similar functional role to cytochrome c_1 in the cytochrome bc_1 complex. However, the two cytochromes have very different structural features. It is an elongated protein, with largely β-sheet secondary structure. The β-sheet structure is very unusual compared with other c-type cytochromes, which are largely α-helical. Cytochrome f has a single transmembrane helical segment near the C-terminal end of the protein, which anchors the globular domain to

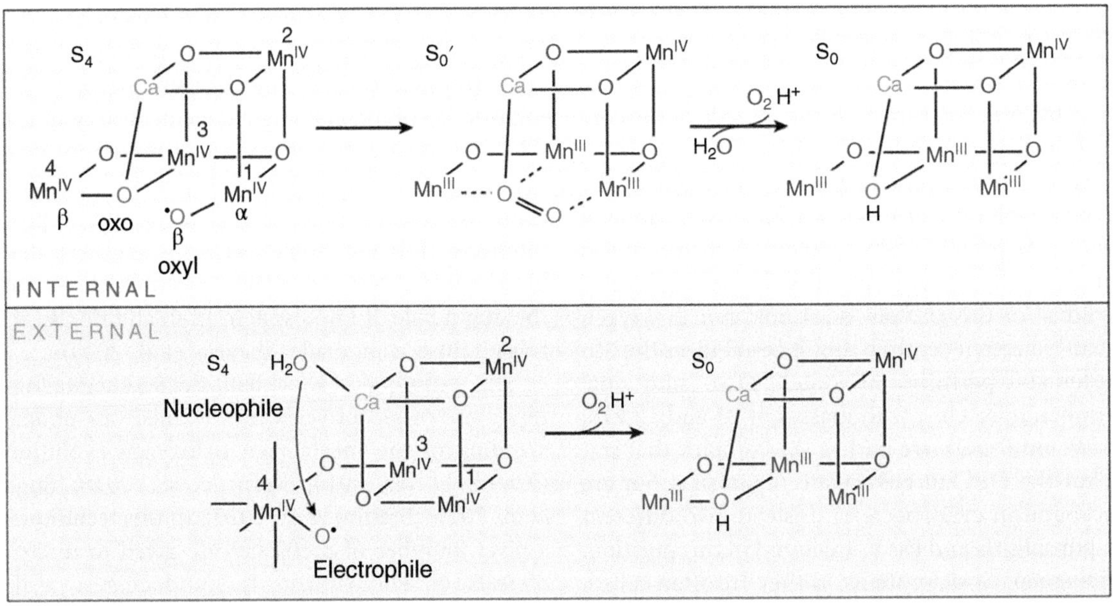

Figure 7.7   Two proposed mechanisms for the formation of O_2 from the state S_4. The two oxygens derived from the substrate waters are colored in green. In the top mechanism, O–O bond formation occurs internally, inside the cubane cage. This mechanism proceeds via an oxo-oxyl coupling. In the bottom mechanism, the O–O bond formation instead occurs externally at the dangler (Mn4) ion. *Source: Cox et al. (2020)/Annual Reviews.*

(a)

(b)

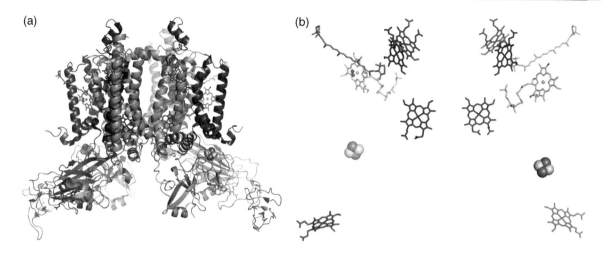

Figure 7.8   Structure of the dimeric cytochrome $b_6 f$ complex from the cyanobacterium *Nostoc* sp. PCC7120. (a) The structure of the complex with the luminal side at the bottom and the stromal side at the top. Subunits are color-coded cytochrome b (green), subunit 4 (purple), ISP (cyan), and cytochrome f (orange). Other small subunits have been omitted for clarity. (b) Cofactors of the complex. Hemes b (blue), heme c_n (brown), heme f, (red), chlorophyll (green), β carotene (orange), the ISP Fe–S center is shown in space filling representation. *Source:* Produced from PDB file 2ZT9 using Pymol.

Table 7.2   Subunit structure of the cytochrome $b_6 f$ complex

Subunit name	Gene	Gene location[a]	Mass (kDa)[b]	Cofactors	Function
PetA (cyt f)	petA	C	32	Heme c	Electron transfer
PetB (cyt b_6)	petB	C	24	Protoheme	Electron transfer
PetC (Rieske Fe–S protein)	PetC	N	19	Fe–S cluster	Electron transfer
PetD (subunit IV)	petD	C	18		Part of cyt b
PetG	petG	C	4		?
PetL[c]	petL	C	3		?
PetM	PetM	N	4		?
PetN	petN	C	3		

[a] Gene location applies only to eukaryotic organisms.
[b] Mass is actual mass based on gene sequence.
[c] Found only in eukaryotic organisms.

the luminal side of the thylakoid membrane. In some species, this segment is easily cleaved from the globular portion that contains the covalently attached heme group. Cytochrome f is also unusual in that the N-terminal amino group is one of the ligands to the Fe in the heme. It donates electrons to plastocyanin (or, in some organisms, the soluble cytochrome c_6).

The Rieske Fe–S protein (ISP) in the cytochrome $b_6 f$ complex consists of two domains: an N-terminal transmembrane helical region that anchors the protein to the membrane and a soluble domain located on the luminal side of the thylakoid membrane that contains the Fe–S redox cofactor (Carrell *et al.*, 1997). The soluble domain is largely β-sheet in secondary structure and is further divided into two subdomains. One of the subdomains contains the Fe–S cofactor and is structurally almost identical to the corresponding part of the ISP from the cytochrome bc_1 complex, while the other subdomain is very different. This is thought to reflect the different reaction partners (cytochrome c_1 vs. cytochrome f) of the ISP in the two types of complexes.

Another important difference between the cytochrome b_6f and the cytochrome bc_1 complex is the cytochrome b portion of the complex. In the cytochrome bc_1 complex, cytochrome b is an integral membrane protein with eight transmembrane helices, as discussed above. However, cytochrome b_6 is much smaller, with only four transmembrane helices predicted. Another subunit of the cytochrome b_6f complex, called subunit IV, exhibits sequence similarity to the C-terminal half of the cytochrome b in the bc_1 complex. Subunit IV contains three predicted transmembrane helices; the eighth transmembrane helix of the cytochrome bc_1 complex is missing. A surprising difference between the cytochrome b_6f and bc_1 complexes is the presence of an additional heme group, called heme c_n in the cytochrome b_6f complex. This heme has a single thioether linkage to the protein instead of two linkages found in almost every other c-type cytochrome. This was first discovered in the cytochrome b_6f complex isolated from the green alga *Chlamydomonas reinhardtii* (Stroebel *et al.*, 2003), but is also found in the complex from cyanobacteria. The function of this additional heme is not understood, but is perhaps involved in cyclic electron flow and may also be an entry point for electrons from other sources. This would change the mechanism from the strict Q-cycle found in the cyt bc_1 complex to a more complicated and potentially flexible one to adapt to the ever-changing environment experienced by oxygenic photosynthetic organisms (Sarewicz *et al.*, 2021).

In addition to the subunits described above, the cytochrome b_6f complex contains some other protein subunits not found in the cytochrome bc_1 complex (Table 7.2). The cytochrome b_6f complex surprisingly contains one molecule each of chlorophyll and β carotene, although it is not known if these pigments serve an important functional role. The cytochrome b_6f complex is also dimeric *in vivo*, like the cytochrome bc_1 complex, although it can easily become monomeric and inactive once isolated (Baniulis *et al.*, 2008).

7.7   Plastocyanin donates electrons to Photosystem I

Cytochrome f donates electrons to the small copper protein **plastocyanin**, which in turn reduces oxidized P700$^+$ in Photosystem I (Gross, 1996; Sigfridsson, 1998; Höhner *et al.*, 2020). Plastocyanin is located in the thylakoid lumen, where it is freely diffusible in the luminal space. Recall that the lateral distribution of the photosystems and the cytochrome b_6f complex is such that, at least in some cases, the mobile electron carrier that links the widely separated photosystems is plastocyanin, while in others it is the membrane-associated plastoquinone. In agreement with this idea, under certain conditions plastocyanin can be observed to migrate between the grana and stroma regions of the thylakoid membrane (Haehnel *et al.*, 1989).

The structure of plastocyanin is shown in Fig. 7.9. It has a largely β-sheet secondary structure with a single Cu atom in a distorted tetrahedral environment. There are two electron transfer pathways that have been proposed to be how electrons can enter and exit plastocyanin. One pathway is through the hydrophobic patch near the Cu ion, and the other is through an acidic (negatively charged) patch on the side of the protein.

In order to carry out its function, plastocyanin must bind to cytochrome f, receive an electron, dissociate, diffuse to Photosystem I, dock to a specific reaction site, donate an electron, and then dissociate again. It thus interacts with two very different electron carriers, each of which may impose a distinct set of constraints on its function. A number of biochemical and kinetic studies of both wild-type and mutant proteins indicate that, in eukaryotes, the acidic patch of plastocyanin binds to Photosystem I via the PsaF subunit, which is located on the luminal side of the Photosystem I complex. However, electron transfer to the Photosystem I pigment dimer P700 (see below) appears to take

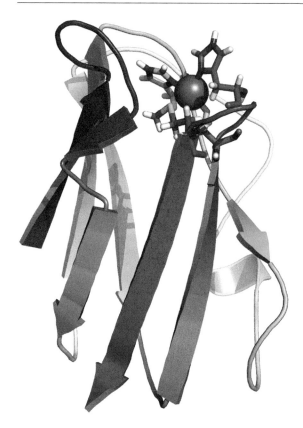

Figure 7.9 Structure of plastocyanin from poplar. The Cu ion and its ligands, two histidines, cysteine and methionine are shown, as well as tyrosine 83, which is implicated in binding and electron transfer. The color-coding runs from blue at the N terminus to red at the C terminus. *Source:* Produced from PDB file 1PLC using Pymol.

place via the hydrophobic patch. In this case, unlike the c-type cytochromes such as cytochrome c_2 discussed earlier, the binding and recognition functions are not located in the same part of the molecule as the electron transfer function. In addition, the luminal helix of the PsaB protein that forms the core of the Photosystem I reaction center is also involved in recognizing the soluble electron carriers that donate to Photosystem I (Kuhlgert *et al.*, 2012).

In many cyanobacteria and green algae, a c-type cytochrome, cytochrome c_6, is found either in addition to, or instead of, plastocyanin (Kerfeld, 1997).

The relative amounts of the two proteins that are present are controlled by copper availability in the growth media (Merchant and Dreyfuss, 1998).

7.8   Photosystem I structure and electron transfer pathway

In contrast to Photosystem II, which operates in a highly oxidizing regime, Photosystem I is much more reducing (Nelson and Yocum, 2006; Golbeck, 2006; Nelson and Junge, 2015). The redox potentials of the early electron acceptors in Photosystem I are approximately –1 V, with the excited state of the photoactive chlorophyll P700 estimated to be –1.26 V (Table A.3). Photosystem I also has a very different pigment complement than either the purple bacterial reaction center or Photosystem II, which have only six to eight pigments associated with the core protein complex. Photosystem I has approximately 100 molecules of chlorophyll and 12–16 β-carotene molecules associated with the core proteins. Most of the chlorophyll molecules function as antenna pigments, so that the entire complex is classed as a fused antenna/reaction center system (Chapter 5).

The Photosystem I reaction center consists of a heterodimeric protein core complex of two 82–83 kDa integral membrane complexes, the PsaA and PsaB proteins, along with about ten additional proteins. Table 7.3 lists the protein components of the Photosystem I reaction center complex and their proposed functions.

The structure of a trimeric Photosystem I complex from a cyanobacterium has been determined using X-ray crystallography (Jordan *et al.*, 2001). The structure of the complete trimeric complex is shown in Fig. 7.10. The structure of the electron transfer cofactors in Photosystem I along with the pathway of electron transfer is shown in Fig. 7.11. Some cyanobacteria instead contain a tetrameric Photosystem I (Li *et al.*, 2019). Eukaryotic Photosystem I complexes are monomeric, and in

Table 7.3   Subunit structure of Photosystem I

Subunit name	Gene name	Gene location[a]	Mass (kDa)[b]	Cofactors	Function
PSI-A (PsaA)	psaA	C	84	Chlorophyll, quinone, β-carotene, Fe–S	Core reaction center of Photosystem I
PSI-B (PsaB)	psaB	C	83	Chlorophyll, quinone, β-carotene, Fe–S	Core reaction center of Photosystem I
PSI-C	psaC	C	9	Fe–S	Fe–S$_A$ and Fe–S$_B$
PSI-D	PsaD	N	18		Ferredoxin docking
PSI-E	PsaE	N	10		Role in cyclic electron transport
PSI-F	PsaF	N	17		Plastocyanin docking
PSI-G[c]	PsaG	N	11		Role in Q_A binding
PSI-H[c]	PsaH	N	11		Interaction with LHCII
PSI-O	PsaO	N	16		Interaction with LHCII
PSI-I	psaI	C	4		Required for PsaL incorporation.
PSI-J	psaJ	C	5		Required for PsaF incorporation.
PSI-K	PsaK	N	9		Role in docking of LHCI
PSI-L	PsaL	N	18		Trimer formation in cyanobacteria
PSI-M[d]	psaM		3		Trimer formation in cyanobacteria
PSI-N[c]	PsaN	N	10		Plastocyanin docking
PSI-X[d]	psaX		3	Chlorophyll	?
LHC-I[c] (LHCI-730)	Lhca1	N	22	Chlorophyll, carotenoid	Antenna function
LHC-I[c] (LHCI-680)	Lhca2	N	23	Chlorophyll, carotenoid	Antenna function
LHC-I[c] (LHCI-680)	Lhca3	N	25	Chlorophyll, carotenoid	Antenna function
LHC-I[c] (LHCI-730)	Lhca4	N	21	Chlorophyll, carotenoid	Antenna function
–	Lhca5	N	23	Chlorophyll, carotenoid	Required for PSI-LHCI-NDH1 supercomplex, substoichiometric.
–	Lhca6	N	23	Chlorophyll, carotenoid	Required for PSI-LHCI-NDH1 supercomplex, substoichiometric.

[a] Gene location applies only to eukaryotic organisms. C, chloroplast; N, nucleus.
[b] Mass is actual mass based on gene sequence.
[c] Found only in eukaryotic organisms.
[d] Found only in cyanobacteria.

addition contain LHCI antenna complexes that are not found in cyanobacteria (Amunts *et al.*, 2007). The structure of a eukaryotic Photosystem I complex, including the LHCI antenna proteins, is shown in Fig. 7.12.

The antenna pigments intrinsic to the core of Photosystem I surround a central group of electron transfer cofactors. The electron transfer cofactors form two potential electron transfer chains, similar to the case in purple bacteria and Photosystem II. A dimer of chlorophylls near the luminal side is the special pair P700. Four additional chlorophyll molecules and two phylloquinone molecules are also present. A feature unique to Photosystem I and

related photosystems is the set of three iron-sulfur (Fe–S) clusters that function as early electron acceptors (Golbeck, 1999). The structures of the Fe–S clusters in some ferredoxin-type proteins are shown in Fig. 6.12. One of the Fe–S clusters, called F_X, is held between the halves of the core reaction center complex, coordinated to cysteine residues from each subunit. The other two Fe–S clusters, called F_A and F_B, are bound to a small protein subunit, the PsaC protein. The three Fe–S centers have distinct EPR spectra, which have been characterized in detail.

The electron transfer processes in Photosystem I begin with excitation of the antenna pigments. Energy transfer delivers excitations to the core

(a)   (b)

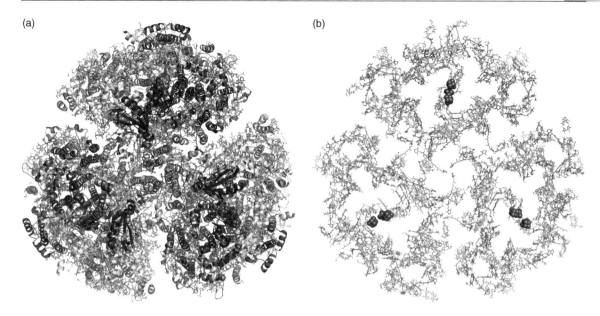

Figure 7.10 Structure of Photosystem I reaction center complex from the thermophilic cyanobacterium *Synechococcus elongates*. (a) Complete structure including protein subunits and cofactors. The structure is shown from the stromal side of the membrane. Subunit color-coding, PsaA (cyan), PsaB (magenta), PsaC (brown), PsaD (blue), PsaE (yellow), PsaF (red), PsaL (salmon), other subunits (gray). Chlorophylls are green, carotenoids are orange, and quinones are yellow. The Fe-S centers are shown as space-filling spheres in magenta. (b) Cofactors only, from the same orientation as (a). Color-coding of cofactors is the same as (a). *Source:* Produced from PDB file 1JB0 using Pymol.

electron transfer cofactors (Melkozernov *et al.*, 2006). For many years, the electron transfer was thought to take place from the excited P700 dimer, in a similar manner as electron transfer takes place in the purple bacterial reaction centers. However, it has been proposed that the primary electron transfer takes place from one of the accessory chlorophylls and the positive charge then rapidly migrates to P700 (Holzwarth *et al.*, 2006; Müller *et al.*, 2010). These two mechanisms lead to the same final products and are difficult to distinguish experimentally. The primary electron transfer produces a state in which P700 is oxidized, and another chlorophyll (called A_0) is reduced. The electron then moves in ~25 ps to one of the quinones (called A_1), and from there to F_X in ~200 ns.

The electron is next transferred from F_X to the F_A center or the F_B center. These centers are contained in the PsaC protein subunit of Photosystem I. The redox potential of F_B (−580 mV), is somewhat more negative than that of F_A (−520 mV), so the initial view was that the reaction sequence $F_X \rightarrow F_B \rightarrow F_A$ was the most likely, based on the decreasing redox gradient that is observed with many electron transport chains. However, considerable evidence now suggests that the correct order is $F_X \rightarrow F_A \rightarrow F_B$ (Golbeck, 1999). Uphill electron transfer reactions are fully consistent with everything that is known about the principles of electron transfer (discussed in Chapter 6), as long as the donor and the acceptor are in close proximity and the uphill reaction is followed by downhill transfers that make the overall process favorable. Recall that a much more uphill transfer from cytochrome c_2 to the tetraheme cytochrome is observed in *Blastochloris viridis* reaction centers.

An electron is transferred from the Photosystem I complex to the soluble protein ferredoxin, and

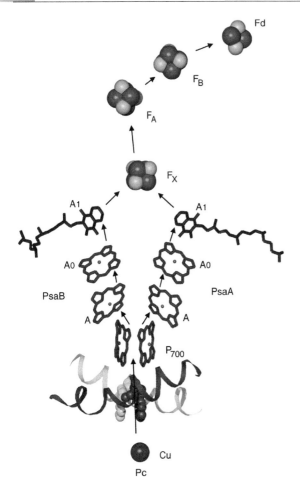

Figure 7.11 Structural model of the pathway for light-induced electron transport from plastocyanin to ferredoxin in photosystem I. Chls (blue), quinones (black), the copper atom of plastocyanin (P_c) (blue), and Fe (red balls) and S (green balls) of the three Fe_4S_4 clusters and the Fd Fe_2S_2 are depicted. Two tryptophan residues (light-blue and light-pink space-filling structures) implicated in electron transport from plastocyanin to P700 are also shown in the context of their secondary structural environment. *Source:* Nelson and Yocum (2006). Reproduced with permission of Annual Reviews.

ultimately NADP$^+$ is reduced as described below. The PsaD and PsaE proteins are located on the stromal side of the complex and are involved with the processes that take place there, either in helping to dock and position the proteins correctly or in the assembly of the complex.

P700$^+$ is re-reduced in most organisms by electron transfer from the blue copper protein plastocyanin. The PsaF protein is implicated in plastocyanin binding, although the core heterodimer PsaA and PsaB complex is also involved in this process. Plastocyanin oxidation kinetics are complex and depend on whether or not the plastocyanin is bound to the Photosystem I complex prior to P700 oxidation (Santabarbara *et al.*, 2009).

There is an interesting difference between the electron transfer pathway in Photosystem I and what is observed in both the purple bacterial reaction center and Photosystem II. In these two latter complexes, the available evidence overwhelmingly supports the view that electron transfer proceeds down only one of the two apparent pathways that are observed in the structure. While Photosystem I also has a heterodimeric core protein arrangement, it is now clear that both potential pathways are utilized to a significant degree, so that an electron produced by P700 oxidation can go down either pathway (Guergova-Kuras *et al.*, 2001). The possible functional significance of this bidirectional electron flow is not clear, as the two pathways intersect at the Fx Fe–S center.

Figure 7.13 summarizes the energies and kinetics of the electron transfer processes in the Photosystem I reaction center. This information has been derived from a large number of studies, using a range of techniques.

7.9   Ferredoxin and ferredoxin-NADP reductase complete the noncyclic electron transport chain

The final steps in the noncyclic electron transfer process are from the bound Fe–S centers in the Photosystem I reaction center to NADP$^+$. A small soluble protein electron carrier, **ferredoxin** (Fd), along with a second protein, **ferredoxin-NADP reductase** (FNR), carry out this function. Ferredoxins are a large class of Fe–S proteins found

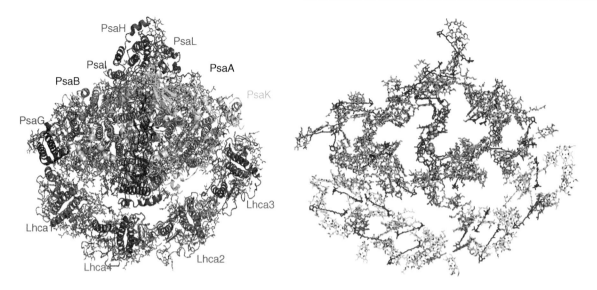

Figure 7.12   Structure of monomeric Photosystem I from pea. Left. View from the stromal side of the membrane showing the protein subunits of the core and the light-harvesting complexes. Right. Pigment organization in PSI-LHCI. The central pigments of the internal electron transport chain are colored red, chlorophylls of the core antenna green, chlorophyll *a* in LHCI in cyan, and chlorophyll *b* in magenta. Carotenoids, which are distributed throughout the complex, are colored in blue and lipids in key connecting points and conserved positions in the core, in orange. *Source:* Taken from Mazor *et al.* (2015). Reproduced with permission of *eLife*.

in all organisms (Knaff, 1996; Schoepp *et al.*, 1999). The ferredoxins that function in oxygenic photosynthesis form a subclass that contains two Fe and two acid-labile sulfide groups, with a cofactor geometry shown in Fig. 6.12. Four cysteine residues coordinate the two Fe atoms, which are bridged by two sulfur atoms. These sulfurs are released upon acid treatment and are therefore called acid-labile sulfides. Ferredoxin accepts a single electron, which is shared between the two Fe atoms. It has an extremely negative reduction potential, −430 mV *vs.* NHE. This potential is such that it can easily be reduced by the F_A and F_B clusters that are associated with the PsaC protein, one of the subunits of the Photosystem I reaction center. The center that reduces ferredoxin, F_B, has a reduction potential of −550 mV. Reduced ferredoxin can, in turn, reduce NADP⁺, which has a reduction potential of −320 mV. Ferredoxin binds via electrostatic interactions to the PsaC protein on the stromal side of the Photosystem I complex, mediated by two other

Photosystem I reaction center subunits, the PsaD and PsaE proteins. In cyanobacteria that are subjected to low Fe levels or high salt stress, Fd is replaced by flavodoxin, a flavin-containing protein that contains no Fe (Hagemann *et al.*, 1999).

The final step in the light-driven noncyclic electron transport sequence is the reduction of NADP⁺, mediated by FNR (Knaff, 1996; Arakaki *et al.*, 1997). The structure of NADP is shown in Fig. 7.14. Unlike quinones and FAD, NADP is always a two-electron carrier.

FNR is a 35–45 kDa protein that accepts electrons one at a time from ferredoxin, yet carries out the two-electron reduction of NADP⁺ to NADPH. The reduction of NADP⁺ to NADPH involves addition of two electrons and one proton and is thus formally described as a hydride (H⁻) transfer. The overall reaction catalyzed by FNR is given in Eq. (7.4):

$$2Fd_{red} + NADP^+ + H^+ \rightarrow 2Fd_{ox} + NADPH \quad (7.4)$$

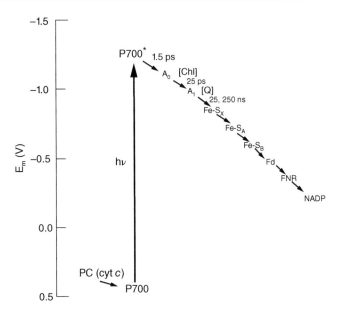

Figure 7.13   Energy-kinetic diagram of photochemistry and early electron transfer reactions in Photosystem I.

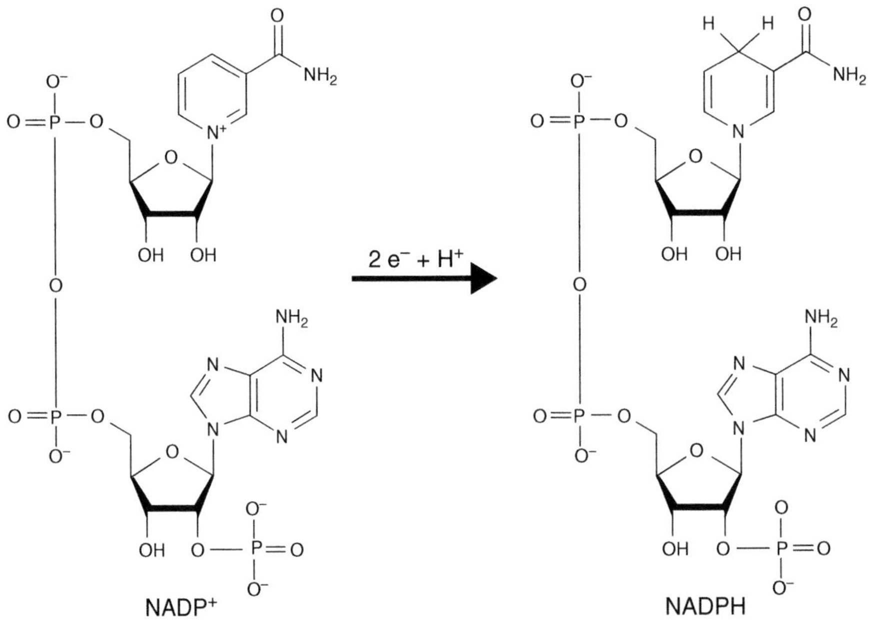

Figure 7.14   The structure of NADP. The redox changes are localized to the nicotinamide group at the top of the molecule. NAD has the same structure except for the phosphate group attached to the ribose.

FNR thus serves as a one-electron to two-electron converter, in that it converts the intrinsically one-electron-at-a-time chemistry of the reaction center into the two-electrons-at-a-time chemistry characteristic of most chemical reactions. In order for this to work, the FNR enzyme must have a functional group that can accept one electron from Fd and hold it while a second Fd provides a second electron. This cofactor is flavin adenine dinucleotide (FAD), whose structure is shown in Fig. 7.15. The redox-active part of the FAD cofactor is the isoalloxazine ring, which can exist in the fully oxidized (quinone), partially reduced (semiquinone), and fully reduced (hydroquinone) forms, as also shown in Fig. 7.15. The reduction potentials of the first and second reductions of FAD in FNR are approximately −370 mV and −270 mV, giving a two-electron potential of −320 mV, which is roughly isoenergetic with $NADP^+$ reduction.

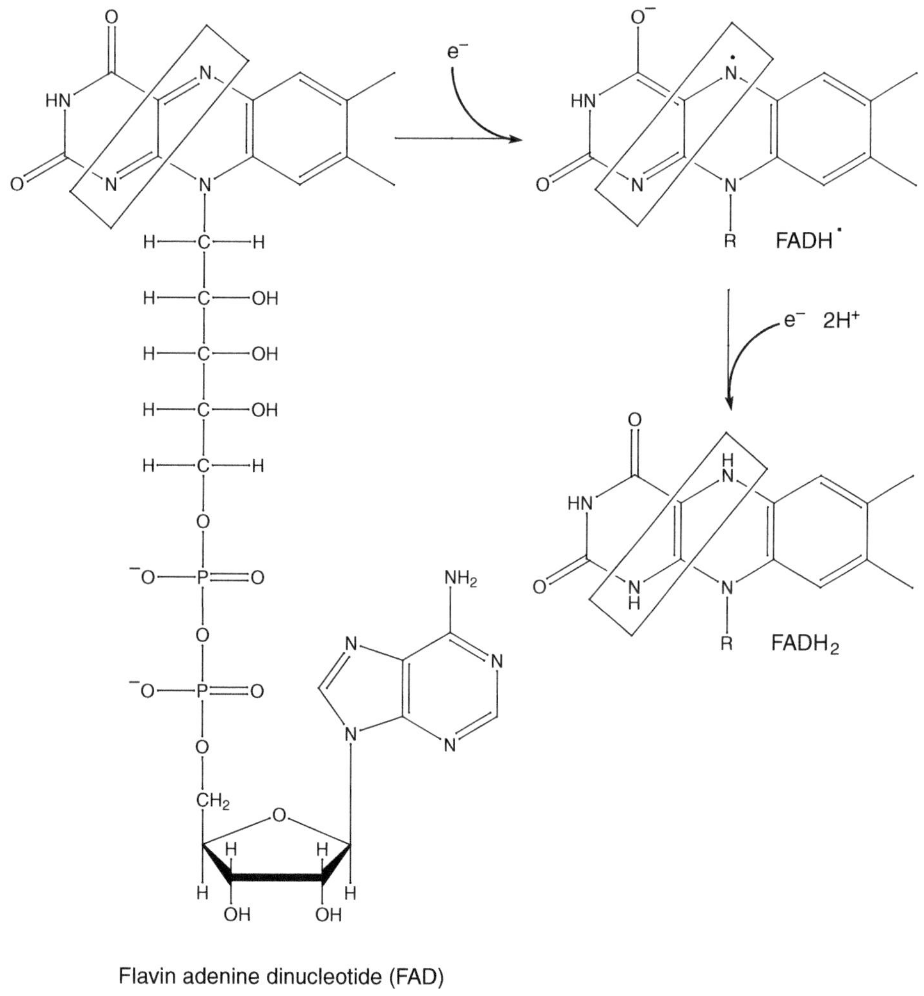

Flavin adenine dinucleotide (FAD)

Figure 7.15   Structure of flavin adenine dinucleotide (FAD), the cofactor for ferredoxin-NADP reductase. The oxidized quinone (FAD), partially reduced semiquinone (FADH•), and fully reduced hydroquinone (FADH2) forms are shown. The redox-active portion of the isoalloxazine ring is boxed.

The structures of ferredoxin and FNR have been solved by X-ray crystallography for a number of organisms. FNR consists of two domains, one that binds FAD and the other containing the $NADP^+$ binding site. Ferredoxin binds in a cleft between the two domains, with the Fe–S cofactor in close proximity to the FAD (Karplus *et al.*, 1991). The structure of a complex between ferredoxin and FNR is shown in Fig. 7.16 (Kurisu *et al.*, 2001).

7.9.1 Other roles of ferredoxin and cyclic electron transfer around Photosystem I

In addition to the major function of reducing $NADP^+$ via FNR, ferredoxin serves as a reductant for a large number of other chloroplast processes, including reduction of nitrate to ammonia, synthesis of the amino acid glutamate, and reductive

regulation via thioredoxin of the ATP synthase enzyme and several of the enzymes involved in the Calvin-Benson cycle (Knaff, 1996). Reduced Fd is easily autooxidized by molecular oxygen to form superoxide, O_2^-, which eventually forms H_2O_2, in a process called the Mehler reaction. Reduced Fd can also reduce monodehydroascorbate to ascorbate, which is an important part of the protection system against active oxygen species, as discussed in Chapter 10. Finally, reduced Fd can catalyze a cycle around Photosystem I, in which electrons are eventually returned to P700 via the cytochrome b_6f complex (see below). It is thus a central player in a number of biosynthetic and regulatory processes. A schematic picture of the different roles of ferredoxin in oxygenic photosynthesis is shown in Fig. 7.17.

Ferredoxin-catalyzed cyclic electron flow around Photosystem I was first observed by Arnon and coworkers in 1954 (Arnon, 1984). It is readily activated if electron flow from Photosystem II is cut off, for example with the inhibitor DCMU, especially if an artificial electron transfer cofactor is included in the sample mixture. But establishing whether or not cyclic electron flow is physiologically important in intact photosynthetic organisms has been difficult. Arnon's original experiments

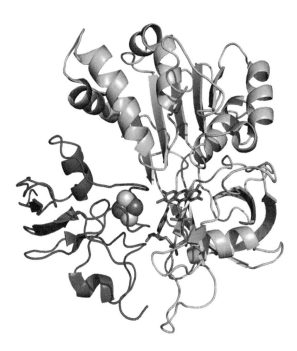

Figure 7.16   Structure of a complex between ferredoxin (Fd) and ferredoxin-NADP reductase (FNR) from maize. Fd is in red and FNR is in green. The Fe–S cofactor of Fd and FAD cofactor of FNR are shown. *Source: Produced from PDB file 1GAQ using Pymol.*

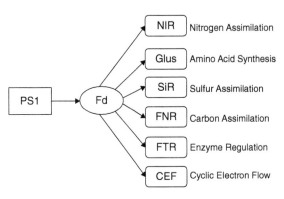

Figure 7.17   Roles of reduced ferredoxin (Fd) in various cellular processes. Photosystem I (PSI) reduces ferredoxin (Fd), which in turn reduces other ferredoxin-dependent enzymes: NiR, nitrate reductase; GluS, glutamate synthase; SiR, sulfite reductase; FNR, ferredoxin-NADP reductase; FTR, ferredoxin-thioredoxin reductase; CEF, cyclic electron flow.

involved measuring ATP formation, so this process came to be known as cyclic photophosphorylation. This requires that the electron flow be coupled in some manner to energy conservation, which we now know requires generation of a proton motive force, as will be discussed in Chapter 8.

It is now clear that there are multiple distinct pathways for cyclic electron flow around Photosystem I (Bendall and Manasse, 1995; Corneille *et al.*, 1998; Shikanai, 2007; Johnson, 2011; Nawrocki *et al.*, 2019), although there is much uncertainty about the relative importance of the various pathways. One pathway for cyclic electron flow involves an NAD(P)H dehydrogenase enzyme, similar to Complex 1 of the mitochondrial respiratory electron transport chain. Despite its name, this complex appears to use reduced Fd to reduce plastoquinone, which is then reoxidized by the cytochrome b_6f complex. Another pathway involves a protein called PGR5. When the gene for this protein is inhibited, cyclic electron flow is blocked. A protein supercomplex that catalyzes cyclic electron flow has been described in the green alga *Chlamydomonas reinhardtii* (Iwai *et al.*, 2010), although this complex has not been observed in higher plants. Finally, the direct oxidation of reduced ferredoxin by the cytochrome b_6f complex, which may be the elusive pathway of ferredoxin-quinone reductase (FQR), may be an important pathway for cyclic electron flow. The difficulties of resolving the different pathways, coupled with the technical challenges of measuring cyclic electron flow, where there is no net oxidation or reduction of redox species, has made this perhaps the most poorly understood aspect of the photosynthetic electron transport chain. Evidence suggests that cyclic electron flow is inhibited by ATP, which has the effect of regulating the amount of cyclic flow to provide enough ATP but not an excess (Fisher *et al.*, 2019).

Is cyclic electron flow required in the normal functioning of the photosynthetic electron transport chain? The amount of cyclic electron flow in the normal functioning of thylakoids is uncertain (Fork and Herbert, 1993; Kramer and Evans, 2011). The requirement for 3 ATP per CO_2 fixed in the Calvin-Benson cycle strongly suggests that some amount of cyclic electron flow is necessary.

The significant amount of cyclic electron flow in the bundle sheath chloroplasts of those plants that utilize the C4 carbon metabolism pathway is well documented. This is discussed in Chapter 9.

References

Allen, J. F. and Forsberg, J. (2001) Molecular recognition in thylakoid structure and function. *Trends in Plant Science* 6: 317–326.

Amunts, A., Drory, O., and Nelson, N. (2007) The structure of a plant photosystem I supercomplex at 3.4 Å resolution. *Nature* 447: 58–63.

Anderson, J. M. (2002) Changing concepts about the distribution of Photosystems I and II between granaappressed and stroma exposed thylakoid membranes. *Photosynthesis Research* 73: 157–164.

Arakaki, A. K., Ceccarelli, E. A., and Carrillo, N. (1997) Plant-type ferredoxin-NADP+ reductases: A basal structural framework and a multiplicity of functions. *FASEB Journal* 11: 133–140.

Arnon, D. I. (1984) The discovery of photosynthetic phosphorylation. *Trends in Biochemical Sciences* 9: 258–262.

Baniulis, D., Yamashita, E., Zhang, H., Hasan, S. S., and Cramer, W. A. (2008) Structure–function of the cytochrome b_6f complex. *Photochemistry and Photobiology* 84: 1349–1358.

Bendall, D. S. and Manasse, R. S. (1995) Cyclic photophosphorylation and electron transport. *Biochimica et Biophysica Acta* 1229: 23–38.

Berthold, D. A., Babcock, G. T., and Yocum, C. F. (1981) A highly resolved, oxygen-evolving photosystemII preparation from spinach thylakoidmembranes. *FEBS Letters* 134: 231–234.

Blankenship, R. E., Babcock, G. T., Warden, J. T., and Sauer, K. (1975) Observation of a new EPR transient in chloroplasts that may reflect the electron donor to Photosystem II at room temperature. *FEBS Letters* 51: 287–293.

Bricker, T. M., Roose, J. L., Fagerlund, R. D., Frankel, L. K., and Eaton-Rye, J. J. (2012) The extrinsic proteins of Photosystem II. *Biochimica et Biophysica Acta* 1817: 121–142.

Carrell, C. J., Zhang, H. M., Cramer, W. A., and Smith, J. L. (1997) Biological identity and diversity in photosynthesis and respiration: structure of the lumenside domain of the chloroplast Rieske protein. *Structure* 5: 1613–1625.

Corneille, S., Cournac, L., Guedney, G., Havaux, M., and Peltier, G. (1998) Reduction of the plastoquinone pool by exogenous NADH and NADPH in higher plant chloroplasts. Characterization of a NAD(P)H-plastoquinone oxidoreductase activity. *Biochimica et Biophysica Acta* 1363: 59–69.

Cox, N., Pantazis, D. A., and Lubitz, W. (2020) Current understanding of the mechanism of water oxidation in Photosystem II and Its relation to XFEL data. *Annual Review of Biochemistry* 89: 795–820.

Cox, N. and Messinger, J. (2013) Reflections on substrate water and dioxygen formation. *Biochimica et Biophysica Acta* 1827: 1020–1030.

Cramer, W. A. and Kallas, T., (eds.) (2016) *Cytochrome Complexes: Evolution, Structures, Energy Transduction and Signaling*. Dordrecht: Springer.

Cramer, W. A., Hasan, S. S., and Yamashita, E. (2011) The Q cycle of cytochrome bc complexes: A structure perspective. *Biochimica et Biophysica Acta* 1807: 788–802.

Debus, R. J. (1992) The manganese and calcium ions of photosynthetic oxygen evolution. *Biochimica et Biophysica Acta* 1102: 269–352.

Evans, J. R. (1987) The dependence of quantum yield on wavelength and growth irradiance. *Australian Journal of Plant Physiology* 14: 69–79.

Ferreira, K. N., Iverson, T. M., Maghlaoui, K., Barber, J., and Iwata, S. (2004) Architecture of the photosynthetic oxygen evolving center. *Science* 303: 1831–1838.

Fisher, N., Bricker, T. M., and Kramer, D. M. (2019) Regulation of photosynthetic cyclic electron flow pathways by adenylate T status in higher plant chloroplasts. *Biochimica et Biophysica Acta* 1860: 148081.

Fork, D. C. and Herbert, S. K. (1993) Electron transport and photophosphorylation by photosystem I in vivo in plants and cyanobacteria. *Photosynthsis Research* 36: 149–168.

Fromme, P., (ed.) (2008) *Photosynthetic Protein Complexes*. Weinheim: Wiley-Blackwell.

Golbeck, J. H. (1999) A comparative analysis of the spin state distribution of in vitro and in vivo mutants of PsaC. A biochemical argument for the sequence of electron transfer as FX→FA→FB→Ferredoxin. *Photosynthsis Research* 61: 107–144.

Golbeck, J. H., (ed.) (2006) *Photosystem I: The Light-Driven Plastocyanin: Ferredoxin Oxidoreductase*. Dordrecht: Springer.

Gross, E. L. (1996) Plastocyanin: Structure, location, diffusion and electron transfer mechanism. In: D. R. Ort and C. F. Yocum, (eds.) *Oxygenic Photosynthesis:*

The Light Reactions. Dordrecht: Kluwer Academic Publishers, pp. 413–429.

Guergova-Kuras, M., Boudreaux, B., Joliot, A., Joliot, P., and Redding, K. (2001) Evidence for two active branches for electron transfer in photosystem I. *Proceedings of the National Academy of Sciences USA* 98: 4437–4442.

Haehnel, W., Ratajczak, R., and Robenek, H. (1989) Lateral distribution and diffusion of plastocyanin in chloroplast thylakoids. *Journal of Cell Biology* 108: 1397–1405.

Hagemann, M., Jeanjean, R., Fulda, S., Havaux, M., Joset, F., and Erdmann, N. (1999) Flavodoxin accumulation contributes to enhanced cyclic electron flow around photosystem I in salt-stressed cells of *Synechocystis* sp. strain PCC 6803. *Physiologia Plantarum* 105: 670–678.

Hammes-Schiffer, S. (2018) Controlling electrons and protons through theory: Molecular electrocatalysts to nanoparticles. *Accounts of Chemical Research* 51: 1975–1983.

Hasan, S. S., Yamashita, E., Baniulis, D., and Cramer, W. A. (2013) Quinone-dependent proton transfer pathways in the photosynthetic cytochrome b_6f complex. *Proceedings of the National Academy of Sciences USA* 110: 4297–4302.

Hoganson, C. W. and Babcock, G. T. (1997) A metallo-radical mechanism for the generation of oxygen from water in photosynthesis. *Science* 277: 1953–1956.

Höhner, R., Pribil, M., Herbstová, M., Lopez, L. S., Kunz, H. H., Li, M., Wood, M., Svoboda, V., Puthiyaveetil, S., Leister, D., and Kirchhoff, H. (2020) Plastocyanin is the long-range electron carrier between photosystem II and photosystem I in plants. *Proceedings of the National Academy of Sciences USA* 117: 15354–15362.

Holzwarth, A. R., Müller, M. G., Niklas, J., and Lubitz, W. (2006) Ultrafast transient absorption studies on photosystem I reaction centers from *Chlamydomonas reinhardtii*. 2. Mutations around the P700 reaction center chlorophylls provide new insight into the nature of the primary electron donor. *Biophysical Journal* 90: 552–565.

Ibrahim, M., Fransson, T., Chatterjee, R., Cheah, M. H., Hussein, R., Lassalle, L., Sutherlin, K. D., Young, I. D., Fuller, F. D., Gul, S., Kim, I.-S., Simon, P. S., de Lichtenberg, C., Chernev, P., Bogacz, I., Pham, C. C., Orville, A. M., Saichek, N., Northen, T., Batyuk, A., Carbajo, S., Alonso-Mori, R., Tono, K., Owada, S., Bhowmick, A., Bolotovsky, R., Mendez, D., Moriarty, N. W., Holton, J. M., Dobbek, H., Brewster, A. S., Adams, P. D., Sauter, N. K., Bergmann, U., Zouni, A., Messinger, J., Kern, J., Yachandra, V. K., and Yano, J. (2020) Untangling the sequence of events during the $S_2 \rightarrow S_3$ transition in photosystem II and implications for the water oxidation mechanism. *Proceedings of the National Academy of Sciences USA* 117: 12624–12634.

Iwai, M., Takizawa, K., Tokutsu, R., Okamuro, A., Takahashi, Y., and Minagawa, J. (2010) Isolation of the elusive supercomplex that drives cyclic electron flow in photosynthesis. *Nature* 464: 1210–1213.

Johnson, G. N. (2011) Physiology of PSI cyclic electron transport in higher plants. *Biochimica et Biophysica Acta* 1807: 384–389.

Jordan, P., Fromme, P., Witt, H. T., Klukas, O., Saenger, W., and Krauss, N. (2001) Three-dimensional structure of cyanobacterial photosystem I at 2.5 Å resolution. *Nature* 411: 909–917.

Kana, R. (2013) Mobility of photosynthetic proteins. *Photosynthesis Research* 116: 465–479.

Karplus, P. A., Daniels, M. J., and Herriott, J. R. (1991) Atomic structure of ferredoxin-NADP$^+$ reductase: Prototype for a structurally novel flavoenzyme family. *Science* 251: 60–66.

Kern, J., Chatterjee, R., Young, I. D., Fuller, F. D., Lassalle, L., Ibrahim, M., Sheraz Gul, S., Fransson, T., Brewster, A. S., Alonso-Mori, R., Hussein, R., Zhang, M., Douthit, L., de Lichtenberg, C., Cheah, M. H., Shevela, D., Wersig, J., Seuffert, I., Sokaras, D., Pastor, E., Weninger, C., Kroll, T., Sierra, R. G., Aller, P., Butryn, A., Orville, A. M., Liang, M., Batyuk, A., Koglin, J. E., Carbajo, S., Boutet, S., Moriarty, N. W., Holton, J. M., Dobbek, H., Adams, P. D., Bergmann, U., Sauter1, N. K., Zouni, A., Messinger, J., Yano, J., and Yachandra, V. K. (2018) Structures of the intermediates of Kok's photosynthetic water oxidation clock. *Nature* 563: 421–25.

Kerfeld, C. A. (1997) Structural comparison of cytochrome c_2 and cytochrome c_6. *Photosynthesis Research* 54: 81–98.

Kirchhoff, H., Hall, C., Wood, M., Herbstová, M., Tsabari, O., Nevo, R., Charuvi, D., Shimoni, E., and Reich, Z. (2011) Dynamic control of protein diffusion within the thylakoid lumen. *Proceedings of the National Academy of Sciences USA* 108: 20248–20253.

Kirchhoff, H., Li, M., and Puthiyaveetil, S. (2017) Sublocalization of cytochrome $b_6 f$ complexes in photosynthetic membranes. *Trends in Plant Science* 22: 574–582.

Knaff, D. B. (1996) Ferredoxin and ferredoxin-dependent enzymes. In: *Oxygenic Photosynthesis: The Light Reactions* D. R. Ort and C. F. Yocum (eds.), Dordrecht: Kluwer Academic Publishers, pp. 333–361.

Kok, B., Forbush, B., and McGloin, M. (1970) Cooperation of charges in photosynthetic O2 evolution. I. A linear four step mechanism. *Photochemistry and Photobiology* 11: 457–475.

Kouřil, R., Dekker, J. P., and Boekema, E. J. (2012) Supramolecular organization of photosystem II in green plants. *Biochimica et Biophysica Acta* 1817: 2–12.

Kramer, D.M. and Evans, J. R. (2011) The importance of energy balance in improving photosynthetic productivity. *Plant Physiology* 155: 70–78.

Kuhlgert, S., Drepper, F., Fufezan, C., Sommer, F., and Hippler, M. (2012) Residues PsaB Asp612 and PsaB Glu613 of Photosystem I confer pH-dependent binding of plastocyanin and cytochrome c_6. *Biochemistry* 51: 7297–7303.

Kupitz, C., Basu, S., Grotjohann, I., Fromme, R., Zatsepin, N. A., Rendek, K. N., Hunter, M. S., Shoeman, R. L., White, T. A., Wang, D. J., James, D., Yang, J. H., Cobb, D. E. Reeder, B., Sierra, R. G., Liu, H. G., Barty, A., Aquila, A. L., Deponte, D., Kirian, R. A., Bari, S., Bergkamp, J. J., Beyerlein, K. R., Bogan, M. J., Caleman, C., Chao, T. C., Conrad, C. E., Davis, K. M., Fleckenstein, H., Galli, L., Hau-Riege, S. P., Kassemeyer, S., Laksmono, H., Liang, M. N., Lomb, L., Marchesini, S., Martin, A. V., Messerschmidt, M., Milathianaki, D., Nass, K., Ros, A., Roy-Chowdhury, S., Schmidt, K., Seibert, M., Steinbrener, J., Stellato, F., Yan, L. F., Yoon, C, Moore, T. A., Moore, A. L., Pushkar, Y., Williams, G. J., Boutet, S., Doak, R. B., Weierstall, U., Frank, M., Chapman, H. N., Spence, J. C. H., and Fromme, P. (2014) Serial time-resolved crystallography of photosystem II using a femtosecond X-ray laser. *Nature* 513: 261–265.

Kurisu, G., Kusonoki, M., Katoh, E., Yamazaki, T., Teshima, K., Onda, Y., Kimata-Ariga, Y., and Hase, T. (2001) Structure of the complex between ferredoxin and ferredoxin–NADP$^+$ reductase. *Nature Structural Biology* 8: 117–121.

Li, M., Calteau, A., Semchonok, D. A., Witt, T. A., Nguyen, J. T., Sassoon, N., Boekema, E. J., Whitelegge, J., Gugger, M., and Bruce, B. D. (2019) Physiological and evolutionary implications of tetrameric photosystem I in cyanobacteria. *Nature Plants* 5: 1309–1319.

Linke, K. and Ho, F. M. (2013) Water in Photosystem II: Structural, functional and mechanistic considerations. *Biochimica et Biophysica Acta* 1837: 14–32.

McEvoy, J. P. and Brudvig, G. W. (2006) Water-splitting chemistry of photosystem II. *Chemical Reviews* 106: 4455–4483.

Melkozernov, A. N., Barber, J., and Blankenship, R. E. (2006) Light-harvesting in photosystem I supercomplexes. *Biochemistry* 45: 331–345.

Merchant, S. and Dreyfuss, B.W. (1998) Posttranslational assembly of photosynthetic metalloproteins. *Annual*

Review of Plant Physiology and Plant Molecular Biology 49: 25–51.

Michel, H. and Deisenhofer, H. (1988) Relevance of the photosynthetic reaction center from purple bacteria to the structure of photosystem II. *Biochemistry* 27: 1–7.

Müh, F., Glöckner, C., Hellmich, J., and Zouni, A. (2012) Light-induced quinone reduction in photosystem II. *Biochimica et Biophysica Acta* 1817: 44–65.

Müller, M. G., Slavov, C., Luthra, R., Redding, K. E., and Holzwarth, A. R. (2010) Independent initiation of primary electron transfer in the two branches of the photosystem I reaction center. *Proceedings of the National Academy of Sciences USA* 107: 4123–4128.

Nagy, G., Ünnep, R., Zsiros, O., Tokutsu, R., Takizawa, K., Porcar, L., Moyetg, L., Petroutsos, D., Garab, G., Finazzi, G., and Minagawa, J. (2014) Chloroplast remodeling during state transitions in *Chlamydomonas reinhardtii* as revealed by noninvasive techniques in vivo. *Proceedings of the National Academy of Sciences USA* 111: 5042–5047.

Nanba, O. and Satoh, K. (1987) Isolation of a photosystem II reaction center consisting of D-1 and D-2 polypeptides and cytochrome *b*-559. *Proceedings of the National Academy of Sciences USA* 84: 109–112.

Nawrocki, W. J., Bailleul, B., Picot, D., Cardol, P., Rappaport, F., Wollman, F. A., and Joliot, P. (2019) The mechanism of cyclic electron flow. *Biochimica et Biophysica Acta* 1860: 433–438.

Nelson, N. and Junge, W. (2015) Structure and Energy Transfer in Photosystems of Oxygenic Photosynthesis. *Annual Review of Biochemistry* 84: 659–83.

Nelson, N. and Yocum, C. Y. (2006) Structure and Function of Photosystems I and II. *Annual Review of Plant Biology* 57: 521–565.

Nevo, R., Charuvi, D., Tsabari, O., and Reich, Z. (2012) Composition, architecture and dynamics of the photosynthetic apparatus in higher plants. *Plant Journal* 70: 157–176.

Ort, D. R. and Yocum, C. F. (eds.) (1996) *Oxygenic Photosynthesis: The Light Reactions.* Dordrecht: Kluwer Academic Publishers.

Pagliano, C., Saracco, G., and Barber, J. (2013) Structural, functional and auxialary proteins of photosystem II. *Photosynthesis Research* 116: 167–88.

Rhee, K. H., Morriss, E. P., Barber, J., and Kühlbrandt, W. (1998) Three-dimensional structure of the plant photosystem II reaction centre at 8 angstrom resolution. *Nature* 396: 283–286.

Santabarbara, S., Redding, K. E., and Rappaport, F. (2009) Temperature dependence of the reduction of P700$^+$ by tightly bound plastocyanin in vivo. *Biochemistry* 48: 10457–10466.

Sarewicz, M., Pintscher, S., Pietras, R., Borek, A., Bujnowicz, Ł., Hanke, G., Cramer, W., Finazzi, G., and Osyczka, A. (2021) Catalytic reactions and energy conservation in the cytochrome bc_1 and $b_6 f$ complexes of energy-transducing membranes. *Chemical Reviews*, 121: 2020-2108.

Schoepp, B., Brugna, M., Lebrun, E., and Nitschke, W. (1999) Iron–sulfur centers involved in photosynthetic light reactions. *Advances in Inorganic Chemistry* 47: 335–360.

Shen, J. R. (2015) The structure of Photosystem II and the mechanism of water oxidation in photosynthesis. *Annual Review of Plant Biology* 66: 23–48.

Sheng, X., Watanabe, A., Li, A., Kim, E, Song, C. H., Murata, K., Song, D. F., Minagawa, J., and Liu, Z. F. (2019) Structural insight into light harvesting for photosystem II in green algae. *Nature Plants* 5: 1320–1330.

Shevela, D., Eaton-Rye, J. J., Shen, J.-R., and Govindjee (2012) Photosystem II and the unique role of bicarbonate: A historical perspective. *Biochimica et Biophysica Acta* 1817: 1134–1151.

Shevela, D., Do, H.-N., Fantuzzi, A., Rutherford, A. W. and Messinger, J. (2020) Bicarbonate-mediated CO_2 formation on both sides of Photosystem II. *Biochemistry* 59: 2442–2449.

Shi, L-X., Hall, M., Funk, C., and Schröder, W. P. (2012) Photosystem II, a growing complex: Updates on newly discovered components and low molecular mass proteins. *Biochimica et Biophysica Acta* 1817: 13–25.

Shikanai, T. (2007) Cyclic electron transport around Photosystem I: Genetic approaches. *Annual Review of Plant Biology* 58: 199–217.

Shinopoulos, K. E. and Brudvig, G. W. (2012) Cytochrome *b*559 and cyclic electron transfer within photosystem II. *Biochimica et Biophysica Acta* 1817: 66–75.

Siegbahn, P. E. M (2013) Water oxidation mechanism in photosystem II, including oxidations, proton release pathways, O–O bond formation and O_2 release. *Biochimica et Biophysica Acta* 1827: 1003–1019

Sigfridsson, K. (1998) Plastocyanin, an electron-transfer protein. *Photosynthsis Research* 57: 1–28.

Skillman, J. B. (2008) Quantum yield variation across the three pathways of photosynthesis: not yet out of the dark. *Journal of Experimental Botany* 59: 1647–1661.

Stroebel, D., Choquet, Y., Popot, J.-L., and Picot, D. (2003) An atypical haem in the cytochrome b_6f complex. *Nature* 426: 413–418.

Styring, S., Sjöholm, S., and Mamedov, F. (2012) Two tyrosines that changed the world: Interfacing the oxidizing power of photochemistry to water splitting in photosystem II. *Biochimica et Biophysica Acta* 1817: 76–87.

Suga, M., Fakita, F., Yamashita, K., Nakajima, Y., Ueno, G., Li, H., Yamane, T., Hirata, K., Umena, Y., Yonekura, S., Yu, L.-J., Murakami, H., Nomura, T., Kimura, T., Kubo, M., Baba, S., Kumasaka, T., Tono, K., Yabashi, M., Isobe, H., Yamaguchi, K., Yamamoto, M., Ago, H., and Shen, J.-R. (2019) An oxyl/oxo mechanism for oxygen-oxygen coupling in PSII revealed by an x-ray free-electron laser. *Science* 366: 334–338.

Umena, Y., Kawakami, K., Shen, J.-R., and Kamiya, N. (2011) Crystal structure of oxygen-evolving photosystem II at a resolution of 1.9 Å. *Nature* 473: 55–60.

van Bezouwen L. S., Caffarri, S., Kale, R. S., Kouřil, R., Thunnissen, A.-M. W. H., Oostergetel, G. T., and Boekema, E. J. (2017) Subunit and chlorophyll organization of the plant photosystem II supercomplex. *Nature Plants* 3: 17080.

Vinyard, D. J. and Brudvig, G. W. (2017) Progress toward a molecular mechanism of water oxidation in Photosystem II. *Annual Review of Physical Chemistry* 68: 101–116.

Vinyard, D. J., Ananyev, G. M., and Dismukes, G. C. (2013) Photosystem II: The Reaction Center of Oxygenic Photosynthesis. *Annual Review of Biochemistry* 82: 577–606.

Wydrzynski, T. J. and Satoh, K. (eds.) (2005) *Photosystem II: The Light-Driven Water: Plastoquinone Oxidoreductase.* Dordrecht: Springer.

Yano, J., Kern, J., Sauer, K., Latimer, M. J., Pushkar, Y., Biesiadka, J., Loll, B., Saenger, W., Messinger, J., Zouni, A., and Yachandra, V. K. (2006) Where water is oxidized to dioxygen: Structure of the photosynthetic Mn_4Ca cluster. *Science* 314: 821–825.

Yano, J. and Yachandra, V. (2014) Mn_4Ca cluster in photosynthesis: Where and how water is oxidized to dioxygen. *Chemical Reviews* 114: 4175–4205.

Zouni, A., Witt, H. T., Kern, J., Fromme, P., Krauss, N., Saenger, W., and Orth, P. (2001) Crystal structure of photosystem II from *Synechococcus elongatus* at 3.8 Å resolution. *Nature* 409: 739–743.

Chapter 8

Chemiosmotic coupling and ATP synthesis

A significant portion of the energy of photons that drive photosynthesis is stored in the form of chemical energy in adenosine triphosphate (ATP). ATP is the universal energy currency of living systems and is used to power many cellular processes. In this chapter, we will focus on the mechanisms by which photon energy is stored in ATP, including the Mitchell chemiosmotic theory and the details of the structure and function of the ATP synthase enzyme.

8.1 Chemical aspects of ATP and the phosphoanhydride bonds

The chemical structure of ATP is given in Fig. 8.1. It is a nucleotide, consisting of an adenine base, the sugar ribose, and three phosphate moieties. The bonds between the phosphate groups are called **phosphoanhydride bonds** and are the bonds that are made and broken when ATP is synthesized and broken down. The related molecules with two and one phosphate groups are adenosine diphosphate (ADP) and adenosine monophosphate (AMP).

The phosphoanhydride bonds are often called "high-energy" phosphate bonds and are commonly represented by a ~ symbol. Sometimes, it is even suggested that these bonds are unstable, like little powder kegs ready to explode. But this is misleading, as the phosphoanhydride bonds are perfectly normal chemical bonds. They represent a potential energy minimum, such as is shown in Fig. A.10. A typical O–P chemical bond has a bond dissociation energy of about $500 \, \text{kJ} \, \text{mol}^{-1}$, which is the energy needed to break the bond. ATP is thus a very stable chemical species in the solid form and can be stored for years without decomposition. Even in solution, ATP is relatively stable, because the activation energy for hydrolysis is high, so the spontaneous rate of hydrolysis is slow. Why then is this bond usually referred to as high-energy? The reason is that the chemical reaction in which the phosphoanhydride bonds are hydrolyzed by water has a substantial negative standard-state free-energy change, making it a spontaneous process.

The chemical reaction by which ATP is converted to ADP and phosphate is shown in Eq. (8.1).

$$ATP^{4-} + H_2O \rightarrow ADP^{3-} + HPO_4^{2-} + H^+ \quad (8.1)$$

Molecular Mechanisms of Photosynthesis, Third Edition. Robert E. Blankenship.
© 2021 Robert E. Blankenship 2021 by John Wiley & Sons Ltd.
Companion website: https://www.wiley.com/go/blankenship/molecularphotosynthesis3e

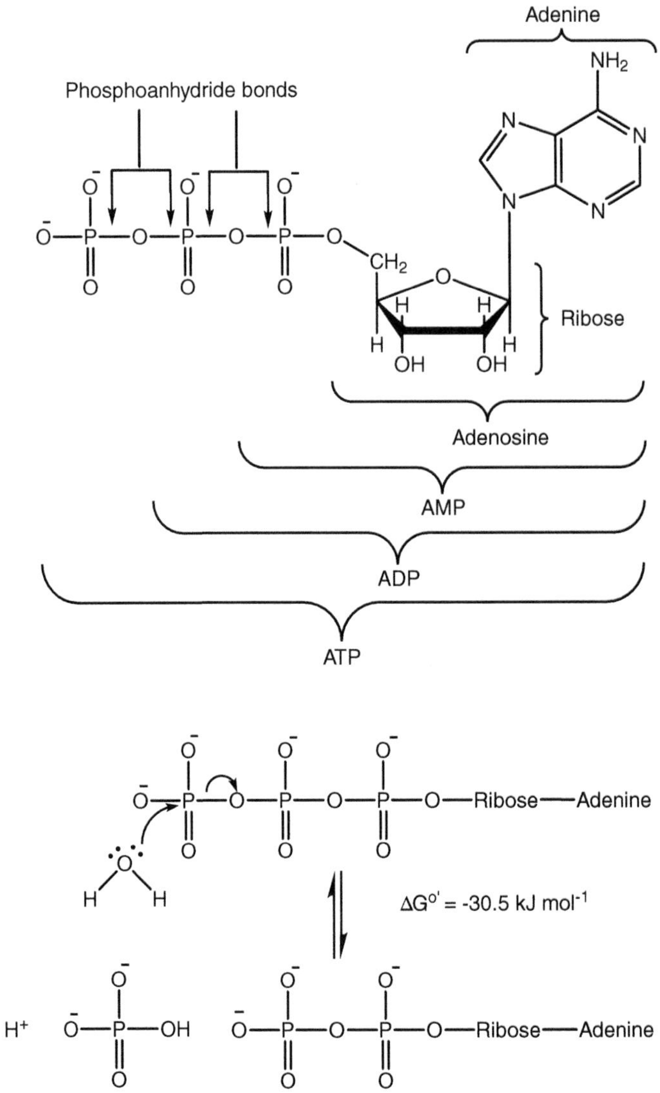

Figure 8.1   Chemical structure of ATP. The structures of ADP and AMP are shorter by one and two phosphate groups, respectively. The hydrolysis of ATP by water to form ADP, P_i, and H^+ is shown in the bottom part of the figure.

This is a hydrolysis reaction, in which water attacks the phosphoanhydride bond, resulting in ADP and inorganic phosphate (often abbreviated P_i). This reaction is also shown in Fig. 8.1. The important point is that numerous bonds are broken and made in the hydrolysis process. The overall free-energy change of the reaction reports on the net change in all these bond energies and does not reflect just the phosphoanhydride bond. The amount of energy released when the new bonds are formed is more than the energy required to break the original bonds, so the overall free-energy change is negative, and the process is spontaneous. Several molecular factors contribute to the negative free energy of hydrolysis of phosphoanhydride bonds. These factors include the extra resonance stabilization of the product phosphate group compared with the phosphates in the ATP molecule, a lower amount of electrostatic repulsion of the products ADP and P_i compared with the more highly charged ATP, and solvation effects, which also favor the products compared with the reactants. While the free energy of hydrolysis of ATP is a useful benchmark, it is important to realize that the actual hydrolysis reaction rarely actually takes place, as that would dissipate the energy as heat and not allow it to be used for cellular processes. Instead, the process is coupled to another energy-requiring reaction so that the energy that would have been released by hydrolysis is used instead to drive the endergonic reaction. The use of coupled reactions is very common in biochemical pathways and provides a mechanism to channel energy in productive ways (Nelson and Cox, 2017).

The standard state free energy of the ATP hydrolysis reaction is $-30.5\,kJ\,mol^{-1}$. This represents the free-energy change when all reactants are at 1M concentrations and at a pH of 7. Of course, the concentrations of the species in Eq. (8.1) in the chloroplast (or any cell or organelle) are not the standard state values, so the actual free-energy change will be quite different from the standard state value. The actual free-energy change of the ATP hydrolysis reaction is given by Eq. (8.2), which is the form of Eq. (A.15) that applies to this particular reaction. The water and H^+ terms have already been incorporated into the standard state expression, so do not appear in Eq. (8.2), although a correction will be needed for pH values other than 7.

$$\Delta G = \Delta G^{o\prime} + RT \ln \frac{[ADP][P_i]}{[ATP]} \qquad (8.2)$$

Most cells maintain the concentrations of ATP, ADP, and P_i in a very narrow range. ATP and P_i are at much higher concentrations than ADP. Typical values are 2.5 mM for ATP, 0.25 mM for ADP, and 2.0 mM for P_i. Using Eq. (8.2), the actual free-energy change for ATP hydrolysis at 298 K and pH 7 is calculated to be $-52\,kJ\,mol^{-1}$. The free energy needed for the synthesis of ATP under these conditions is thus $+52\,kJ\,mol^{-1}$.

8.2   Historical perspective on ATP synthesis

ATP was first isolated in 1929 from muscle tissues. Later work showed that the production of ATP was associated with cellular respiration and the consumption of oxygen. In 1940, Fritz Lipmann proposed that ATP is the universal energy currency in cells. Albert Lehninger and Eugene Kennedy showed in 1948 that the mitochondrion was the site of oxidative phosphorylation, the process in which electron flow from NADH to oxygen is coupled to the synthesis of ATP (Nicholls and Ferguson, 2013; Nelson and Cox, 2017).

Photosynthetic phosphorylation was discovered in the 1950s by Daniel Arnon and coworkers (Arnon, 1984). At the time, the announcement that chloroplasts could make ATP was greeted with considerable skepticism. Phosphorylation was thought to be the exclusive province of the mitochondria, which oxidize NADH to NAD^+ to make ATP during aerobic respiration. The possibility that chloroplasts could make ATP at the same time as they were reducing $NADP^+$ to NADPH was considered to be beyond the pale.

The development of the understanding of how ATP is made in both chloroplasts and mitochondria has a colorful and contentious history (Racker, 1976; Gilbert and Mulkay, 1984; Nicholls and Ferguson, 2013; Prebble, 2002). In the 1930s, the pathway of glycolysis was established, followed by other pathways of intermediary metabolism. Glycolysis produces ATP by a process called substrate-level phosphorylation, in which ATP is formed by a soluble enzyme acting on a series of chemical substrates in the cytoplasm of the cell. An example involves one of the reactions in glycolysis, in which 1,3-bisphosphoglycerate reacts with ADP to form ATP and 3-phosphoglycerate, catalyzed by the enzyme phosphoglycerate kinase (Nelson and Cox, 2017).

When scientists first tried to decipher the mechanism used by mitochondria and chloroplasts to make ATP, the substrate-level phosphorylation reactions were considered as models. Great efforts were expended to isolate phosphorylated intermediates, but these efforts were unsuccessful. Other theories were put forward, in which it was proposed that the process was driven somehow by enzyme conformational changes. However, understanding of the way in which ATP was made was not progressing, and a famous assessment of the situation by Efriam Racker, a prominent worker in the area, summed up the field: "Anyone who is not thoroughly confused just doesn't understand the problem." Not a single scientist doing research on this subject proposed that a membrane might somehow be involved in the phosphorylation process. What was needed was an entirely new approach. Although it was not apparent at the time, simply continuing the same unproductive approaches was never going to prove fruitful.

Finally, in 1961, a British biochemist named Peter Mitchell proposed a phosphorylation mechanism in which free energy, stored as both a pH gradient and an electrical potential across a membrane that surrounded an enclosed space, was transduced into the high-energy phosphoanhydride bond in ATP. Initially, other scientists ignored Mitchell, because he was speaking an entirely new language that didn't mean much to them. Later, he was harshly ridiculed. However, as Mitchell began to produce evidence in support of his theory, more and more scientists slowly began to pay attention.

The essence of the **Mitchell chemiosmotic hypothesis**, as it came to be known, is that electron carriers are arranged in a vectorial manner in an intrinsically asymmetric biological membrane (Mitchell, 1979). As electrons are transferred from one carrier to the next, so protons are transported from one side of the membrane to the other in a series of linked reactions, generating a pH difference between the interior and exterior spaces that the membrane encloses. The membrane is intrinsically impermeable to protons and most other charged species. The proton transfer and other processes also generate electrical potential differences across the membrane, and it is the sum of these two effects, collectively called the **proton motive force**, that provides the energy source for ATP synthesis. Mitchell provided a natural explanation for the mechanism of action of uncouplers, a class of compounds that inhibited ATP synthesis in intact

systems. These compounds are all lipophilic weak acids, such as dinitrophenol, DNP. He proposed that these compounds dissipated the proton gradient by being capable of diffusing through the membrane in both the protonated and deprotonated forms. Mitchell didn't specify the chemical mechanism by which the proton motive force was transduced into ATP, but instead focused on the role of the membrane and the linked proton and electron transfer processes. He ultimately convinced most skeptics and received the Nobel prize in Chemistry in 1978. His ideas are now universally accepted and provide the cornerstone of our understanding of the mechanism of ATP formation in photosynthetic and respiratory energy conservation, as well as a multitude of other membrane-linked cellular processes, such as active transport.

In 1966, dramatic evidence in favor of the chemiosmotic hypothesis was obtained by André Jagendorf and his coworkers, in a classic experiment that was elegant in its simplicity and revealing in its result (Fig. 8.2) (Jagendorf, 1967, 1998). They suspended chloroplast thylakoids in a pH 4 buffer, which permeated the membrane and caused the interior, as well as the exterior, of the thylakoid to equilibrate at this acidic pH. They then rapidly injected the suspension into a pH 8 buffer solution, thereby creating a pH difference of four units across the thylakoid membrane, with the inside acidic relative to the outside. They found that large amounts of ATP were formed from ADP and P_i by this process, without any light input or electron transport. This result supported the predictions of the chemiosmotic hypothesis. Largely because of the Jagendorf experiment, scientists working in photosynthesis were early converts to Mitchell's ideas.

8.3   Quantitative formulation of proton motive force

Mitchell proposed that the total energy available for ATP synthesis, which he called the **proton motive force** (Δp), is the sum of a proton chemical potential and a transmembrane electrical potential

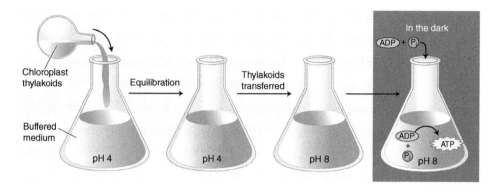

Figure 8.2   Acid–base phosphorylation experiment carried out by Jagendorf and coworkers. Thylakoids were suspended in a pH 4 buffer and allowed to equilibrate. They were then rapidly transferred to a pH 8 buffer, and ADP and P_i were added. ATP was formed, in agreement with the predictions of the Mitchell chemiosmotic mechanism. *Source:* Taiz *et al.* (2018)/Oxford University Press.

(Lowe and Jones, 1984; Nicholls and Ferguson, 2013). These two components of the proton motive force from the outside of the membrane to the inside are given by Eq. (8.3):

$$\Delta p = \Delta \Psi - 59 \Delta pH \qquad (8.3)$$

The units of Δp are mV. In Eq. (8.3), $\Delta \Psi$ is the transmembrane electrical potential (in mV) and ΔpH (or $pH_i - pH_o$) is the pH difference across the membrane. The constant of proportionality (at 25 °C) is 59 mV per pH unit, so a transmembrane pH difference of one unit is equivalent to a membrane potential of 59 mV. Figure 8.3 illustrates the definition of the proton motive force.

We can easily justify Eq. (8.3) by referring to the fundamental equations of thermodynamics. The chemical potential of a particular species is given by Eq. (A.11). If we include an additional term for electrical work in this equation, Eq. (8.4) is the result, where z_j is the charge on the species, F is the Faraday constant, and Ψ_j is the electrical potential (the subscript j is used to avoid confusion with the i (inside) and o (outside) nomenclature used below).

$$\mu_j = \mu_j^\circ + RT \, ln a_j + z_j F \psi_j \qquad (8.4)$$

Because we are most interested in the chemical potential for hydrogen ions, we will rewrite this equation for this species, resulting in Eq. (8.5):

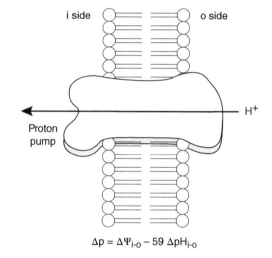

$$\Delta p = \Delta \Psi_{i\text{-}o} - 59 \, \Delta pH_{i\text{-}o}$$

Figure 8.3   Schematic picture of a membrane-enclosed vesicle and the proton motive force generated by coupled electron and proton transport.

$$\mu_{H^+} = RT \, ln \left[H^+ \right] + F \psi_{H^+} \qquad (8.5)$$

For hydrogen ions, the μ° term is zero by definition, and z is +1. The reference point for the electrical potential is not important, because only the differences in these potentials will be considered later. The chemical potential difference for the hydrogen ion between the inside and outside of a membrane vesicle is given by Eqs. (8.6) and (8.7):

$$\Delta\mu_{H^+} = \mu_{H_i^+} - \mu_{H_o^+} \qquad (8.6)$$

$$\Delta p = \frac{\mu_{H^+}}{F} = -\frac{2.3RT}{F}\Delta pH_{i-o} + \Delta\psi_{i-o} \qquad (8.7)$$

Substitution of the numerical values for R, T, and F (see Appendix) and rearrangement leads to Eq. (8.3).

Experiments have established that the two components of the proton motive force are thermodynamically interchangeable (although there may be kinetic differences), so a large ΔE_m, a large ΔpH, or intermediate amounts of both are all equally effective in forming ATP (Hangarter and Good, 1988). Under conditions of steady-state electron transport in isolated thylakoids, the membrane electrical potential is small due to subsequent ion movement across the membrane, so Δp is built almost entirely by ΔpH. In intact plants, the situation may be somewhat different, and the pH gradient may be smaller, with a larger membrane potential (Kramer *et al.*, 1999).

8.4 Nomenclature and cellular location of ATP synthase

The ATP synthase enzyme (often called the coupling factor) is a large multisubunit complex consisting of two subsections, called F_1 and F_o. These names refer to the enzyme from mitochondria that was used for much of the early work. The names are somewhat confusing and need some explanation. The F stands for "fraction," and F_1 refers to the first fraction isolated as an ATP hydrolyzing enzyme (ATPase) after treatment of submitochondrial particles with urea. The subscript on F_o is not zero, but rather a lowercase letter "o" and stands for oligomycin sensitivity. Oligomycin is an inhibitor that specifically blocks proton transfer through this part of the enzyme. Because the F_1 portion of the enzyme is usually assayed in the reverse direction from its normal function of ATP synthesis, it is often called ATPase.

The chloroplast enzyme is by analogy called CF_1 and CF_o and is, in most respects, very similar to the related enzymes from both mitochondria and various bacteria, both photosynthetic and nonphotosynthetic (Böttcher and Gräber, 2000; Strotmann et al. 2000; Junge *et al.*, 2009; Spetzler *et al.*, 2012; Kühlbrandt, 2019). Ironically, CF_o is not sensitive to oligomycin, although it is otherwise generally very similar to F_o. The CF_o portion is located within the membrane, while the CF_1 portion protrudes like a knob on one side of the membrane. The coupling factor is located exclusively in the stromal membranes in chloroplasts, with the F_1 portion sticking out into the stromal space (Fig. 8.4).

In purple photosynthetic bacteria, the enzyme is located in the intracytoplasmic membrane, with the F_1 portion sticking into the cytoplasm (Gromet-Elhanan, 1995; Feniouk and Junge, 2009). When chromatophore membranes are prepared, the process of breaking and resealing causes the cytoplasmic side to become the outside and the periplasmic side to become the inside. (See Fig. 6.1 for a schematic picture of how this occurs.) This process reverses the topological orientation of the chromatophore relative to the intact cell. The result is that protons are pumped in the opposite direction compared with intact cells, so light irradiation causes a pH decrease in the external medium. In chloroplasts, light causes a pH increase in the external medium (and therefore an acidification in the thylakoid lumen). However, this opposite sense of the pH change does not signify a fundamental difference in these systems, but is only a result of the process of preparation of chromatophores. In fact, the F_1–F_o complexes from all organisms and organelles are highly similar, and many hybrids with parts of their systems from one origin and other parts from a separate origin are fully functional.

8.5 Structure of ATP synthase

ATP synthase enzymes from all cells have a similar chemical composition and structure, although there are some differences. In all enzymes, there are

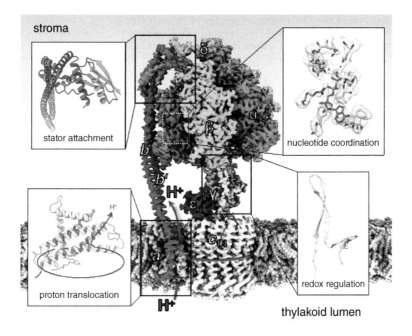

Figure 8.4 Structure of the chloroplast ATP synthase enzyme. Subunits α and β contain the nucleotide-binding sites with resolved nucleotides, Mg^{2+}, and water molecules. Subunit δ joins CF_1 to the membrane-embedded motor via the peripheral stalk (b, b') that positions subunit a against the rotor ring. The electrochemical proton gradient drives ring rotation (arrow). The central stalk ($\gamma\varepsilon$) transmits torque to CF_1. The redox regulator blocks rotation in the dark. *Source:* Hahn *et al.* (2018)/American Association for the Advancement of Science.

three copies each of the α and β subunits and one copy of each of the other F_1 subunits γ, δ, and ε. However, the δ subunit of the mitochondrial enzyme is homologous to the ε subunit of the enzymes from other sources, and the mitochondrial ε subunit is unrelated to any subunit in the bacterial or chloroplast enzymes.

In the F_o portion of the bacterial and mitochondrial enzymes, there is one copy of the a subunit, two copies of the b subunit (or one each of the very similar b and b' subunits in purple bacteria and cyanobacteria), and 8–15 copies of the c subunit. The stoichiometry of the c subunit is different for different organisms. The mechanistic implications of this variable stoichiometry are discussed below. The mitochondrial F_o complex contains five additional subunits not found in the other enzymes. The F_o subunits of the chloroplast enzyme have a different nomenclature, in that the homolog of the a subunit is known as subunit IV, the c subunits are

known as subunit III, and the b subunits are not identical but, rather, are two related but distinct subunits known as I and II (Böttcher and Gräber, 2000). The subunit composition of the chloroplast ATP synthase enzyme is given in Table 8.1.

In 1994, a landmark in ATP synthase research took place with the determination of the structure of the F_1 enzyme from bovine heart mitochondria by X-ray crystallography (Abrahams *et al.*, 1994). These structural studies have been extended to include more of the complex (Rees *et al.*, 2009). While there are some significant differences between the chloroplast and mitochondrial enzymes, their basic structure and mechanism of action are almost certainly the same. Therefore, our discussion of the structure and mechanism of the coupling factor will be largely based on the principles elucidated from the structure of the mitochondrial enzyme. More recent structural studies by

Table 8.1   Subunit structure of the ATP synthase complex of chloroplasts

Subunit name	Number of copies	Gene	Gene location[a]	Mass (kDa)[b]	Function
CF$_1$ subunits					
α	3	atpA	C	55	Noncatalytic nucleotide binding
β	3	atpB	C	52	Catalytic nucleotide binding
γ	1	atpC	N	35	Part of rotor, central stalk, thiol modulation
δ	1	atpD	C	21	Part of stator, attachment to $\alpha_3\beta_3$
ε	1	atpE	N	15	Part of rotor, central stalk
CF$_0$ subunits					
I (b)	1	atpF	C	17	Part of stator, side stalk
II (b)	1	atpG	N	16	Part of stator, side stalk
III (c)	8–15[c]	atpH	C	8	Part of rotor, proton carrier
IV (a)	1	atpI	C	25	Part of stator, proton channel

[a] Gene location applies only to eukaryotic organisms.
[b] Mass is actual mass-based on gene sequence.
[c] Stoichiometry depends on species.

both X-ray crystallography and NMR spectroscopy have revealed the structure of much of the F$_0$ portion of the enzyme (Rastogi and Girvin, 1999; Stock et al., 1999, 2000; Pogoryelov et al., 2009; Watt et al., 2010).

The structure of ATP synthase enzyme is shown in Figs. 8.4 and 8.5. The F$_1$ portion of the coupling factor contains the three α and β subunits, alternating like sections of an orange. The γ subunit consists of an exceptionally long, bent α helical segment, which is found in the central core of the $\alpha_3\beta_3$ complex. The bottom portion of the γ subunit extends into the F$_0$ portion of the enzyme. The δ and ε subunits interact with the F$_0$ portion of the enzyme. The two b subunits form a claw that interacts with the δ subunit on one end and the integral membrane a subunit on the other. The b subunits serve to lock the F$_1$ portion of the enzyme to the membrane-bound a subunit in the F$_0$ portion. The a subunit is predicted to have five or six membrane-spanning helices. The c subunits are arranged in a ring, and each one adopts a hairpin-like structure in the membrane, with two membrane-spanning helices (Rastogi and Girvin, 1999; Stock et al., 1999). The ring of c subunits is tightly associated with the γ subunit, but is not tightly linked to the rest of the complex. These subunits form a rotor that can

move independently of the rest of the structure, which is termed the stator. The terms **rotor** and **stator** are derived from the portions of an electric motor, in which the stator is the stationary part, and the rotor spins in the center. This already suggests the motor-like nature of the mechanism of action of this remarkable enzyme, which is described below.

The CF$_1$ enzyme from chloroplasts is unique among F-type ATPases in that its γ subunit contains a pair of cysteine residues that can be present in either the oxidized disulfide form or the reduced sulfhydryl form. In the oxidized disulfide form the enzyme is inactive, whereas it is fully active in the reduced form (Nalin and McCarty, 1984; Yang et al., 2020).

A soluble protein, thioredoxin, is reduced by ferredoxin and in turn reduces the enzyme, thereby activating it. Thioredoxin is generally in the reduced form only when electron transport through Photosystem I is taking place, so the ATP synthase is activated only when it is needed to produce ATP. This regulatory mechanism prevents it from running backward whenever light-driven electron transport is interrupted. Thioredoxin also regulates many enzymes of the carbon fixation cycle, as discussed in Chapter 9.

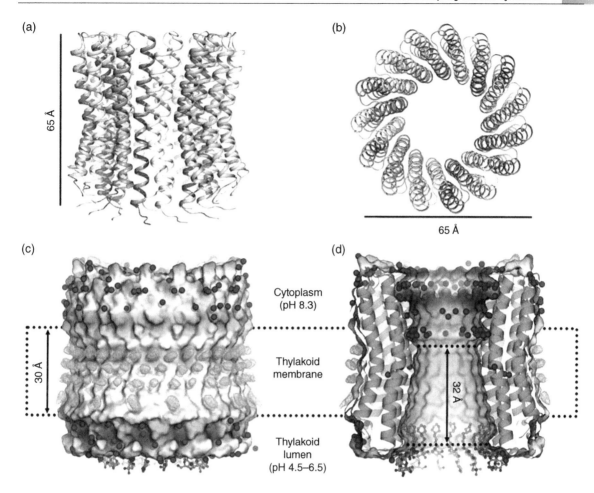

Figure 8.5   Structure of the F_o portion of the F_1F_o ATP synthase enzyme from the cyanobacterium *Spirulina platensis*. (a) and (b) show side and top views of the F_o complex. (c) shows a side view of the electrostatic surface potential and (d) shows a longitudinal section showing the interior surface of the ring. *Source:* Pogoryelov *et al.* (2009)/ Springer Nature.

8.6   The mechanism of chemiosmotic coupling

The basic elements of a molecular-scale mechanism for the action of the ATP synthase are now in place. Although some details of the mechanism outlined below will almost certainly turn out to be wrong, the main aspects of the mechanism are probably correct. The most important aspect of the proposed mechanism is that the rotor portion of the complex, the γ and ε subunits and the ring of

c subunits, rotate with respect to the stator portions of the complex. This rotation was dramatically demonstrated by an ingenious experiment (Noji *et al.*, 1997). The stator section of the F_1 portion of the enzyme was attached to a surface, and a long actin filament was attached to the end of the γ subunit that is normally in contact with the F_o portion of the enzyme. When ATP was added to the complex (which is hydrolyzed in a reaction that is the reverse of the normal reaction in which ATP is made by the enzyme), the actin filaments attached to the γ subunit spun rapidly (Fig. 8.6). Subsequent experiments have established that the ring of c subunits, attached to the γ and ε subunits, also rotates as would be expected if it is part of the rotor (Pänke et al., 2000; Wada et al., 2000). Other experiments have established that the rotary motion takes place in 120-degree jumps (Adachi et al., 2000; Yasuda et al., 2001; Adachi et al., 2007). These and additional biochemical experiments (Fillingame et al., 2000) clearly establish that the ATPase enzyme is a rotary motor.

The **binding change mechanism** for ATP synthesis was proposed by Paul Boyer on the basis of a large number of enzyme kinetic analyses (Boyer, 1997, 2000). This mechanism involves three enzymatic sites, which alternate sequentially from open, loose, and tight binding states (Fig. 8.7). A major feature of the binding change mechanism is the proposal that the main energy-requiring step in ATP synthesis is the release of bound ATP from the enzyme, rather than the formation of ATP itself. The rotary motor model provides a natural mechanism for the alternation of the three enzymatic sites, which are located on the β subunits. As the γ

subunit rotates in the bearing that forms the central channel of the F_1 portion of the enzyme, its bent structure acts like a camshaft and alternately distorts the β subunits in turn, causing the cycling among the open, loose, and tight binding states. This conformational energy is transduced into the ATP phosphoanhydride bond by a process that is not yet well understood.

The final part of the mechanism involves an explanation of how the protons flowing from one side of the membrane to the other can drive the rotation of the rotor portion of the enzyme (Fig. 8.8). Here, attention is focused on the F_o portion of the enzyme, in which the a subunit provides entrance and exit channels for protons, but does not contain a direct connection between the two. A proton that comes into the entrance channel protonates an essential amino acid residue, glutamic acid 61, in the middle of one of the transmembrane helices of one of the 14 c subunits. This causes a conformational change that "ratchets" the c subunits with respect to the a subunit (Rastogi and Girvin, 1999; Junge et al., 2009; Pogoryelov et al., 2009; Junge and

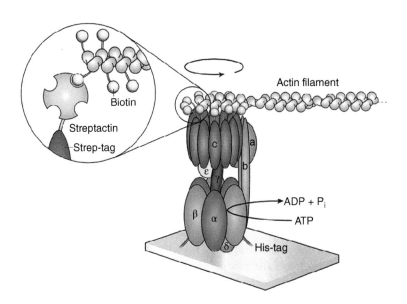

Figure 8.6 Experiment demonstrating that the ATPase is a rotary motor. The F_1 portion of the ATP synthase enzyme was tethered to a glass slide by histidine tags. An actin filament was attached to the γ subunit. Some of the enzymes subsequently carried out ATP-dependent rotation of the actin filaments, which was visualized in a microscope. *Source:* Junge et al. (2009)/Springer Nature.

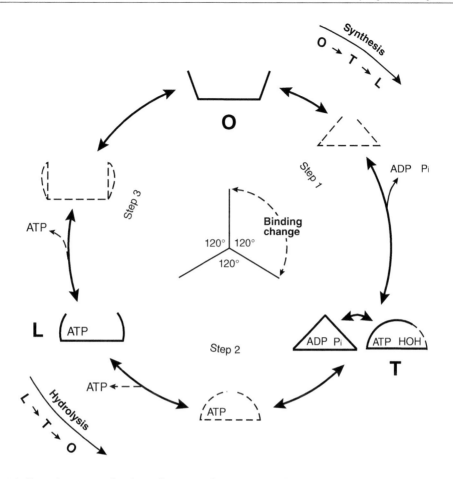

Figure 8.7   Binding change mechanism of ATP synthesis proposed by Boyer. The enzyme has three catalytic nucleotide-binding sites, which cycle between the tight, loose, and open conformations. The conformational changes induced by the rotary motion of the γ subunit cause the enzyme to change its affinity for the nucleotide. *Source:* Boyer (2000)/Elsevier.

Nelson, 2015; Hahn *et al.*, 2018; Kühlbrandt, 2019). The ring of c subunits rotates counterclockwise (viewed from the F_1 side of the membrane), and each of the c subunits in turn picks up a proton, finally discharging it through the exit channel. Clockwise rotation is prevented by the electrostatic interaction of a positively charged arginine residue in the a subunit. The protons neutralize the negative charges on the glutamic acid residues in the c subunits, and the neutral subunits are free to rotate in the hydrophobic membrane. The c subunits each transport one proton during one complete cycle and each of the three β subunits goes through the

conformational sequence described above, with each producing and releasing an ATP molecule.

The stoichiometry of proton pumping and ATP synthesis depends on the number of c subunits in the F_o portion of the enzyme. The number of c subunits in each F_o subunit is not a universal value, but appears to vary from one organism to another, or possibly even as a function of the metabolic state in a single organism. For example, in *E. coli*, the number of c subunits is 12 (Rastogi and Girvin, 1999), while in mitochondria it ranges from 8 to 10 (Stock *et al.*, 1999; Watt *et al.*, 2010). Microscopic analysis of the chloroplast enzyme indicates that it contains 14 c

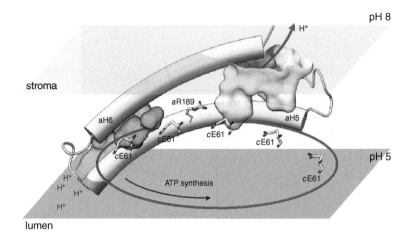

Figure 8.8 Proposed pathway of proton translocation through the F_o portion of the ATP synthase. The long, membrane-intrinsic hairpin of subunit a helices H5 and H6 follows the curvature of the c_{14}-ring, forming the luminal proton entry channel (transparent red) and stromal exit channel (transparent blue). The entry channel conducts protons from the acidic thylakoid lumen to the c-ring glutamate (cE61). After an almost full rotation of the c_{14}-ring, the glutamate encounters the large, hydrophilic exit channel that extends to the stromal membrane surface. *Source:* Hahn *et al.* (2018)/American Association for the Advancement of Science.

subunits in the F_o complex (Seelert *et al.*, 2000), while it is 15 in the cyanobacterium *Spirulina platensis* (Pogoryelov *et al.*, 2009). In each case, one full rotation of the γ subunit with its attached ring of c subunits and *e* subunit leads to the synthesis of three molecules of ATP.

The stoichiometry of H⁺/ATP predicted for any given ATP synthase enzyme is calculated by the number of c subunits in that enzyme divided by three. This conclusion rests on the assumption that each c subunit translocates one H⁺ per cycle, which has not been firmly established. Therefore, in an enzyme that has 12 c subunits, 12 H⁺ are translocated per cycle divided by 3 ATP synthesized per cycle, giving the overall stoichiometry as 4 H⁺/ATP. If the number of c subunits is 14, and all are translocating protons, then the H⁺/ATP stoichiometry is 4.67, while if there are 10 functional c subunits, the H⁺/ATP stoichiometry is 3.33. Recent experimental values of the H⁺/ATP stoichiometry in thylakoids were found to be 4 (Berry and Rumberg, 1996; van Walraven *et al.*, 1996; Kramer *et al.*, 1999; Steigmiller *et al.*, 2008; Petersen *et al.*, 2012). No experimental values as high as 4.67 have been reported. However, these are difficult measurements, which are subject to a number of complicating factors.

There is a fundamental mechanistic implication of the fact that the ratio of H⁺/ATP is not necessarily an integer. A noninteger value means that the coupling of the proton motive force to ATP formation cannot be a direct mechanical gearing but, rather, must take place through a mechanism that might instead be compared to the release of elastic energy stored in a coiled rubber band (Cherepanov *et al.*, 1999; Pänke and Rumberg, 1999; Junge *et al.*, 2009; Martin et al., 2018). Exactly how this works on a molecular level is not yet understood.

There are still many unanswered questions about the mechanism of ATP synthesis by the F_1 ATP synthase, but the combination of the structural and dynamic information and the detailed enzymatic analysis has pointed the way toward a consistent picture, which is now being refined. John Walker, the leader of the X-ray group that determined the structure of the F_1 ATPase, and Paul Boyer, who formulated the binding change mechanism, shared the 1997 Nobel prize in chemistry for their pioneering work.

References

Abrahams, J., Leslie, A. G. W., Lutter, R., and Walker, J. E. (1994) Structure at 2.8 Å resolution of F_1-ATPase from bovine heart mitochondria. *Nature* 370: 621–628.

Adachi, K., Yasuda, R., Noji, H., Itoh, H., Harada, Y., Yoshida, M., and Kinosita, K. (2000) Stepping rotation of F_1-ATPase visualized through angle-resolved single-fluorophore imaging. *Proceedings of the National Academy of Sciences USA* 97: 7243–7247.

Adachi, K., Oiwa, K., Nishizaka, T., Furuike, S., Itoh, H., Yoshida, M., and Kinosita, K. (2007) Coupling of rotation and catalysis in F_1-ATPase revealed by single-molecule imaging and manipulation. *Cell* 130: 309–321.

Arnon, D. I. (1984) The discovery of photosynthetic phosphorylation. *Trends in Biochemical Sciences* 9: 258–262.

Berry, S. and Rumberg, B. (1996) H^+/ATP coupling ratio at the unmodulated CF_0CF_1-ATP synthase determined by proton flux measurements. *Biochimica et Biophysica Acta* 1276: 51–56.

Böttcher, B. and Gräber, P. (2000) The structure of the H^+-ATP synthase from chloroplasts and its subcomplexes as revealed by electron microscopy. *Biochimica et Biophysica Acta* 1458: 404–416.

Boyer, P. D. (1997) The ATP synthase – A splendid molecular machine. *Annual Review of Biochemistry* 66: 717–749.

Boyer, P. D. (2000) Catalytic site forms and controls in ATP synthase catalysis. *Biochimica et Biophysica Acta* 1458: 252–262.

Cherepanov, D. A., Mulkidjanian, A. Y., and Junge, W. (1999) Transient accumulation of elastic energy in proton translocating ATP synthase. *FEBS Letters* 449: 1–6.

Feniouk, B. A. and Junge, W. (2009) Proton translocation and ATP synthesis by the F_0F_1-ATPase of purple bacteria. In: C. N. Hunter, F. Daldal, M. C. Thurnauer, C. N. Hunter, F. Daldal, M. C. Thurnauer, and J. T. Beatty, (eds.) *The Purple Photosynthetic Bacteria*. Dordrecht: Springer, pp. 475–493.

Fillingame, R. H., Jiang, W., Dmitriev, O. Y., and Jones, P. C. (2000) Structural interpretations of F_0 rotary function in the *Escherichia coli* F_1F_0 ATP synthase. *Biochimica et Biophysica Acta* 1458: 387–403.

Gilbert, G. N. and Mulkay, M. (1984) *Opening Pandora's Box*. Cambridge: Cambridge University Press.

Gromet-Elhanan, Z. (1995) The proton-translocating F_0F_1 ATP synthase-ATPase complex. In: R. E. Blankenship, M. T. Madigan, and C. E. Bauer, (eds.) *Anoxygenic Photosynthetic Bacteria*. Dordrecht: Kluwer Academic Publishers, pp. 807–830.

Hahn, A., Vonck, J., Mills, D. J., Meier, T., and Kühlbrandt, W. (2018) Structure, mechanism, and regulation of the chloroplast ATP synthase. *Science* 360: eaat4318.

Hangarter, R. P. and Good, N. E. (1988) Active-transport, ion movements, and pH changes. 2. Changes of pH and ATP synthesis. *Photosynthesis Research* 19: 237–250.

Jagendorf, A. T. (1967) Acid-based transitions and phosphorylation by chloroplasts. *Federation Proceedings Federation of American Society of Experimental Biology* 26: 1361–1369.

Jagendorf, A. T. (1998) Chance, luck and photosynthesis research: An inside story. *Photosynthesis Research* 57: 217–227.

Junge, W. and Nelson, N. (2015) ATP synthase. *Annual Review of Biochemistry* 84: 631–657.

Junge, W., Sielaff, H., and Engelbrecht, S. (2009) Torque generation and elastic power transmission in the rotary F_0F_1-ATPase. *Nature* 459: 364–370.

Kramer, D. M., Sacksteder, C. A., and Cruz, J. A. (1999) How acidic is the lumen? *Photosynthesis Research* 60:151–163.

Kühlbrandt, W. (2019) Structure and mechanisms of F-type ATP synthases. *Annual Review of Biochemistry* 88: 515–549.

Lowe, A. G. and Jones, M. N. (1984) Proton motive force – What price Δp? *Trends in Biochemical Sciences* 12: 234–237.

Martin, J. L., Ishmukhametov, R., Spetzler, D., Hornung, T., and Frasch, W. D. (2018) Elastic coupling power stroke mechanism of the F1-ATPase molecular motor. *Proceedings of the National Academy of Sciences USA* 115: 5750–5755.

Mitchell, P. (1979) Keilin's respiratory chain concept and its chemiosmotic consequences. *Science* 206: 1148–1159.

Nalin, C. M. and McCarty, R. E. (1984) Role of a disulfide bond in the γ subunit in the activation of the ATPase of chloroplast coupling factor. 1. *Journal of Biological Chemistry* 259: 7275–7280.

Nelson, D. L. and Cox, M. M. (2017) *Lehninger Principles of Biochemistry*, 7th Edn., New York: W. H. Freeman.

Nicholls, D. G. and Ferguson, S. J. (2013) *Bioenergetics 4*. London: Academic Press.

Noji, H., Yasuda, R., Yoshida, M., and Kinoshita, K. (1997) Direct observation of the rotation of the F_1-ATPase. *Nature* 386: 299–302.

Pänke, O. and Rumberg, B. (1999) Kinetic modeling of rotary CF_0F_1-ATP synthase: Storage of elastic energy during energy transduction. *Biochimica et Biophysica Acta* 1412: 118–128.

Pänke, O., Gumbiowski, K., Junge, W., and Engelbrecht, S. (2000) F-ATPase: Specific observation of the rotating c subunit oligomer of EF_oEF_1. *FEBS Letters* 472: 34–38.

Petersen, J., Foerster, K., Turina, P., and Graber, P. (2012) Comparison of the H^+/ATP ratios of the H^+-ATP synthases from yeast and from chloroplast. *Proceedings of the National Academy of Sciences USA* 109: 11150–11155.

Pogoryelov, D., Yildiz, O., Faraldo-Gomez, J. D., and Meier, T. (2009) High-resolution structure of the rotor ring of a proton-dependent ATP synthase. *Nature Structural and Molecular Biology* 16: 1068–1073.

Prebble, J. (2002) Peter Mitchell and the ox phos wars. *Trends in Biochemical Sciences* 27: 209–212.

Racker, E. (1976) *A New Look at Mechanisms in Bioenergetics.* New York: Academic Press.

Rastogi, V. K. and Girvin, M. E. (1999) Structural changes linked to proton translocation by subunit c of the ATP synthase. *Nature* 402: 263–268.

Rees, D. M., Leslie, A. G. W., and Walker, J. E. (2009) The structure of the membrane extrinsic region of bovine ATP synthase. *Proceedings of the National Academy of Sciences USA* 106: 21597–21601.

Seelert, H., Poetsch, A., Dencher, N. A., Engel, A., Stahlberg, H., and Muller, D. J. (2000) Proton-powered turbine of a plant motor. *Nature* 405: 418–419.

Spetzler, D., Ishmukhametov, R., Hornung, T., Martin, J., York, J., Jin-Day, L., and Frasch, W. D. (2012) Energy transduction by the two molecular motors of the F_1F_o ATP synthase. In: J. J. Eaton-Rye, B. C. Tripathy, and T. D. Sharkey, (eds.) *Photosynthesis, Plastid Biology, Energy Conversion and Carbon Assimilation.* Dordrecht: Springer.

Steigmiller, S., Turina, P., and Gräber, P. (2008) The thermodynamic H^+/ATP ratios of the H^+-ATP synthases from chloroplasts and *Escherichia coli. Proceedings of the National Academy of Sciences USA* 105: 3745–3750.

Stock, D., Leslie, A. G. W., and Walker, J. E. (1999) Molecular architecture of the rotary motor in ATP synthase. *Science* 286: 1700–1705.

Stock, D., Gibbons, C., Arechaga, I., Leslie, A. G. W., and Walker, J. E. (2000) The rotary mechanism of ATP synthesis. *Current Opinion in Structural Biology* 10: 672–679.

Strotmann, H., Shavit, N., and Leu, S. (2000) Assembly and function of the chloroplast ATP synthase. In: J.-D. Rochaix, M. Golschmidt-Clermont, and S. Merchant, (eds.) *The Molecular Biology of Chloroplasts and Mitochondria of Chlamydomonas.* Dordrecht: Kluwer Academic Publishing, pp. 477–500.

Taiz, L., Zeiger, E., Møller, I. M., and Murphy, A. (2018) *Fundamentals of Plant Physiology.* Sunderland, MA: Sinauer Associates.

Van Walraven, H. S., Strotmann, H., Schwarz, O., and Rumberg, B. (1996) The H^+/ATP coupling ratio of the ATP synthase from thiol-modulated chloroplasts and two cyanobacterial strains is four. *FEBS Letters* 379: 309–313.

Wada, Y., Sambongi, Y., and Futai, M. (2000) Biological nano motor, ATP synthase F_oF_1: From catalysis to *gec*10–12 subunit assembly rotation. *Biochimica et Biophysica Acta* 1459: 499–505.

Watt, I. N., Montgomery, M. G., Runswick, M. J., Leslie, A. G. W., and Walker, J. E. (2010) Bioenergetic cost of making an adenosine triphosphate molecule in animal mitochondria. *Proceedings of the National Academy of Sciences USA* 107: 16823–16827.

Yang, J.-H., Williams, D., Kandiah, E., Fromme, P., and Chiu, P.-L. (2020) Structural basis of redox modulation on chloroplast ATP synthase. *Communications Biology* 3: 482.

Yasuda, R., Noji, H., Yoshida, M., Kinosita, K., and Itoh, H. (2001) Resolution of distinct rotational substeps by submillisecond kinetic analysis of F-1-ATPase. *Nature* 410: 898–904.

Chapter 9

Carbon metabolism

Light-induced electron transport in oxygenic photosynthetic organisms generates ATP and NADPH, which are high-energy compounds of intermediate stability. They are stable against the sorts of rapid loss processes that we have discussed in earlier chapters. However, they are not suitable for long-term storage of energy, such as building plant biomass, or for storage in seeds, tubers, or fruits. For these, it is necessary to convert the energy into a more stable and compact form. Most plants produce sugars or more complex carbohydrates such as starch for long-term energy storage, although some plants produce large quantities of proteins or oils. Some of these conversions are carried out within the chloroplast, while in other cases they take place in the cell cytoplasm from building blocks that are exported from the chloroplast. In this chapter, we will discuss the major processes that are involved in the fixation of CO_2 and its conversion into sugars. Most of our discussion will apply to higher plant systems, although in some places the differences from algae, cyanobacteria, and anoxygenic bacteria will be pointed out.

9.1 The Calvin–Benson cycle is the primary photosynthetic carbon fixation pathway

The metabolic pathway that incorporates carbon into plants by reduction of CO_2 to sugars is known as the Calvin–Benson cycle, after the American chemists Melvin Calvin and Andrew Benson who, along with their associates, worked out the pathway in a brilliant series of experiments in the late 1940s and 1950s (Calvin, 1989, 1992; Benson, 2002). It is also sometimes called the reductive pentose phosphate, or RPP, cycle. Calvin's group utilized two new techniques to assist in deciphering this complex set of biochemical reactions: two-dimensional paper chromatography and radioactive tracers. This was the first use of radioactive tracers in biochemistry to elucidate a metabolic pathway. In honor of this achievement, Calvin was awarded the 1961 Nobel prize for chemistry. An interesting account of the history of the use of radioactive tracers, including Calvin's research and work that preceded it, can be found in the book by Creager (2013).

The chromatography methods permitted the separation of dozens of compounds found in the cells of the organism that they utilized, the green alga *Chlorella pyrenoidosa*. The radioactive tracers were fed to the culture using $^{14}CO_2$, and the culture was incubated in the light for varying periods of time and then rapidly quenched by injection into boiling alcohol. Prior to the addition of radioactivity, the cells were illuminated in a device called a "lollipop," which provided even illumination and easy manipulation (Fig. 9.1). This methodology permitted measurements to be made within seconds after $^{14}CO_2$ addition. With this experimental

Figure 9.1 Lollipop apparatus used by Melvin Calvin and Andrew Benson in studies of the path of carbon in photosynthesis. The lollipop is the oval glass device in the center. It held the algal suspension that was illuminated for varying periods and then quenched by quickly draining it into the alcohol-containing flask below. *Source:* Figure courtesy of the Lawrence Berkeley National Laboratory.

system, the first compounds formed upon $^{14}CO_2$ incorporation were the only ones labeled in an experiment of short duration, while longer experiments generated a more complex set of products (Fig. 9.2).

The first compound labeled with $^{14}CO_2$ was 3-phosphoglyceric acid (PGA), followed by sugar phosphates and, later, amino acids, organic acids, aldehydes, and ketones. The addition of a single carbon atom to a precursor to give a three-carbon product suggested that the precursor molecule might be a two-carbon compound. However, no such two-carbon precursor was ever identified. The finding that five-carbon sugar phosphates were present led to the proposal that the precursor might be a five-carbon compound, and that the resulting six-carbon compound was unstable and rapidly broke down into two three-carbon compounds. This was the key insight, and the basic elements of the cycle were elucidated. The five-carbon precursor was established to be ribulose 1,5-bisphosphate (RuBP). Note that neither the ATP nor the NADPH produced by the electron transport chain is utilized in the reaction that converts RuBP and CO_2 to PGA, which is called carboxylation. The details of this reaction are considered below. An account of the discovery of the Calvin–Benson cycle has been given by Sharkey (2019).

9.1.1 The three phases of the Calvin–Benson cycle are carboxylation, reduction, and regeneration

The Calvin–Benson cycle is a complex series of chemical reactions that can seem intimidating at first. However, the overall cycle can be broken down into three phases, which are more easily understood (Heldt and Piechulla, 2010; Nelson and Cox, 2017; Taiz *et al.*, 2018). The three phases of **carboxylation**, **reduction**, and **regeneration** are schematically illustrated in Fig. 9.3. The entire cycle is shown in Fig. 9.4, and the reactions are summarized in Table 9.1. The carboxylation step generates

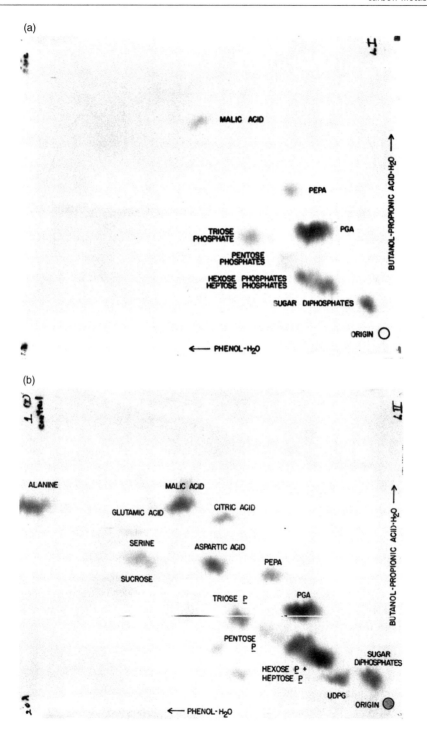

Figure 9.2   Autoradiograms of two-dimensional paper chromatograms used by Calvin and Benson to track the fate of radiolabeled $^{14}CO_2$. The top panel (a) shows a short-duration labeling experiment of two seconds, in which the majority of the label is found in 3-phosphoglyceric acid (PGA). The bottom panel (b) shows a long exposure of 60 seconds, in which the radioactivity is found in many more compounds. *Source:* Figure courtesy of the Lawrence Berkeley National Laboratory.

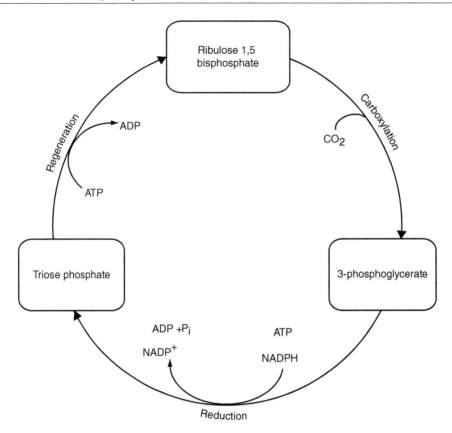

Figure 9.3   Three stages of the Calvin–Benson cycle: carboxylation, reduction, and regeneration. Most of the ATP and all the NADPH are used in the reduction phase, and some ATP is used in the regeneration phase. The product that is drawn off for storage is triose phosphate, a collective term for glyceraldehyde 3-phosphate and dihydroxyacetone phosphate.

PGA. The reduction phase uses the NADPH and some of the ATP produced by electron transport to reduce the PGA to triose phosphate. Most of the triose phosphate is used to regenerate RuBP in a complex set of reactions, while some is drawn off for starch synthesis in the chloroplast and sucrose synthesis in the cytoplasm.

Most of the reactions that make up the Calvin–Benson cycle are identical to metabolic reactions that are well-known in nonphotosynthetic tissues, including gluconeogenesis and the oxidative pentose phosphate cycle (Nelson and Cox, 2017). The enzymes that catalyze these steps are closely related to the ones that act in the nonphotosynthetic reactions, but are specialized for their role in the photosynthetic process, often in their method of regulation. The two unique reactions in the Calvin–Benson cycle that are not found in metabolic processes in most cells are the carboxylation reaction and the final step in the regeneration of RuBP.

9.1.2   Rubisco is the carboxylation enzyme

A remarkable enzyme called ribulose-bisphosphate carboxylase/oxygenase, or rubisco, carries out the carboxylation step in carbon fixation. Rubisco is a complex enzyme that is highly regulated and exhibits two distinct activities: carboxylation and

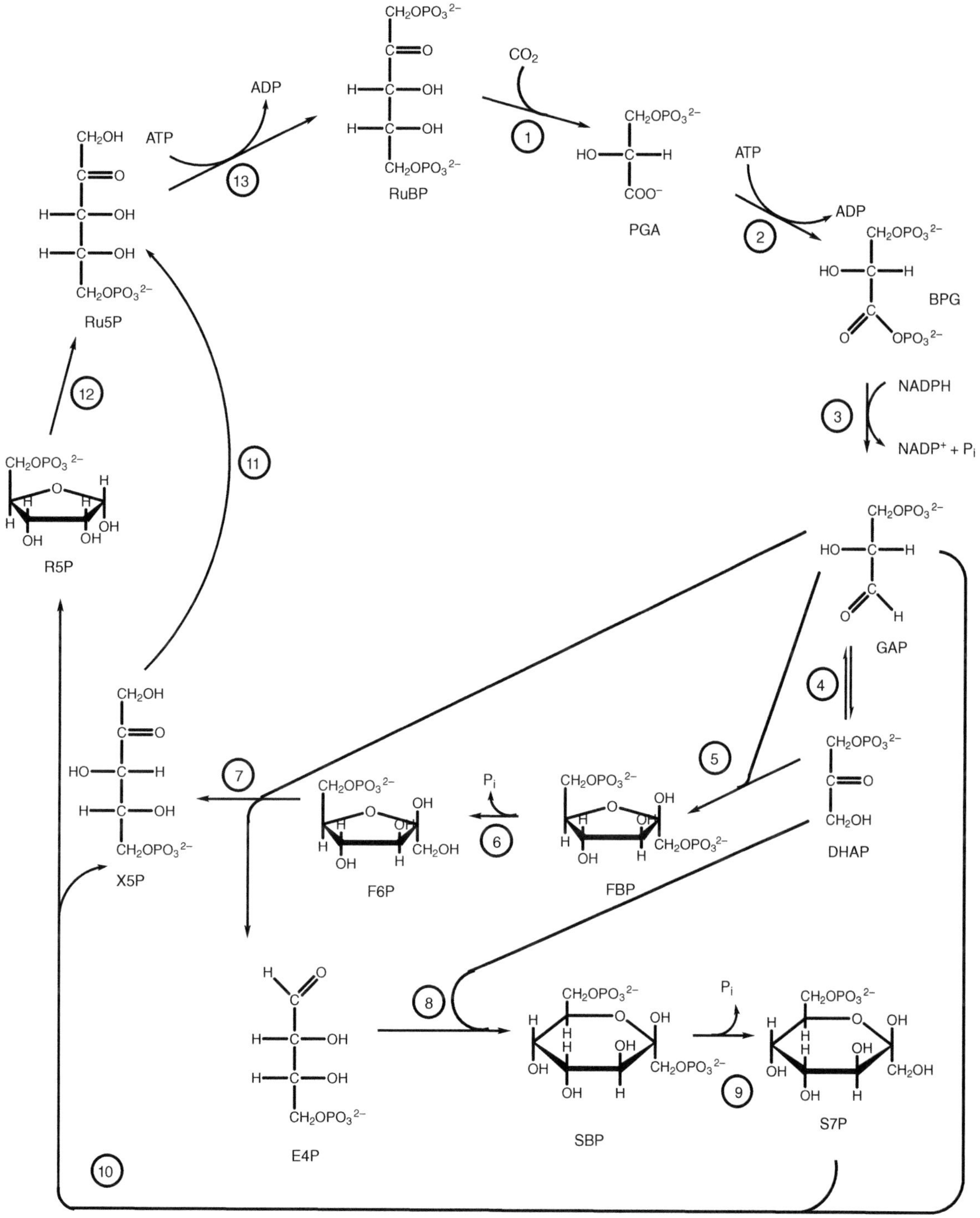

Figure 9.4 Calvin–Benson cycle for reduction of CO_2 in photosynthesis. The individual reactions and the enzymes that catalyze them are given in Table 9.1.

Table 9.1   Chemical reactions of the Calvin–Benson cycle.

Table 9.1   (Continued)

(8) E4P + DHAP →(aldolase)→ SBP

(9) SBP + H₂O →(sedoheptulose 1,7-bisphosphate phosphatase)→ S7P + HOPO₃²⁻

(10) S7P + GAP →(transketolase)→ R5P + X5P

(11) X5P →(ribulose 5-phosphate epimerase)→ Ru5P

(12) R5P →(ribose 5-phosphate isomerase)→ Ru5P

(13) Ru5P + ATP →(phosphoribulokinase)→ RuBP + ADP

oxygenation. Oxygenation is a wasteful activity and is discussed in more detail below.

Rubisco from higher plants and most photosynthetic bacteria consists of eight copies each of large (L) subunits of mass 51–58 kDa and small (S) subunits of mass 12–18 kDa, giving an L_8S_8 quaternary structure (Fig. 9.5) (Hartman and Harpel, 1994; Roy and Andrews, 2000; Spreitzer and Salvucci, 2002; Bracher *et al.*, 2017). The L subunits contain the catalytic site, which is shared between two L subunits, so the minimal active complex is an L_2 dimer. Some purple photosynthetic bacteria contain two distinct types of rubisco complexes: both an L_2 dimer and an L_8S_8 complex. The function of the S subunit is not well understood, but it may stabilize the larger L_8S_8 complex or in some way increase the catalytic efficiency (Spreitzer, 1999). In the majority of eukaryotic photosynthetic organisms, the S subunits are nuclear-encoded, whereas the L subunits are chloroplast-encoded (some exceptions to this general rule will be discussed in Chapter 12). The coordination of the import and assembly of the two subunits is discussed in Chapter 10. Chaperonins, large protein-folding complexes, are essential for rubisco assembly.

The energetics of carboxylation are very strongly exergonic ($\Delta G^{\circ\prime} = -35\,kJ\,mol^{-1}$), so carboxylation is essentially an irreversible reaction. The mechanism of the reaction is thought to involve the spontaneous formation of an enediol form of RuBP. The enediol formation is followed by carboxylation, to give an unstable intermediate, 2-carboxy-3-ketoarabinitol-1,5-bisphosphate, followed by its spontaneous breakdown into two molecules of PGA, one of which contains the carbon atom that was derived from atmospheric CO_2 (Fig. 9.6).

Before rubisco can become active catalytically, a lysine residue in the active site must be modified by **carbamylation**. In this process, a CO_2 molecule (not one that will be fixed into photosynthate) reacts with the ε-amino group of a specific lysine in the L subunit, as shown in Fig. 9.7. A Mg^{2+} ion coordinates to the carbamylated residue to generate the active form of the enzyme. The carbamate is thought to act as a general base, facilitating the formation of the enediol intermediate (Cleland *et al.*, 1998). RuBP binds very

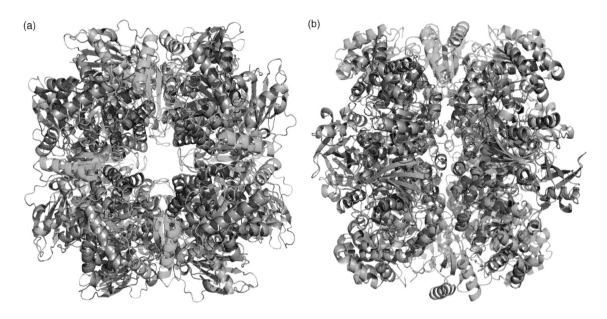

(a)   (b)

Figure 9.5   Structure of spinach rubisco. (a) Model of the L_8S_8 active complex, with the S4 caps at the top and bottom of the L_8 central portion. (b) Side view of the L_8S_8 complex. L subunits are colored salmon and cyan, and S subunits are colored green. Structure from PDB file 1AUS, displayed using PyMol.

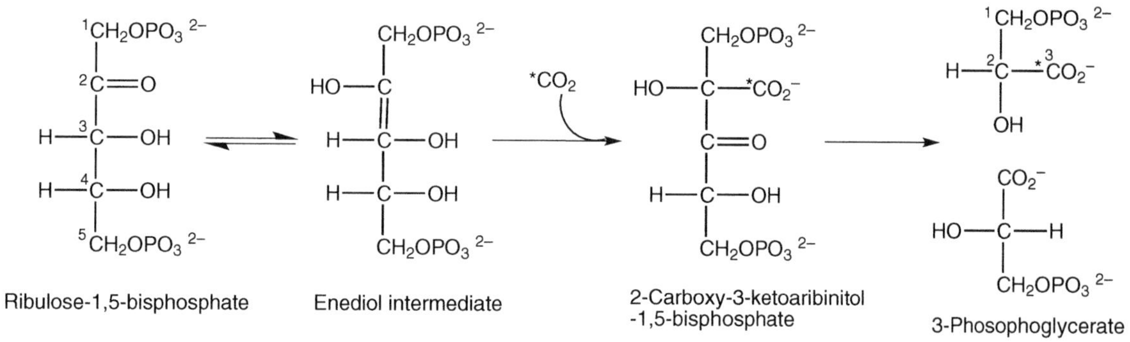

$^1CH_2OPO_3{}^{2-}$ $CH_2OPO_3{}^{2-}$ $CH_2OPO_3{}^{2-}$ $^1CH_2OPO_3{}^{2-}$

Ribulose-1,5-bisphosphate   Enediol intermediate   2-Carboxy-3-ketoaribinitol-1,5-bisphosphate   3-Phosophoglycerate

Figure 9.6 Chemical steps of carboxylation in rubisco. RuBP binds to the active site. Enediol formation takes place, promoted by the carbamate group on an active site lysine. CO_2 is then added to form the 2-carboxy-3-ketoarabinitol-1,5-bisphosphate intermediate, which is then hydrolyzed to form two molecules of PGA. The newly added C atom is in the three position of one of the PGA molecules.

tightly to rubisco, changing the conformation of rubisco in a way that effectively blocks access to the active site carbamylated group. If RuBP binds before carbamylation takes place, the enzyme becomes "stuck" in an inactive state. Another enzyme, **rubisco activase**, promotes carbamylation of rubisco, probably by altering the structure of rubisco in a way that weakens the binding of RuBP to the nonactivated

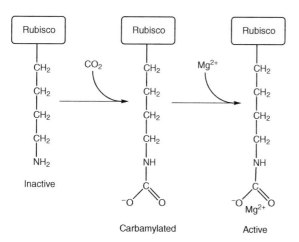

Figure 9.7 Carbamylation of lysine in rubisco. A CO_2 molecule reacts with the ε-amino group of a lysine residue in the active site in rubisco, which then binds Mg^{2+} to form the active enzyme. The negatively charged carbamylate group acts as a general base in the formation of the enediol shown in Fig. 9.6.

enzyme. Because this process requires ATP, the activity of rubisco changes with light intensity, so the rate of carboxylation is coordinately regulated with electron transport activity. In addition to RuBP, rubisco activase also facilitates the removal of other compounds that bind to the active site of rubisco, including carboxyarabinitol-1-phosphate, a naturally occurring inhibitor, as well as xylulose 1,5-bisphosphate, an unreactive isomer of RuBP that can form on the enzyme and inhibit its activity.

The most important alternate reaction that rubisco carries out involves the addition of O_2 instead of CO_2 and is therefore called **oxygenation**, which leads to photorespiration (see below). Oxygenation is a significant drain on a plant's resources, because in most plants it takes place approximately a third of the time instead of carboxylation. This process is shown in Fig. 9.8. Isotope effect evidence suggests that the oxygenation process begins with electron transfer to form superoxide, O_2^-, which subsequently attacks the substrate RUBP (Bathellier et al., 2020).

Because CO_2 and O_2 are competitive substrates, the ratio of carboxylation to oxygenation reactions is determined by the catalytic rates and affinities of rubisco for the two substrates, weighted by the concentrations of O_2 and CO_2 in air. Table 9.2 lists the relevant kinetic parameters for rubisco. The enzyme turnover number k_{cat} is the number of molecules of the substrate that can be processed by

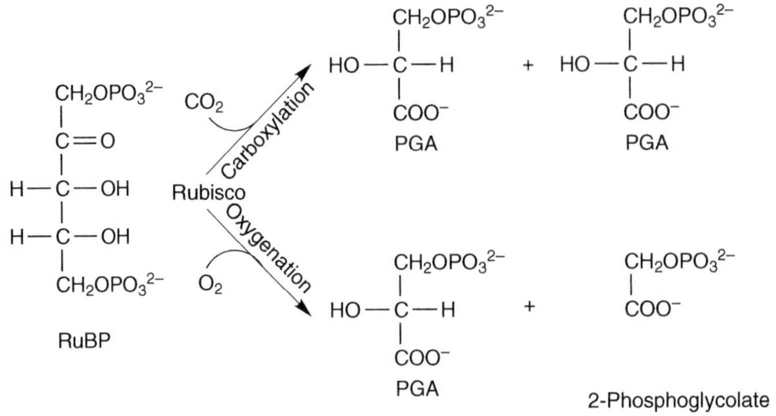

Figure 9.8   Carboxylation and oxygenation reactions of rubisco. RuBP reacts with CO_2 to form two molecules of PGA. Alternatively, RuBP reacts with O_2 to form one molecule of PGA and one molecule of 2-phosphoglycolate. The 2-phosphoglycolate is recovered by a complex series of reactions as described in Fig. 9.11.

Table 9.2   Enzymatic and physical parameters for rubisco.

(a) Enzyme kinetic constants for rubisco

$k_{cat}^{CO_2}$ (s^{-1})	$k_{cat}^{O_2}$ (s^{-1})	$K_M^{CO_2}$ (μM)	$K_M^{O_2}$ (μM)	$k_{cat}^{CO_2} / K_M^{CO_2}$ (s^{-1}M^{-1})	$k_{cat}^{O_2} / K_M^{O_2}$ (s^{-1}M^{-1})
3.3	2.4	9	535	3.7×10^5	4.5×10^3

Values are for 25 °C (Woodrow and Berry, 1988).

(b) Carbon dioxide and oxygen concentration in water vs. temperature

Temperature (°C)	$[CO_2]$ (μM)	$[O_2]$ (μM)	$[CO_2]/[O_2]$
5	21.9	401	0.0515
15	15.7	320	0.0462
25	11.7	265	0.0416
35	9.1	228	0.0376

the enzyme at saturating substrate concentration. The K_M constant, or Michaelis constant, is the concentration of substrate at which the enzyme shows half the rate that it does at saturating substrate concentrations. The specificity factor, τ, is the kinetic preference of the enzyme for CO_2 over O_2. To obtain the specificity factor for a given substrate, it is first necessary to determine the ratio of k_{cat} and K_M for that substrate. The significance of k_{cat}/K_M in enzyme kinetics is that it is equal to the effective second-order rate constant for the reaction of the enzyme with the substrate to form a product and is considered to be a measure of the

catalytic efficiency of the enzyme (Fersht, 1999). The specificity factor τ is then obtained by dividing the k_{cat}/K_M values for carboxylation by the corresponding values for oxygenation:

$$\tau = \frac{k_{cat}^{CO_2}}{K_M^{CO_2}} \Bigg/ \frac{k_{cat}^{O_2}}{K_M^{O_2}} \qquad (9.1)$$

The actual rate of carboxylation compared with oxygenation at given concentrations of CO_2 and O_2 can then easily be calculated from Eq. (9.2), using the values of k_{cat} and K_M given in Table 9.2 for both the carboxylation and the oxygenation reactions, from which a specificity factor of about 80 is determined for higher plant rubiscos.

$$\frac{v_{CO_2}}{v_{O_2}} = \tau \frac{[CO_2]}{[O_2]} \quad (9.2)$$

Although this number indicates a high preference for CO_2 over O_2, higher plant rubiscos have significant oxygenation activity because of the much higher concentration of O_2 compared with CO_2 in the atmosphere (21% *vs.* 0.035%). For example, using Eq. (9.2) with the concentrations of CO_2 and O_2 in the air gives the result that the rate of carboxylation is only about three times the rate of oxygenation under normal physiological conditions. The temperature dependences of the solubilities of CO_2 and O_2 in water are also different, with the result that, at higher temperatures, the ratio of $[CO_2]/[O_2]$ decreases, thereby increasing the relative importance of oxygenation compared with carboxylation.

Rubisco is a relatively sluggish enzyme, with a turnover number (k_{cat}) of only a few CO_2 molecules fixed per second for each enzyme molecule, even at saturating concentrations of CO_2 (Table 9.2). This means that in order to have a high rate of CO_2 fixation, a high concentration of rubisco is needed. Remarkably, under typical conditions, the concentration of rubisco in a chloroplast (~4 mM) is as much as 1000 times higher than the concentration of one substrate, CO_2, and is comparable with the concentration of the other substrate, RuBP. Consequently, as much as 50% of the soluble protein in a leaf is rubisco. It is very likely that rubisco is the most abundant protein on Earth (Bar-On and Milo, 2019).

Rubiscos from the more primitive cyanobacteria and the dimeric enzyme from anoxygenic photosynthetic bacteria have much lower specificity factors than do higher plant rubiscos, ~35 and 10, respectively (Tabita, 1999). This is not a liability in environments with low oxygen contents, such as anaerobic areas, where many anoxygenic photosynthetic bacteria live. Cyanobacteria contain a carbon-concentrating mechanism that greatly increases the concentration of CO_2, thereby

avoiding oxygenation (see below). The early Earth is thought to have been essentially anaerobic when rubisco was first selected by evolution as the carboxylation enzyme (Whitney *et al.*, 2011). As oxygen levels increased as a result of oxygenic photosynthesis, evolutionary selection pressure improved the performance of the enzyme. However, it is apparently not possible to avoid the oxygenation activity completely, and the specificity factor may be limited by intrinsic chemical factors. So, rubisco may well be an example of an enzyme that evolved under a specific set of environmental conditions where it worked well, but became less efficient as changes in environmental conditions outpaced its rate of evolution. Although rubisco is no longer a highly efficient enzyme, it is apparently too great a jump for plants to change to another primary carboxylation system. Rubisco has thus been optimized as much as possible, but still remains a major source of energy loss in most plants. Efforts to improve rubisco by genetic engineering have not yet been successful (Spreitzer and Salvucci, 2002). A few rubisco enzymes, such as the one found in the thermophilic alga *Cyanidium calderium*, have remarkably high specificity factors of more than 200 (Tabita, 1999; Roy and Andrews, 2000), so there may still be room for improvement that has not yet been realized. However, there appears to be a tradeoff in the kinetic properties of rubisco so that enzymes with higher rates have lower specificities and *vice versa* (Tcherkez *et al.*, 2006; Whitney *et al.*, 2011).

Some strategies have been employed by evolution to avoid the oxygenation reaction by increasing the CO_2 concentration where rubisco is located and are discussed later in this chapter. However, these mechanisms also have an energetic cost.

9.1.3 Details of the other Calvin–Benson cycle reactions

Following carboxylation, the next steps in the Calvin–Benson cycle are a phosphorylation step, followed by the reduction of the PGA formed in the

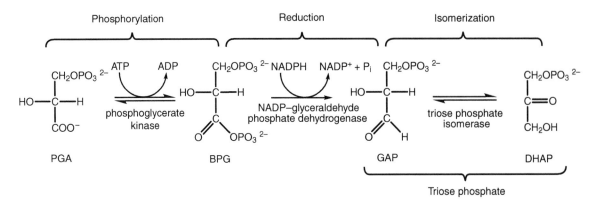

Figure 9.9   Reduction phase of the Calvin–Benson cycle. PGA is phosphorylated to form BPG, which is then reduced to form GAP. GAP is then rapidly equilibrated to form a pool together with its isomer, DHAP. GAP and DHAP are collectively known as triose phosphate.

carboxylation step (Heldt and Piechula, 2010; Taiz *et al.*, 2018). Collectively, they are called the **reduction phase** (Fig. 9.9). Chemically, these reactions are nearly identical to steps of glycolysis (and the reverse reaction, gluconeogenesis) and are catalyzed by enzymes that are evolutionarily related to the cytosolic glycolytic enzymes. The phosphorylation step converts PGA to 1,3-bisphosphoglycerate (BPG) using the enzyme **phosphoglycerate kinase** and ATP generated by the light-dependent reactions.

The next step is the reduction of BPG to glyceraldehyde 3-phosphate (GAP) using NADPH and the enzyme **NADP-glyceraldehyde phosphate dehydrogenase**. A phosphate group is also released as part of this reaction. The mechanism of the reaction involves a thioester bond formation with a Cys residue of the enzyme with the release of phosphate, followed by the reduction with NADPH, and finally the release of the aldehyde product. The overall reaction in this phase of the Calvin–Benson cycle is the reduction of a carboxylic acid to an aldehyde. The high-energy thioester intermediate is formed at the expense of the high-energy phosphate from ATP.

The enzyme **triose phosphate isomerase** catalyzes the rapid equilibrium between GAP and dihydroxyacetone phosphate (DHAP), through a 1,2 enediol intermediate. The rapid equilibrium

between these two sugars ensures that they react as a pool, and collectively they are known as **triose phosphate**. The triose phosphate pool is the starting point for the final phase of the Calvin–Benson cycle, the regeneration of RuBP, as well as the point of export of reduced product from the chloroplast. This latter process is described below.

The **regeneration** phase of the Calvin–Benson cycle is a complex series of reactions in which the three-carbon triose phosphates are regenerated into the five-carbon sugar ribulose 5-phosphate (Ru5P) (Fig. 9.4). There are four-, five-, six-, and seven-carbon intermediates, erythrose 4-phosphate (E4P), xylulose 5-phosphate (X5P), ribose 5-phosphate (R5P), fructose 1,6-bisphosphate (FBP), fructose 6-phosphate (F6P), sedoheptulose 1,7-bisphosphate (SBP), and sedoheptulose 7-phosphate (S7P). The regeneration phase looks formidable, but is really just a reshuffling of the carbon skeletons to produce the needed five-carbon substrate for carboxylation. This entire series of reactions is virtually identical to the pentose phosphate cycle in nonphotosynthetic metabolism, and the chloroplast enzymes are very similar to the cytosolic ones. Regulatory aspects that differ for the chloroplast enzymes and the cytosolic ones are discussed below.

The final step in the regeneration phase involves the phosphorylation of Ru5P to give RuBP using the enzyme phosphoribulokinase (Fig. 9.4). This enzyme is one of two enzymes (the other is rubisco) that are unique to photosynthetic carbon fixation. The RuBP made in this reaction is ready to undergo another cycle of carboxylation, reduction, and regeneration.

9.1.4 The stoichiometry and efficiency of the Calvin–Benson cycle reactions

The overall stoichiometry of the Calvin–Benson cycle from CO_2 to triose phosphate is given in Eq. (9.3).

$$3\ CO_2 + 9\ ATP^{4-} + 6\ NADPH + 5\ H_2O \rightarrow$$
$$DHAP2^- + 9\ ADP^{3-} + 8\ P_i^{2-} + 6\ NADP^+ + 3\ H^+$$
$$(9.3)$$

This includes the three ATP and two NADPH required for assimilation of each of the three CO_2 molecules needed to form one molecule of triose phosphate, including the regeneration of the same amount of RuBP as consumed. As discussed in Chapter 7, modern values indicate that this is close to the amount of these reactants provided by the light reactions for each O_2 produced. The triose phosphate either can be drawn off by export from the chloroplast, as described below, or can be used for starch synthesis within the chloroplast. Alternatively, it can be reinvested to form additional molecules of RuBP and thereby increase the metabolite pool available. If this carbon is used to generate more RuBP, the overall reaction is given by Eq. (9.4).

$$5\ CO_2 + 9\ H_2O + 16\ ATP^{4-} + 10\ NADPH \rightarrow$$
$$RuBP^{4-} + 14\ P_i^{2-} + 6\ H^+\ 16\ ADP^{3-} + 10\ NADP^+$$
$$(9.4)$$

The net result is that the cycle acts on its own to increase the substrate level, which is termed **autocatalytic**. This is in contrast to the case with the TCA cycle in aerobic metabolism. The TCA cycle is a catalytic cycle, in that the normal operation of the cycle does not change the concentrations of the metabolite pools.

Note that both Eqs. (9.3) and (9.4) are balanced chemical reactions that correctly describe the Calvin–Benson cycle. They differ in whether the triose phosphate is used to generate additional RuBP or is drawn off. This can be varied depending on the needs of the plant. When the plant has been in darkness, the RuBP level is relatively low, so most of the carbon is reinvested to increase the capacity of the cycle. When the level of cycle intermediates is sufficient, then carbon is drawn off at the triose phosphate level.

How efficient is the Calvin–Benson cycle in terms of energy balance? The overall energy efficiency of photosynthesis under ideal conditions (low-intensity red light and high CO_2) is calculated to be 27% (see Chapter 13). This includes losses due to energy trapping through primary photochemistry, secondary electron transfer processes, ATP production, and carbon fixation, but does not include losses due to light saturation effects and photorespiration. It is possible to isolate the efficiency of just the Calvin–Benson cycle by considering the thermodynamics of the overall process calculated in Chapter 13 compared with the thermodynamics of the light reactions. The total standard state free energy stored for each CO_2 fixed and O_2 evolved is +479.5 kJ mol^{-1} (Chapter 13). The energy stored in the light reactions is the redox energy from the reduction of NADPH and oxidation of H_2O plus the phosphate bond energy from ATP formation. Using the redox potentials given in Table A2 along with Eq. (A25), the redox energy input is found to be +440 kJ mol^{-1} for each CO_2 fixed and O_2 evolved. The phosphate bond energy is +31 kJ mol^{-1} for each ATP formed, for a total of +93 kJ mol^{-1} for three ATP molecules. Thus, the total energy input to the Calvin–Benson cycle is +533 kJ mol^{-1} for each CO_2 fixed, of which +479.5 kJ mol^{-1} is recovered, for an overall efficiency of 90%. Of this energy input, approximately 80% comes from redox energy (440/533), and 20% from phosphate bond energy (93/533). Of course, these are standard state thermodynamic values, and the actual values, including the real concentrations of the various chemical species, may give somewhat different values. Nevertheless, this calculation gives us an idea of the relative efficiencies of the light reactions compared with the dark reactions and also the relative free-energy contributions from redox and phosphate sources.

9.1.5   The Calvin–Benson cycle is regulated by disulfide reduction of key enzymes and ion movements

As discussed above, most of the enzymes of the Calvin–Benson cycle are closely related to enzymes that catalyze the same reactions in other cellular compartments. However, several of the Calvin–Benson cycle enzymes are activated by the reduction of disulfide bonds between cysteine residues to generate the sulfhydryl form of cysteine. This reduction is mediated by thioredoxin, a small redox-active protein that also contains a pair of cysteine residues that can be either oxidized or reduced (Buchanan, 1991; Jacquot *et al.*, 1997; Buchanan and Balmer, 2005). Thioredoxin is reduced by an enzyme, ferredoxin–thioredoxin oxidoreductase, which is in turn reduced by ferredoxin produced in the light reactions. This reductive cascade is shown schematically in Fig. 9.10. The **ferredoxin–thioredoxin reductase** (FTR) has an unusual flat and thin shape (Dai *et al.*, 2000). Ferredoxin binds on one side and transfers electrons to an Fe–S cluster which then transfers them to thioredoxin, which binds on the other side. A mixed disulfide between FTR and thioredoxin is an intermediate. For the enzymes that are regulated by thiol reduction mediated by thioredoxin, the active site of the oxidized enzyme is held in an unfavorable geometry by the disulfide bond, but the reduced form is free to assume a conformation that permits the normal enzymatic activity.

The enzymes that are activated by thioredoxin action are the enzyme that reduces BPG to GAP, NADP-glyceraldehyde phosphate dehydrogenase, and three enzymes of the regeneration phase: fructose 1,6-bisphosphate phosphatase, sedoheptulose

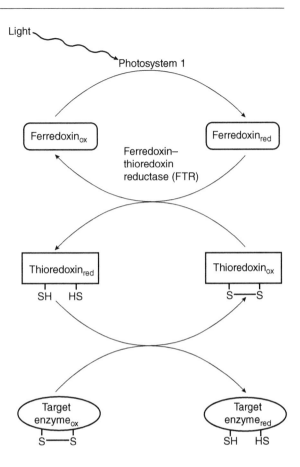

Figure 9.10   Thioredoxin-mediated redox control of enzyme activity. Ferredoxin is reduced by electrons from photosystem 1 and in turn, reduces ferredoxin–thioredoxin reductase (FTR). FTR reduces thioredoxin to the sulfhydryl form, and thioredoxin reduces the disulfide of the target enzyme, in most cases activating its enzymatic activity.

1,7-bisphosphate phosphatase, and phosphoribulokinase. In addition, rubisco activase, the enzyme that regulates the activation state of rubisco, is itself activated by thioredoxin.

One enzyme of the chloroplast, glucose 6-phosphate dehydrogenase, is inhibited in the reduced sulfhydryl form and activated in the oxidized disulfide form, just the opposite of the pattern described above. This enzyme is not part of the Calvin–Benson cycle, but instead is the entry point for the oxidative pentose phosphate cycle that is utilized to make the ribose necessary for DNA and RNA synthesis using glucose 6-phosphate to generate ribose 5-phosphate and CO_2, along with the reduction of NADPH. If this oxidative pathway, which releases CO_2 and NADPH instead of utilizing them, as does the Calvin–Benson cycle, were also to be active during photosynthesis, it would constitute a futile cycle. The two pathways would feed substrates to each other, with the net result being the wasting of ATP. This is prevented by the opposite sense of the disulfide regulation of glucose 6-phosphate dehydrogenase, compared with the Calvin–Benson cycle enzymes. This ensures that the two cycles are not simultaneously active.

The chloroplast ATP synthase is also reductively activated by disulfide reduction of the γ subunit (Mills, 1996). Upon darkening, the chloroplast becomes less reducing, and the disulfide is formed, rendering the enzyme inactive. This control mechanism prevents the ATP that was synthesized in the light from being hydrolyzed in the dark by the enzyme running backward.

In addition to the reductive thioredoxin regulation of the Calvin–Benson cycle enzymes, light-driven ion movements activate several enzymes. As discussed in Chapter 8, electron transport causes the acidification of the thylakoid lumen and therefore the alkalization of the stroma, typically raising the pH from 7 to 8. A light-dependent increase of stromal Mg^{2+} concentration is coupled to the H^+ efflux. The pH optimum of several of the Calvin–Benson cycle enzymes is in the alkaline range, so they are activated by this pH change. In addition, the enhanced Mg^{2+} concentration stimulates the activity of some of the enzymes. Finally, the levels of metabolites also regulate some of the Calvin–Benson cycle enzymes, so that their specific activity is enhanced when their substrates are present in higher concentrations (Heldt and Piechulla, 2010).

The regulation of the Calvin–Benson cycle is thus a multifaceted phenomenon, which includes covalent modification by thioredoxin and rubisco carbamylation, plus noncovalent modification by ion and metabolite levels, and protein–protein interactions between rubisco and rubisco activase. These mechanisms act in concert to keep the cycle in balance and ensure maximum efficiency.

9.2   Photorespiration is a wasteful competitive process to carboxylation

Oxygenation is an unwanted side reaction in which rubisco reacts with oxygen instead of CO_2. The products of the oxygenation reaction are 2-phosphoglycolate and PGA (Fig. 9.8). The 2-phosphoglycolate inhibits the Calvin–Benson cycle and must be metabolized to PGA for eventual recovery by the Calvin–Benson cycle. The overall process, including the oxygenation and recovery, is called **photorespiration**, because O_2 is taken up in a light-dependent manner, and CO_2 is released (Oliver, 1998; Douce and Heldt, 2000; Foyer et al., 2009; Heldt and Piechulla, 2010). The recovery is a remarkably elaborate process, involving three organelles: the chloroplast, mitochondrion, and peroxisome. The overall process is shown in Fig. 9.11, and the chemical reactions are summarized in Table 9.3.

The photorespiratory cycle is initiated in the chloroplast with the hydrolysis of 2-phosphoglycolate to glycolate, which leaves the chloroplast via a specific glycolate–glycerate exchange transporter present in the inner chloroplast envelope membrane. The glycolate then enters another organelle, the peroxisome, by way of passive diffusion through a nonspecific pore. In the peroxisome, glycolate is oxidized to glyoxylate using O_2 and the glycolate oxidase enzyme. This reaction forms H_2O_2 as a product, which is rapidly broken down by the high levels of catalase present in the peroxisome. The glyoxylate is metabolized in two different ways. In one reaction, it reacts with glutamate to form α-ketoglutarate and glycine in a reaction catalyzed by the enzyme glutamate–glycolate aminotransferase. The α-ketoglutarate formed in this reaction is imported back into the chloroplast, where it is converted back into glutamate and then sent back to the peroxisome. This reaction involves reduction by ferredoxin and transamination from glutamine, which is itself made from glutamate. This process serves to reassimilate the NH_3 that is produced in the mitochondrion, as described in the next paragraph. The other glyoxalate reaction involves

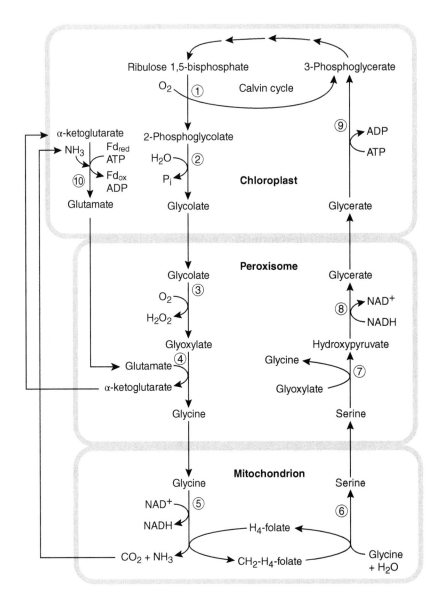

Figure 9.11   Photorespiratory cycle. The 2-phosphoglycolate formed in the oxygenation reaction is converted to glycolate, exported from the chloroplast, and is imported into the peroxisome, where it is metabolized into glycine. The glycine is exported from the peroxisome and taken up by the mitochondrion, where two molecules are combined and decarboxylated to form one molecule of serine. The serine is then transported to the peroxisome, where it is converted to glycerate and then reimported into the chloroplast and phosphorylated to form PGA. The individual reactions are given in Table 9.3.

Table 9.3   Chemical reactions involved in photorespiration.

(1) RuBP + O_2 $\xrightarrow{\text{rubisco}}$ 2-phosphoglycolate + PGA + 2 H^+

(2) 2-phosphoglycolate + H_2O $\xrightarrow{\text{phosphoglycolate phosphatase}}$ glycolate + $HOPO_3^{2-}$

(3) glycolate $\xrightarrow{\text{glycolate oxidase}}$ glyoxalate + H_2O_2

(4) glyoxalate + glutamate $\xrightarrow{\text{glyoxalate:glutamate aminotransferase}}$ glycine + α-ketoglutarate

(5) glycine + NAD^+ + H_4-folate $\xrightarrow{\text{glycine decarboxylase}}$ NADH + H^+ + CO_2 + NH_3 + CH_2-H_4-folate

(6) + CH_2-H_4-folate + H_2O + glycine $\xrightarrow{\text{serine hydroxymethyl transferase}}$ serine + H_4-folate

(7) serine + glyoxalate $\xrightarrow{\text{serine aminotransferase}}$ hydroxypyruvate + glycine

(8) hydroxypyruvate + NADH + H^+ $\xrightarrow{\text{hydroxypyruvate reductase}}$ glycerate + NAD^+

(9) glycerate + ATP $\xrightarrow{\text{glycerate kinase}}$ PGA + ADP + H^+

(10) α-ketoglutarate + ATP + NH_3 + 2 Fd_{red} $\xrightarrow[\text{glutamate synthase}]{\text{glutamine synthase}}$ glutamate + ADP + $HOPO_3^{2-}$ + 2 Fd_{ox}

reaction with serine and is also described below. Both H_2O_2 and glyoxalate are toxic compounds, so their localization in the peroxisome is beneficial.

The glycine produced in photorespiration is released from the peroxisome and taken up by the mitochondrion. For every two glycine molecules imported, a CO_2 and an NH_3 are released, and an NAD^+ is reduced to NADH by the large enzyme complex glycine decarboxylase. The remaining carbon atom is added to the other glycine molecule to form serine by the enzyme serine hydroxymethyl-transferase, which uses folate as a cofactor. The serine exported from the mitochondrion reenters the peroxisome, where it reacts with glyoxalate in a transamination reaction to form glycine and hydroxypyruvate. The hydroxypyruvate is reduced by NADH to form glycerate, which leaves the peroxisome and is translocated into the chloroplast. The final step is the phosphorylation of the glycerate to form PGA, which is one of the metabolites in the Calvin–Benson cycle.

Overall, for every two molecules of RuBP that react via oxygenation to form two molecules of 2-phosphoglycolate and two molecules of PGA, one CO_2 is lost. The other three carbon atoms in the two 2-phosphoglycolate molecules are recycled to form one additional PGA molecule. The energetic cost of photorespiration might not appear to be too high when looking only at Fig. 9.11, in which only one ATP is hydrolyzed during the entire cycle (NADH is both oxidized and reduced, so there is no net change). However, in order to get back to where we started (which is how one should calculate the energetic cost), it is necessary to regenerate two RuBP molecules via the Calvin–Benson cycle. The overall cost of photorespiration is five ATP and three NADPH for every oxygenation event. This is a significant energetic cost, which can add up to approximately 50% of the amount of energy expended on carboxylation (Heldt and Piechulla, 2010).

Is photorespiration an evolutionary artifact that the plant is unable to avoid, or does it perhaps serve a useful purpose? The answer to this question is not yet clear. C4 plants (see below) avoid photorespiration almost entirely by concentrating CO_2, so it is clearly not essential in all conditions. Also, other pathways exist for the synthesis of glycine and serine. Photorespiration may prevent overreduction of thylakoid components under conditions of high light and low CO_2. This may occur when the stomata are closed at midday to prevent water loss.

9.3 The C4 carbon cycle minimizes photorespiration

Certain plants, including agronomically important species such as maize and sugarcane, have a mechanism for circumventing photorespiration by increasing the concentration of CO_2 in the vicinity of rubisco. This is accomplished by a pumping mechanism such that CO_2 is temporarily fixed in one cell type, and the resultant organic compound is transported to another cell type where it is decarboxylated to regenerate CO_2. The CO_2 released by decarboxylation is then fixed in the normal manner by the Calvin–Benson cycle (Hatch, 1988; Furbank et al., 2000). This mechanism is therefore not a replacement for the Calvin–Benson cycle, but an addition to it. Of course, such an active pumping mechanism carries an energetic cost, but under many conditions, the cost is much less than the losses suffered by photorespiration (Hatch, 1988). The C4 pathway concentrates CO_2 by 15-fold or more, from 5 to 70 μM, and thus, according to Eq. (9.2), virtually eliminates photorespiration by increasing the CO_2/O_2 ratio from 0.044 to 0.264.

The C4 mechanism of CO_2 concentration relies in part on the fact that those plants that contain the C4 pathway have a unique cellular differentiation not observed in C3 plants. C4 plants are so named because 4-carbon compounds are the initial products of carboxylation, in contrast to the Calvin–Benson cycle, which incorporates atmospheric CO_2 into 3-carbon compounds and is therefore termed C3 photosynthesis. The anatomy of C3 and C4 plants is shown in Fig. 9.12. The C4 plant has a

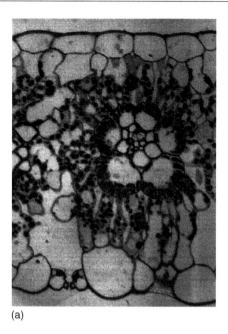

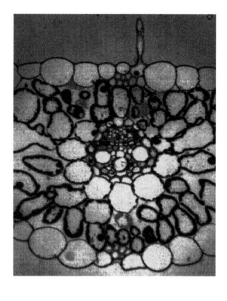

(a)   (b)

Figure 9.12   Anatomical differences between C3 and C4 plants. (a) A C4 plant, *Zea maize*, corn, which contains both mesophyll cells and bundle sheath cells. (b) A C3 plant, *Triticum aestivum*, wheat, which contains only mesophyll cells. *Source:* Figure courtesy of Angela Franceschi.

characteristic circle of cells surrounding the vascular bundle. These cells are called bundle sheath cells and are described as having a *Kranz* (German for "wreath") anatomy. The other type of cell is termed mesophyll cells and appear generally similar to spongy mesophyll cells in the C3 plant (although there are important metabolic differences). The cells are connected by a large number of highly specific pores called **plasmodesmata**, which permit molecules of molecular mass up to about 1000 Daltons to pass between the cells. Therefore, plasmodesmata allow the free exchange of metabolites between cells, but generally exclude the transfer of macromolecules such as enzymes.

The primary fixation of CO_2 takes place in the mesophyll cells by the action of an enzyme phosphoenolpyruvate (PEP) carboxylase. It catalyzes the reaction of HCO_3^- with PEP to form oxaloacetate (Fig. 9.13 and Table 9.4). Because the substrate of PEP carboxylase is HCO_3^-, and not CO_2, O_2 is a very poor substrate, and the enzyme does not have oxygenation activity. The HCO_3^- is formed from CO_2 and H_2O by the action of the enzyme carbonic anhydrase, which dramatically speeds up the equilibrium between these two species. The carboxylation reaction is efficient and essentially irreversible. The oxaloacetate formed in this reaction is then either reduced to malate or transaminated to form aspartate, which is exported from the mesophyll cell to the bundle sheath cell and then decarboxylated. The Calvin–Benson cycle enzymes are found in the bundle sheath cells, wherein the CO_2 released is fixed in the normal manner.

There are three variants of C4 photosynthesis that are found in different classes of plants. All three types utilize PEP carboxylase as the primary carboxylation enzyme in the mesophyll cells and decarboxylate a 4-carbon compound in the bundle sheath cells. They differ in the mode and location of C4 acid decarboxylation. We will limit our

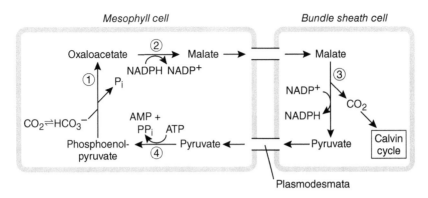

Figure 9.13 C4 pathway found in the NADP-malic enzyme type of C4 photosynthesis. Initial carboxylation is carried out on phosphoenolpyruvate (PEP) and HCO_3^- by the PEP carboxylase enzyme. The reduced product, malate, is exported from the mesophyll cells and enters the bundle sheath cells, where it is decarboxylated to form CO_2 and pyruvate. The CO_2 is used as a substrate for the Calvin–Benson cycle with rubisco, and the pyruvate is returned to the mesophyll cell, where it is activated to form PEP. The net result of the C4 cycle is to concentrate CO_2 in the bundle sheath cells, where it is fixed under conditions of low O_2 and high CO_2, thus minimizing the oxygenation reaction in rubisco. Individually numbered reactions are detailed in Table 9.4.

discussion to the type of C4 photosynthesis found in maize and sugarcane, which is known as the NADP-malic enzyme type.

In NADP-malic enzyme type C4 photosynthesis, the carboxylation takes place in the cytoplasm of the mesophyll cells, and the oxaloacetate formed is translocated into the chloroplast of the mesophyll cells, where it is reduced to malate by NADPH, using the enzyme NADP-malate dehydrogenase. The malate is then exported from the mesophyll chloroplast and enters the bundle sheath cells through the plasmodesmata and is finally imported into the chloroplast of the bundle sheath cells. The import of oxaloacetate and export of malate take place by specific transporters, while the transport through the plasmodesmata occurs by passive diffusion. The malate is then decarboxylated to form pyruvate in the bundle sheath chloroplast, with the production of NADPH in a reaction catalyzed by NADP-malic enzyme. The pyruvate is exported from the bundle sheath chloroplast, diffuses through the plasmodesmata, and is then imported into the mesophyll chloroplast.

The final step in the cycle is the phosphorylation of pyruvate to form phosphoenolpyruvate. The reaction that forms phosphoenolpyruvate (PEP) is given in Eq. (9.5), where PPi is pyrophosphate.

$$\text{pyruvate} + \text{ATP} + P_i \rightarrow \text{PEP} + PP_i + \text{AMP} \quad (9.5)$$

The enzyme that catalyzes this reaction is pyruvate–phosphate dikinase. This reaction has some interesting thermodynamic and mechanistic features. The thermodynamics of the formation of phosphoenolpyruvate from pyruvate and phosphate are very strongly uphill, $\Delta G^{o\prime} = +62\,\text{kJ mol}^{-1}$. This is about twice the energy that the hydrolysis of ATP provides. The thermodynamics of the reaction are therefore very unfavorable, even when coupled to ATP hydrolysis. The net free-energy change of Eq. (9.5) is $\Delta G^{o\prime} = +31\,\text{kJ mol}^{-1}$. In order to drive the reaction to completion, it is directly coupled to the hydrolysis of the PP_i to form $2P_i$ by a pyrophosphatase enzyme. This latter reaction is energetically favorable, with $\Delta G^{o\prime} = -33\,\text{kJ mol}^{-1}$. The overall thermodynamics for the coupled reactions are therefore slightly favorable. This reaction takes place in the chloroplast of the mesophyll cell. The phosphoenolpyruvate is then exported from the chloroplast to the cytosol by way of a specific transporter, in exchange for a phosphate, and is ready for another cycle of carboxylation.

The mechanism of pyruvate–phosphate dikinase involves first the transfer of a phosphate from ATP to P_i to form ADP and PP_i. Next, a second phosphate is transferred from ADP to an enzyme histidine residue, to form a phosphohistidine and AMP.

Table 9.4   Chemical reactions involved in C4 and CAM photosynthesis.

(1) PEP $+$ HCO$_3^-$ $\xrightarrow{\text{PEP carboxylase}}$ oxaloacetate $+$ HOPO$_3^{2-}$

(2) oxaloacetate $+$ NADPH $+$ H$^+$ $\xrightarrow{\text{NADP malate dehydrogenase}}$ malate $+$ NADP$^+$

(3) malate $+$ NADP$^+$ $\xrightarrow{\text{NADP malic enzyme}}$ pyruvate $+$ CO$_2$ $+$ NADPH $+$ H$^+$

(4) pyruvate $+$ HOPO$_3^{2-}$ $+$ ATP $\xrightarrow{\text{Pyruvate-phosphate dikinase}}$ PEP $+$ AMP $+$ H$_2$P$_2$O$_7^{2-}$

The phosphohistidine then phosphorylates pyruvate, to form phosphoenolpyruvate.

It takes energy to drive the C4 cycle, because the net result is to transport a CO_2 molecule from a low concentration to a high concentration. All the redox steps balance exactly, so there is no net expenditure of redox energy. This is as expected, because there is no overall redox change in the cycle. However, the equivalent of two molecules of ATP is hydrolyzed during the cycle for each CO_2 pumped. This energy expenditure is in addition to the three ATP molecules needed to drive the Calvin–Benson cycle. Under ideal conditions and at moderate temperatures with sufficient water availability, this extra energetic cost overshadows the losses due to photorespiration and, therefore, under these conditions, C4 plants are not as competitive as C3 plants. However, at high temperatures, where the CO_2 concentration in the water of the plant tissues is low, C4 plants have a distinct advantage over C3 plants. This effect is amplified under water stress conditions. Because C4 plants are more efficient at scavenging CO_2, they do not need to keep their stomates open as much as C3 plants do, and therefore they suffer less water loss. For all these reasons, C4 plants are much more common in hot, dry climates.

An interesting aspect of the C4 cycle of the type found in maize and sugarcane is that the bundle sheath cells do not evolve oxygen or generate NADPH from water, because of a lack of Photosystem II activity. NADPH produced by the oxidative decarboxylation of malate (malic enzyme reaction) is only half the amount needed for reducing the PGA produced by rubisco. The PGA that is produced by rubisco in the bundle sheath must therefore be partly reduced in the mesophyll chloroplasts using NADPH and ATP made during photosynthesis. The triose phosphate is then imported into the bundle sheath cells, where it is returned to the Calvin–Benson cycle (Hatch, 1988). The PGA is returned to the mesophyll cell, and the cycle repeated. Note that the formation of triose phosphate in the mesophyll cell does not involve fixation of CO_2, but only reduction of existing PGA. This indirect process means that the bundle sheath cells do not need to carry out normal noncyclic electron transfer in order to generate all the metabolites needed to drive the Calvin–Benson cycle. The almost complete absence of Photosystem II from the bundle sheath cells in many C4 plants is remarkable, and the only light-driven electron transfer process occurring there is cyclic electron flow mediated by Photosystem I, which produces ATP. An advantage of this system is that, without Photosystem II, O_2 is not produced in the bundle sheath cells, so oxygenation is further suppressed compared with carboxylation.

The C4 cycle is found in only about 1% of plant species worldwide. However, it is extremely important economically, with C4 plants accounting for many major agricultural crops. It has been estimated that 30% of terrestrial global photosynthesis takes place by C4 plants (Heldt and Piechulla, 2010). Evolutionary aspects of the C4 cycle are discussed in Chapter 12.

9.4   Crassulacean acid metabolism avoids water loss in plants

Plants that live in hot, dry climates are usually severely stressed for water. Any mechanism that can reduce water loss will be of tremendous selective advantage. C4 plants lose less water than C3 plants, because they can open their stomates less fully and still obtain enough CO_2 for photosynthesis because of their higher affinity for CO_2. However, they must still open their stomates to perform

photosynthesis and must do so during the middle of the day, when water is lost very rapidly.

Some plants have a strategy that permits them to carry out photosynthesis during the day and keep the stomates entirely closed. Water loss is thus greatly minimized. They must open their stomates to carry out import of CO_2, but they do so during the night, when temperatures are cooler, and water loss is much reduced. This mechanism is known as **crassulacean acid metabolism (CAM)**, after the family Crassulaceae, in which it was first elucidated (Lüttge, 1998; Cushman *et al.*, 2000). However, CAM is also found in many other families of plants, so this name is something of a misnomer. The acid part of the CAM name is accurate, because the signature of CAM is the production during the night of large amounts of malic (occasionally citric) acid, which is stored in the vacuole and then broken down during the day, releasing CO_2, which is refixed by the normal action of the Calvin–Benson cycle.

Water loss from stomates can be very substantial. CAM plants lose 50–100 g of H_2O for every gram of CO_2 fixed. However, C4 plants lose approximately 250 g of H_2O for every gram of CO_2 fixed, and C3 plants lose up to 500 g (Taiz *et al.*, 2018). So, CAM can cut water loss by a factor of up to ten.

CAM is also found in aquatic plants, which is at first surprising, as these plants are obviously not water-stressed (Keeley, 1998). However, a closer examination reveals the advantage of CAM in these organisms. Aquatic environments are often very low in CO_2 availability. This is especially true in some environments, such as shallow pools, which often experience dramatic daily variations in CO_2 levels due to the action of algae and cyanobacteria, with their efficient CO_2-concentrating mechanism (see below). By adopting CAM, aquatic plants are thus able to take advantage of the higher nighttime levels of CO_2 and thereby gain selective advantage.

The mechanism of CAM is quite similar to that of C4 metabolism, including the primary carboxylation which is carried out by the enzyme PEP carboxylase, using PEP as a substrate. The major conceptual difference between the two pathways is that C4 metabolism is a mechanism that concentrates CO_2 spatially, whereas CAM is a mechanism that concentrates it temporally. The overall CAM cycle is shown in Fig. 9.14, and the reactions are summarized in Table 9.4. First, we will discuss the nighttime cycle and then the daytime cycle.

At night, starch, or in some cases sucrose, is broken down in the cytosol to form triose phosphate and is converted to PEP. Carboxylation by PEP carboxylase produces oxaloacetate, which is then reduced to malate using NADH produced during the starch breakdown. The malate is pumped into the vacuole, where it is stored as the free acid form, malic acid. The transport of malate into the vacuole is driven by ATP hydrolysis, because it is operating against a significant thermodynamic gradient. The ATP hydrolysis is carried out by a V-type ATPase enzyme, which is similar in many ways to the F-type ATP synthase discussed in Chapter 8, although it functions in the opposite direction, pumping H^+ at the expense of ATP hydrolysis. The ATPase pumps hydrogen ions across the membrane, where they react with the malate, which enters via a specific channel, forming malic acid. The malic acid formed cannot leak back through the channel and accumulates, allowing more malate to enter. Considerable quantities of malic acid can be transported in this manner. The pH of the vacuole is about 3, below the pKa for malate/malic acid, ensuring that it is in the free acid form.

During the day, the malate is released from the vacuole and enters the chloroplast. There it is decarboxylated to form CO_2 and pyruvate, with the formation of NADPH. The CO_2 is fixed by rubisco and processed by the Calvin–Benson cycle in the normal way to form triose phosphate and, eventually, starch. The pyruvate is activated to PEP by ATP and the pyruvate–phosphate dikinase, the same enzyme as discussed earlier in the C4 pathway. This reaction produces PEP and AMP as products. The PEP is converted into PGA and also enters the Calvin–Benson cycle, so most of the carbon atoms in malate are eventually stored as starch.

CAM significantly reduces water loss, but there is a downside to the CAM cycle. The amount of malic acid that can be stored in the vacuole, although significant, is not sufficient to generate nearly as much CO_2 as C3 or C4 plants can import

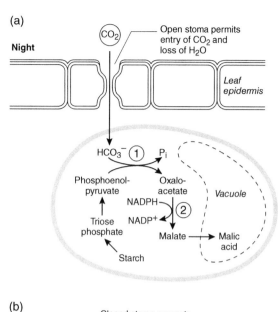

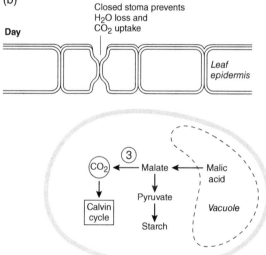

Figure 9.14 Crassulacean acid metabolism avoids water loss by temporally separating gas exchange from photosynthesis. (a) During the night, CO_2 is taken in through the stomates and is fixed by PEP carboxylase to form malate, which is stored in the vacuole as malic acid. (b) During the day, the malic acid is released from the vacuole and decarboxylated to form CO_2, which is fixed by rubisco and the Calvin–Benson cycle. Individually numbered reactions are detailed in Table 9.4.

during the day. The consequence of this is that CAM plants tend to grow very slowly. Some plants have the best of both worlds, in that they carry out normal C3 photosynthesis when water is plentiful, but switch to CAM when water is limiting or when the plant is salt-stressed. Other plants – for example, the giant saguaro cactus that grows in Arizona – are obligate CAM plants and are therefore extremely slow-growing, taking more than 100 years to reach maturity.

9.5   Algae and cyanobacteria actively concentrate CO_2

The final type of CO_2-concentrating mechanism is found in various forms in most cyanobacteria and algae (Kaplan and Reinhold, 1999; Moroney and Somanchi, 1999; Badger and Spalding, 2000; Wang et al., 2011; Kupriyanova et al., 2013). It consists of a combination of pumping mechanisms, the enzyme carbonic anhydrase, and a dense packaging of rubisco, as shown schematically in Fig. 9.15. Recent evidence has implicated the NADH dehydrogenase complex (Complex I) in cyanobacteria in the carbon concentration mechanism (Laughlin et al., 2020; Schuller et al., 2020). The cyanobacterial Complex I contains additional subunits that function as a vectorial carbonic anhydrase. The net effect of the carbon concentration system is to concentrate CO_2 as HCO_3^- so that it cannot simply leak back out through the membrane. Rubisco is very densely packaged in dense cytoplasmic structures called **carboxysomes** in cyanobacteria as shown in Fig. 9.16 (Kerfeld et al., 2018). In algae, the rubisco, along with rubisco activase, is packed into chloroplast complexes called pyrenoid bodies. In both cases, the bodies are almost crystalline rubisco and also contain carbonic anhydrase. This latter enzyme catalyzes the hydration of CO_2 to form HCO_3^- as given in Eq. (9.6).

$$CO_2 + H_2O + HCO_3^- + H^+ \qquad (9.6)$$

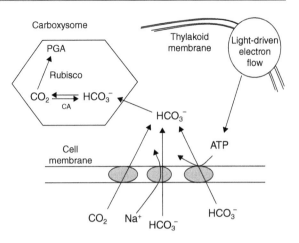

Figure 9.15   Schematic picture of the carbon-concentrating mechanism found in cyanobacteria. CO_2 and HCO_3^- are taken up by active pumps in the cell membrane and deposited in the cytoplasm as HCO_3^-, which is in turn taken into the carboxysome, where rubisco and carbonic anhydrase (CA) are found. The CA converts the HCO_3^- back to CO_2, and rubisco fixes it via the Calvin–Benson cycle.

CO_2 or HCO_3^- is pumped into the cell and, in the case of algae, is also pumped into the chloroplast. In cyanobacteria, there are multiple pumps, an ATP-driven pump, a Na^+/HCO_3^- symport, and a facilitated diffusion transporter for CO_2. A symport is a membrane complex that transports two substrates across the membrane simultaneously, both from the same side of the membrane. The inorganic carbon is stored primarily in the form of HCO_3^-. In contrast to CO_2, which is very permeable through the membrane and cell wall and therefore easily leaks out, HCO_3^- does not leak out of the cell, because the charge makes the membrane relatively impermeable to HCO_3^-.

The carbonic anhydrase in the carboxysome converts HCO_3^- to CO_2, which is immediately fixed by rubisco before it can leak out. Interestingly, there is no carbonic anhydrase in the cytoplasm of the cyanobacteria, and if it is introduced via genetic transformation, the cells are not able to concentrate CO_2. The explanation for this observation is that carbonic anhydrase in the cytoplasm increases the amount of free CO_2 in the cytoplasm, and it is then subject to

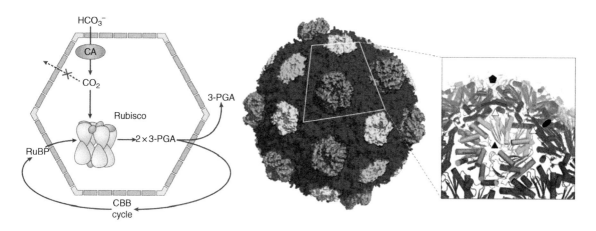

Figure 9.16   Carboxysome structure and function. Left: Schematic model of carboxysome including rubisco and carbonic anhydrase (CA). Middle: X-ray structure of carboxysome. Right: Detail of corner of carboxysome. *Source: Kerfeld et al.* (2018)/Springer Nature.

leakage. The pyrenoid body of the chloroplast also contains carbonic anhydrase with a similar function.

If cells are grown in a high concentration of CO_2, the concentration system is suppressed. If the cells are then transferred to a low concentration of CO_2, they are initially unable to concentrate CO_2, but then develop the ability as the genes that code for the components of the system are induced and the proteins synthesized (Omata *et al.*, 1999).

9.6   Sucrose and starch synthesis

Carbon fixed by the Calvin–Benson cycle is processed for longer-term storage in two distinct forms. One of these forms is starch, which is made and stored in the chloroplast. The other form is sucrose, which is made in the cytoplasm (Huber and Huber, 1996). In both cases, triose phosphate is the beginning point for the conversion. Both processes take place at significant rates and the interplay between them is highly regulated. In both cases, the storage product is a nonreducing oligosaccharide that is not phosphorylated. The monosaccharide glucose is rather easily oxidized and is not suitable as a storage product. In addition, by converting sugar to a polymeric compound such as starch, large quantities of carbon can be stored without increasing the osmotic pressure inside the cells. A phosphorylated compound would rapidly tie up much of the phosphate in the cell and cause many reactions to be inactivated. Figure 9.17 schematically illustrates the overall pathways of starch and sucrose synthesis.

Starch synthesis takes place in the chloroplast. During the day, large quantities of starch are made and deposited in starch granules in the chloroplast stroma. This starch is largely mobilized during the night to supply carbon for a variety of chloroplast processes and so does not represent a long-term storage product but, rather, an intermediate storage product.

Starch synthesis begins with triose phosphate, the principal product of the Calvin–Benson cycle, as shown in Fig. 9.17 and summarized in Table 9.5. The first several reactions are identical to some of the reactions of the regeneration of RuBP, proceeding through fructose 1,6-bisphosphate and fructose 6-phosphate (F6P). The F6P is isomerized to glucose 6-phosphate (G6P) by the enzyme hexose phosphate isomerase, and the G6P is converted to glucose 1-phosphate (G1P) by phosphoglucomutase. G1P is activated with ATP by the

Table 9.5   Chemical reactions involved in starch and sucrose synthesis.

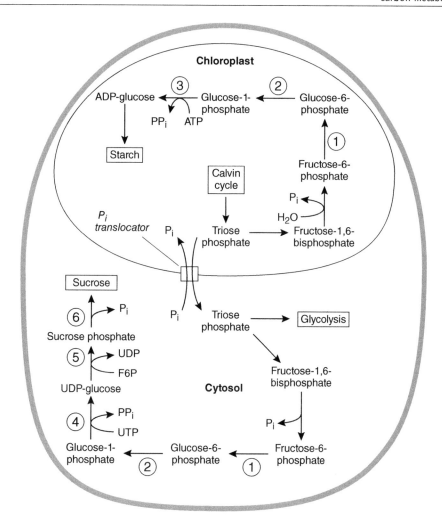

Figure 9.17   Starch and sucrose synthesis reactions provide for long-term storage of photosynthate. Triose phosphate is either converted to starch in the chloroplast or exported from the chloroplast to the cytoplasm, where it is converted to sucrose. The reactions of starch and sucrose synthesis are detailed in Table 9.5.

enzyme ADP-glucose pyrophosphorylase to form the nucleotide-linked sugar ADP-glucose and pyrophosphate. The pyrophosphate is immediately hydrolyzed by pyrophosphatase, which drives the reaction towards forming ADP-glucose. Finally, the glucose moiety of ADP-glucose is linked from carbon 1 to carbon 4 of a growing starch chain by the enzyme starch synthase, forming a 1 → 4 glycosidic link between the sugars and releasing ADP.

Sucrose synthesis takes place in the cytoplasm. Triose phosphate synthesized by the Calvin–Benson cycle is exported from the chloroplast by the action of a triose phosphate–inorganic phosphate translocator in the chloroplast envelope inner membrane. This membrane protein catalyzes an obligate exchange of triose phosphate exit for Pi entry. The strict one-for-one exchange regulates the export of phosphorylated intermediates so that they will not be depleted from the chloroplast.

The reactions of sucrose synthesis are shown in Fig. 9.17 and summarized in Table 9.5. The early reactions are chemically identical to the first steps of starch synthesis, although the enzymes are cytosolic isoforms of the chloroplast enzymes. G1P is also converted into a nucleotide-linked sugar, although the nucleotide that it reacts with is uridine triphosphate (UTP), and the product is UDP-glucose. This is condensed with F6P to form sucrose 6-phosphate by the enzyme sucrose phosphate synthase. Finally, the phosphate is hydrolyzed by sucrose phosphate phosphatase to yield the final product sucrose.

Triose phosphate has three possible fates: starch synthesis, sucrose synthesis, or the regeneration of RuBP. The balance of these three pathways is controlled by a complex regulatory mechanism (Heldt and Piechulla, 2010). If too much triose phosphate is devoted to either starch or sucrose formation, then the pool of Calvin–Benson cycle intermediates will rapidly be depleted, and the Calvin–Benson cycle will collapse catastrophically.

Starch synthesis is regulated by the levels of the primary product of carboxylation by rubisco and PGA, along with the inorganic phosphate concentration. These two compounds are present in large quantities, so an increase in the concentration of PGA is accompanied by a decrease in P_i. Elevated PGA and reduced P_i concentrations have the effect of activating starch synthesis by activating ADP-glucose pyrophosphorylase.

The regulation of sucrose synthesis is primarily at two steps: where fructose 1,6-bisphosphate is converted to fructose 6-phosphate by fructose 1,6-bisphosphate phosphatase and where F6P is condensed with UDP-glucose to form sucrose phosphate by sucrose phosphate synthase. FBPase is inhibited by the powerful metabolic regulator fructose 2,6-phosphate. The concentration of this regulator is in turn controlled by the levels of a number of metabolites, with the result that, until the triose phosphate level reaches a threshold level, the activity of the phosphatase is very low. This prevents triose phosphate from being utilized for sucrose synthesis until a sufficient quantity has been synthesized to regenerate RuBP. Sucrose

phosphate synthase is also regulated by the concentration of cytosolic metabolites, particularly P_i and G6P, which act as allosteric effectors. The sensitivity of SPS to these effectors depends on the phosphorylation status of the enzyme, which changes in part depending on the demand for sucrose synthesis.

9.7   Other carbon fixation pathways in anoxygenic phototrophs

All oxygenic photosynthetic organisms utilize the Calvin–Benson cycle as the mechanism of CO_2 fixation, sometimes with additions such as the C4 pathway or CAM, as described above. However, some anoxygenic phototrophic bacteria can utilize other carbon fixation pathways or in some cases do not appear to be capable of autotrophic carbon fixation at all.

Most purple phototrophic bacteria utilize the Calvin–Benson cycle for carbon fixation in much the same way that it operates in oxygenic organisms (Tabita, 1999). The aerobic anoxygenic phototrophic purple bacteria, however, appear to have lost the ability to do autotrophic carbon fixation and have lost the genes for rubisco and PRK (Swingley et al., 2007; Tang et al., 2009). These organisms live in the highly oxygenated upper layers of the ocean and have access to significant quantities of organic carbon, so would lose much of their fixed CO_2 to photorespiration if they utilized the Calvin–Benson cycle.

The green sulfur bacteria run the tricarboxylic acid (TCA) cycle backward. In most organisms, this cycle operates by using the three-carbon acid pyruvate as substrate and through a series of reactions produce three molecules of CO_2 and reduced compounds such as NADH, which are oxidized by the respiratory electron transport chain. In the green sulfur bacteria, reduced ferredoxin produced by the reaction center (Chapter 6) is used to reduce CO_2 (Buchanan and Arnon, 1990; Fuchs, 2011) (Fig. 9.18a). Flux analysis of *Cb. tepidum*

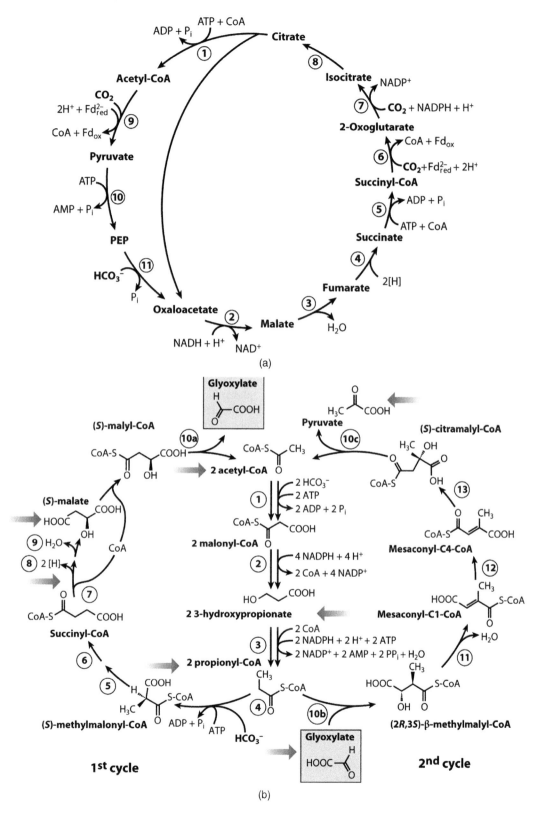

Figure 9.18 Alternative CO_2 fixation pathways in anoxygenic phototrophs. (a) Reverse TCA cycle. (b) Hydroxypropionate cycle. *Source:* Fuchs (2011)/Annual Reviews.

has traced the metabolic pathways in detail and established the flux of carbon in this organism (Feng *et al.*, 2010).

The filamentous anoxygenic phototrophs utilize yet another pathway called the 3-hydroxypropionate cycle. It uses acetyl CoA as substrate and produces the two-carbon acid glyoxylate as an intermediate product and then utilizes a second cycle to produce pyruvate as the final product (Fuchs, 2011) (Fig. 9.18b).

The phototrophic heliobacteria do not appear to be capable of autotrophic carbon fixation (Sattley *et al.*, 2008). They assimilate a variety of organic compounds under strictly anaerobic conditions (Tang *et al.*, 2010).

References

Badger, M. R. and Spalding, M. H. (2000) CO$_2$ acquisition, concentration and fixation in cyanobacteria and algae. In: R. C. Leegood, T. D. Sharkey, and S. von Caemmerer (eds.) *Photosynthesis: Physiology and Metabolism*. Dordrecht: Kluwer Academic Publishers, pp. 369–397.

Bar-On, Y. M. and Milo, R. (2019) The global mass and average rate of rubisco. *Proceedings of the National Academy of Sciences USA* 116: 4738–4743.

Bathellier, C., Yu, L.-J., Farquhar, G. D., *et al.* (2020) Ribulose 1,5-bisphosphate carboxylase/oxygenase activates O$_2$ by electron transfer. *Proceedings of the National Academy of Sciences USA* 117: 24234–24242.

Benson, A. A. (2002) Following the path of carbon in photosynthesis: a personal story. *Photosynthesis Research* 73: 31–49.

Bracher, A., Whitney, S. M., Hartl, F. U., and Hayer-Hartl, M. (2017) Biogenesis and metabolic maintenance of rubisco. *Annual Review of Plant Biology* 68: 29–60.

Buchanan, B. B. (1991) Regulation of CO$_2$ assimilation in oxygenic photosynthesis – The ferredoxin/thioredoxin system – Perspective on its discovery, present status, and future development. *Archives of Biochemistry and Biophysics* 288: 1–9.

Buchanan, B. B. and Arnon, D. I. (1990) A reverse KREBS cycle in photosynthesis: consensus at last. *Photosynthesis Research* 24: 47–53.

Buchanan, B. B. and Balmer, Y. (2005) Redox regulation: A broadening horizon. *Annual Review of Plant Biology* 56: 187–220.

Calvin, M. (1989) Forty years of photosynthesis and related activities. *Photosynthesis Research* 21: 3–16.

Calvin, M. (1992) *Following the Trail of Light: A Scientific Odyssey*. Washington, DC: American Chemical Society.

Cleland, W. W., Andrews, T. J., Gutteridge, S., Hartman, F. C., and Lorimer, G. H. (1998) Mechanism of rubisco: The carbamate as general base. *Chemical Reviews* 98: 549–561.

Creager, A. (2013) *Life Atomic: Radioisotopes as Tools in Science and Medicine*. Chicago: University of Chicago Press.

Cushman, J. C., Taybi, T., and Bohnert, H. J. (2000) Induction of crassulacean acid metabolism – Molecular aspects. In: R. C. Leegood, T. D. Sharkey, and S. von Caemmerer (eds.) *Photosynthesis: Physiology and Metabolism*. Dordrecht: Kluwer Academic Publishers, pp. 551–582.

Dai, S. D., Schwendtmayer, C., Schurmann, P., Ramaswamy, S., and Eklund, H. (2000) Redox signaling in chloroplasts: Cleavage of disulfides by an iron–sulfur cluster. *Science* 287: 655–658.

Douce, R. and Heldt, H.-W. (2000) Photorespiration. In: R. C. Leegood, T. D. Sharkey, and S. von Caemmerer (eds.) *Photosynthesis: Physiology and Metabolism*. Dordrecht: Kluwer Academic Publishers, pp. 115–136.

Feng, Y., Tang, K.-H., Blankenship, R. E., and Tang, Y. J. (2010) Metabolic flux analysis of the mixotrophic metabolisms in the green sulfur bacterium *Chlorobaculum tepidum. Journal of Biological Chemistry* 285: 39544–39550.

Fersht, A. (1999) *Structure and Mechanism in Protein Science*. New York: W. H. Freeman.

Foyer, C. H., Bloom, A., Queval, G., and Noctor, G. (2009) Photorespiratory metabolism: Genes, mutants, energetics and redox signaling. *Annual Review of Plant Biology* 60: 455–484.

Fuchs, G. (2011) Alternative pathways of carbon dioxide fixation: Insights into the early evolution of life? *Annual Review of Microbiology* 65: 631–658.

Furbank, R. T., Hatch, M. D., and Jenkins, C. L. D. (2000) C4 photosynthesis: Mechanism and regulation. In: R. C. Leegood, T. D. Sharkey, and S. von Caemmerer (eds.) *Photosynthesis: Physiology and Metabolism*. Dordrecht: Kluwer Academic Publishers, pp. 435–457.

Hartman, F. C. and Harpel, M. R. (1994) Structure, function, regulation, and assembly of D-ribulose-1,

5-bisphosphatecarboxylase oxygenase. *Annual Review of Biochemistry* 63: 197–234.

Hatch, H. (1988) C4 photosynthesis: A unique blend of modified biochemistry, anatomy and ultrastructure. *Biochimica et Biophysica Acta* 895: 81–106.

Heldt, H.-W. and Piechulla, B. (2010) *Plant Biochemistry*, 4th Edn. Amsterdam: Academic Press.

Huber, S. C. and Huber, J. L. (1996) Role and regulation of sucrose-phosphate synthase in higher plants. *Annual Review of Plant Physiology and Plant Molecular Biology* 47: 431–444.

Jacquot, J.-P., Lancelin, J. M., and Meyer, Y. (1997) Thioredoxins: Structure and function in plant cells. *New Phytologist* 136: 543–570.

Kaplan, A. and Reinhold, L. (1999) CO_2 concentrating mechanisms in photosynthetic microorganisms. *Annual Review of Plant Physiology and Plant Molecular Biology* 50: 539–570.

Keeley, J. E. (1998) CAM photosynthesis in submerged aquatic plants. *Botanical Reviews* 64: 121–175.

Kerfeld, C. A., Aussignargues, C., Zarzycki, J., Cai, F., and Sutter, M. (2018) Bacterial microcompartments. *Nature Reviews Microbiology* 16: 277–290.

Kupriyanova, E. V., Sinatova, M. A., Cho, S. M., Park, Y. I., Los, D., and Pronina, N. A.. (2013) CO_2-concentrating mechanism in cyanobacterial photosynthesis: Organization, physiological role, and evolutionary origin. *Photosynthesis Research* 117: 133–146.

Laughlin, T. G., Savage, D. F., and Davies, K. M. (2020) Recent advances on the structure and function of NDH-1: The complex I of oxygenic photosynthesis. *Biochimica et Biophysica Acta* 1861: 148254.

Lüttge, U. (1998) Crassulacean acid metabolism. In: A. S. Raghavendra (ed.) *Photosynthesis: A Comprehensive Treatise*. Cambridge: Cambridge University Press, pp. 136–149.

Mills, J. D. (1996) The regulation of chloroplast ATP synthase CF_o–CF_1. In: D. R. Ort and C. F. Yocum (eds.) *Oxygenic Photosynthesis: The Light Reactions*. Dordrecht: Kluwer Academic Publishers, pp. 469–492.

Moroney, J. V. and Somanchi, A. (1999) How do algae concentrate CO_2 to increase the efficiency of photosynthetic carbon fixation? *Plant Physiology* 119: 9–16.

Nelson, D. L. and Cox, M. M. (2017) *Lehninger Principles of Biochemistry*, 7th Edn. New York: W. H. Freeman.

Oliver, D. J. (1998) Photorespiration and the C2 cycle. In: A. S. Raghavendra (ed.) *Photosynthesis: A Comprehensive Treatise*. Cambridge: Cambridge University Press, pp. 173–182.

Omata, T., Price, G. D., Badger, M. R., *et al.* (1999) Identification of an ATP-binding cassette transporter involved in bicarbonate uptake in the cyanobacterium *Synechococcus* sp. strain PCC 7942. *Proceedings of the National Academy of Sciences USA* 96: 13571–13576.

Roy, H. and Andrews, T. J. (2000) Rubisco: Assembly and mechanism. In: R. C. Leegood, T. D. Sharkey, and S. von Caemmerer (eds.) *Photosynthesis: Physiology and Metabolism*. Dordrecht: Kluwer Academic Publishers, pp. 53–83.

Sattley, W. M., Madigan, M. T., Swingley, W. D., *et al.* (2008) The genome of *Heliobacterium modesticaldum*, a phototrophic representative of the firmicutes containing the simplest photosynthetic apparatus. *Journal of Bacteriology* 190: 4687–4696.

Schuller, J. M., Saura, P., Thiemann, J., Schuller, S. K., Gamiz-Hernandez, A. P., Kurisu, G., Nowaczyk, M. M., and Kaila, V. R. I. (2020) Redox-coupled proton pumping drives carbon concentration in the photosynthetic complex I. *Nature Communications* 11: 494.

Sharkey, T. D. (2019) Discovery of the canonical Calvin–Benson cycle. *Photosynthesis Research* 140: 235–252.

Spreitzer, R. J. (1999) Questions about the complexity of chloroplast ribulose-1,5-bisphosphate carboxylase/oxygenase. *Photosynthesis Research* 60: 29–42.

Spreitzer, R. J. and Salvucci, M. E. (2002) Rubisco: Interactions, associations and the possibilities of a better enzyme. *Annual Review of Plant Biology* 53: 449–475.

Swingley, W. D., Sadekar, S., Mastrian, S. D., Matthies, H. J., Hao J., Ramos, H., Acharya, C. R., Conrad, A. L., Taylor, H. L., Dejesa, L. C., Shah, M. K., O'Huallachain, M. E., Lince, M. T., Blankenship, R. E., Beatty, J. T. and Touchman, J. W. (2007) The complete genome sequence of *Roseobacter denitrificans* reveals a mixotrophic rather than photosynthetic metabolism. *Journal of Bacteriology* 189: 683–690.

Tabita, F. R. (1999) Microbial ribulose 1,5-bisphosphate carboxylase/oxygenase: A different perspective. *Photosynthesis Research* 60: 1–28.

Taiz, L., Zeiger, E., Møller, I. M., and Murphy, A. (2018) *Fundamentals of Plant Physiology*. Sunderland, MA: Sinauer Associates.

Tang, K.-H., Feng, X., Tang, Y., and Blankenship, R. E. (2009) Carbohydrate metabolism and carbon fixation in *Roseobacter denitrificans* OCh114. *PLoS One* 4: e7233.

Tang, K.-H., Yue, H., and Blankenship, R. E. (2010) Energy metabolism of *Heliobacterium modesticaldum* during phototrophic and chemotrophic growth. *BMC Microbiology* 10: 150.

Tcherkez, G. G. B., Farquhar, G. D., and Andrews, T. J. (2006) Despite slow catalysis and confused substrate specificity, all ribulose bisphosphate carboxylases may be nearly perfectly optimized. *Proceedings of the National Academy of Sciences USA* 103: 7246–7251.

Wang, Y., Duanmu, D., and Spalding, M. H. (2011) Carbon dioxide concentrating mechanism in *Chlamydomonas reinhardtii*: inorganic carbon transport and CO_2 recapture. *Photosynthesis Research* 109: 115–122.

Whitney, S. M., Houtz, R. L., and Alonso, H. (2011) Advancing our understanding and capacity to engineer nature's CO_2-sequestering enzyme, rubisco. *Plant Physiology* 155: 27–35.

Woodrow, J. E. and Berry, J. A. (1988) Enzymatic regulation of photosynthetic CO_2 fixation in C3 plants. *Annual Review of Plant Physiology and Plant Molecular Biology* 39: 533–594.

Chapter 10

Genetics, assembly, and regulation of photosynthetic systems

In Chapters 5, 6 and 7, we have considered the structure and function of the photosynthetic apparatus in both bacterial and eukaryotic organisms. In this chapter, we will explore how the system is assembled and how that assembly is regulated. First, we will consider the genetic organization and regulation in the relatively simple and well-understood anoxygenic phototrophic bacteria. Then, we will move to cyanobacteria, which must regulate and coordinate the expression, assembly, and function of two photosystems working together. We will next consider the much more complex job of coordinating and regulating two distinct genomes in the chloroplasts of eukaryotic photosynthetic organisms. This also involves the import into the chloroplast and processing of the majority of the proteins that make up the chloroplast.

10.1   Gene organization in anoxygenic photosynthetic bacteria

Complete genome sequences have been determined for a large number of anoxygenic phototrophs. The status of genome projects can be found at the Genomes Online (GOLD) database (https://gold.jgi.doe.gov). Genome sequences have been determined, annotated, and published for a number of anoxygenic phototrophs. Much of this information can be found at the US Department of Energy Integrated Microbial Genome website (http://img.jgi.doe.gov/).

The purple photosynthetic bacteria are a well-studied group of anoxygenic phototrophs, whose gene organization and regulation are relatively well understood. Most of the genetic information needed to construct the photosynthetic apparatus in purple bacteria is contained in a photosynthesis gene cluster (PGC), a ~40–50 kb stretch of contiguous DNA located in the circular chromosome of the cell (Fig. 10.1). This collection of genes codes for 38 open reading frames (ORFs) that together largely comprise the additional information needed, beyond normal metabolism, to permit the organism to become photosynthetic (Naylor *et al.*, 1999; Choudhary *et al.*, 2009). Of the DNA in the PGC, 50% is dedicated to coding for the enzymes involved in the later stages of bacteriochlorophyll synthesis, 18% is for carotenoid synthesis, while the rest is composed of

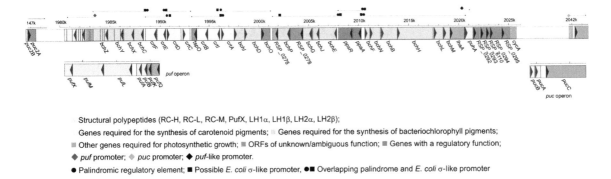

Structural polypeptides (RC-H, RC-L, RC-M, PufX, LH1α, LH1β, LH2α, LH2β);

Genes required for the synthesis of carotenoid pigments;   Genes required for the synthesis of bacteriochlorophyll pigments;

■ Other genes required for photosynthetic growth; ■ ORFs of unknown/ambiguous function; ■ Genes with a regulatory function;

◆ *puf* promoter; ◆ *puc* promoter; ◆ *puf*-like promoter;

● Palindromic regulatory element; ■ Possible *E. coli* σ-like promoter, ●■ Overlapping palindrome and *E. coli* σ-like promoter

Figure 10.1   The photosynthesis gene cluster from *Rhodobacter sphaeroides*. Gene names are indicated below the map. Numbers above the map indicate the position in kilobases. The arrows within the genes indicate the direction of transcription of various genes, and the arrows above the map indicate the genes grouped into operons. *Source:* Naylor *et al.* (1999).

structural genes for the light-harvesting and reaction center proteins (8%), other genes needed for photosynthetic growth (10%), plus a few unknown ORFs and noncoding regions. The early stages of chlorophyll biosynthesis, the steps up to protoporphyrin IX (Chapter 4), are shared with heme biosynthesis, and the *hem* genes that code for these enzymes are located elsewhere on the chromosome, as are the *pet* (photosynthetic electron transport) genes that code for the components of the cytochrome bc_1 complex, which is also shared with the respiratory pathways. Other photosynthesis genes that are not part of the PGC include the *cbb* genes that code for the Calvin–Benson cycle enzymes and the *puc* genes that code for the LH2 antenna complex.

The arrangement of genes in the PGC of *Rb. sphaeroides* is shown in Fig. 10.1 (Naylor *et al.*, 1999). There are several subclusters of genes, which code for bacteriochlorophyll biosynthesis, carotenoid biosynthesis, and the structural proteins of the photosystem. The reaction center L and M proteins and the α and β peptides of the LH1 complex are contained in an operon near the upstream end of the PGC known as *puf* (photosynthetic unit-fixed). The H subunit is located near the other end of the PGC and is coded for by the *puhA* gene. The bacteriochlorophyll and carotenoid biosynthesis

genes are also located in several clusters of genes that are transcribed as **operons**. The LH2 antenna complexes are coded for by the *puc* operon, which is not part of the PGC, but is located approximately 20 kb upstream of the *puhA* gene (Choudhary *et al.*, 2009).

All purple bacteria studied to date have a PGC, although in many cases it has been significantly rearranged, and in some cases, it is split into two subclusters (Swingley *et al.*, 2009). The Gram-positive bacterium *Heliobacillus mobilis* also has a PGC (Xiong *et al.*, 1998; Sattley *et al.*, 2008), although it is significantly different from the one found in the purple bacteria and appears to have been assembled separately. The green sulfur bacteria and filamentous anoxygenic phototrophic bacteria, as well as the cyanobacteria, do not have a PGC. It is not clear why some organisms exhibit extreme clustering of the genetic information needed to carry out photosynthesis, while others do not. In other types of bacteria, such clustering is often found when the genes are laterally transferred as a unit to other organisms. Alternatively, it may be that some aspect of the regulation of the expression of the photosynthetic apparatus requires that the genes be physically close on the chromosome. However, this proximity does not seem to be important for many other organisms, suggesting that this is not the reason.

10.2   Gene expression and regulation of purple photosynthetic bacteria

Purple photosynthetic bacteria are remarkably versatile organisms metabolically and can meet their energy needs by photosynthesis, aerobic and anaerobic respiration, and fermentation. In order to meet the differing needs of these various metabolic lifestyles, the gene expression is highly regulated. In most purple bacteria, the expression of the photosynthetic apparatus is observed only under anaerobic conditions. In addition, the relative amounts of the reaction center and the LH1 and LH2 antenna complexes are modified in response to light intensity. The entire membrane architecture in purple bacteria is fundamentally altered upon expression of the photosynthetic apparatus, with the formation of the invaginated intracytoplasmic membrane system (Chapter 6).

The regulation of the expression of the photosynthetic apparatus is controlled by the interplay of several regulatory circuits that affect transcription of the various operons (Gregor and Klug, 1999; Bauer et al., 2009; Klug and Masuda, 2009; Kumka et al., 2017). As is typical for bacteria, regulation is accomplished largely at the transcriptional level, and many, if not most, genes are organized in operons of several genes that are cotranscribed into a single messenger RNA, which is then translated into protein at the ribosome. In some cases, there are multiple distinct transcripts. Overlapping multigene transcripts form what are called **superoperons**. Although the overall result of the regulatory mechanisms is almost the same in the two purple bacteria that have been well studied, Rb. sphaeroides and Rb. capsulatus, the molecular mechanisms by which this is accomplished are surprisingly different, requiring that the original literature be consulted for precise details of any particular system. The powerful technique of RNA-seq has been employed to give a detailed picture of gene expression patterns under different conditions (Imam et al., 2014; Kumka et al., 2017).

The first level of regulation is the repression of all photosynthesis gene expression under aerobic conditions. When oxygen is present, a repressor is synthesized that inhibits the expression of all the photosynthetic operons. The gene that codes for this repressor, ppsR (also known as crtJ), is located in the middle of the PGC. Its gene product inhibits transcription of all the other operons in the PGC by binding to a conserved palindrome sequence. Under anaerobic conditions, this repression is released, and all the genes in the PGC are transcribed to a moderate degree.

The second level of regulation is the activation of the reaction center and light-harvesting structural protein genes (puf, puh, and puc) under anaerobic conditions. This is regulated by a two-component signal transduction system consisting of a membrane-bound sensor kinase (RegB) and a soluble response regulator (RegA) (Bauer et al., 2009). RegB autophosphorylates in response to the oxygen level and then phosphorylates RegA, which in turn binds to the promoter upstream of the operon and activates transcription. In addition, the expression of the cytochrome c_2 and the Calvin–Benson cycle enzymes are regulated by this system. Ultimately, this system must sense the redox state of the cell.

The third level of transcription regulation is manifested under low light conditions, where the level of expression of the puf and puh operons is increased in dim light (Klug and Masuda, 2009). This is regulated by a light-sensitive circuit that is not very well understood at the molecular level. The circuit includes the hvrA gene product and may also involve blue light photoreceptors and a suite of other light-sensitive complexes such as bacteriophytochrome. Finally, the ratios of the reaction center and the LH1 and LH2 complexes are also modulated by light intensity. This effect is at least partially accomplished post-translationally, although there is also a high light-induced repression of the puc operon.

Although all the details are not yet in hand, the major elements of the light- and oxygen-regulated expression of photosynthesis are evident. Aerobic conditions almost completely repress the expression of photosynthesis genes, which are activated when the cells become anaerobic. In dim light, more copies of the photosystem components are made, and the levels of reaction center and antenna complexes are fine-tuned by light intensity to optimize light collection. Additional levels of regulation involve mRNA stability and the assembly of photosystem protein complexes.

The above discussion applies to those purple bacteria that express photosynthesis only under anaerobic conditions. There is, however, a large group of Aerobic Anoxygenic Phototrophic bacteria (AAP) that have just the opposite pattern of gene expression, in that they express photosynthesis genes and carry out photosynthesis only under aerobic conditions (Yurkov and Beatty, 1998; Yurkov and Csotonyi, 2009). These organisms are something of an enigma, in that they seem to express photosynthetic components only under conditions where it seems that they do not really need them and are unable to carry out photosynthesis under anaerobic conditions, yet they are very numerous, especially in the upper reaches of the oceans (Kolber et al., 2001). An understanding of the possible advantages of this pattern of regulation of gene expression for these organisms has been elusive.

10.3   Gene organization in cyanobacteria

Cyanobacteria have a fundamentally different pattern of gene organization and regulation from that of the anoxygenic purple bacteria. This is probably due to the fact that almost all of them are obligately aerobic species and live largely by photoautotrophic metabolism. This changes the constraints affecting the gene regulation and expression. Generally, cyanobacteria exhibit regulatory mechanisms that are intermediate between the very different patterns found in the anoxygenic bacteria (Section 10.2) and chloroplasts (Section 10.5) (Wollman et al., 1999).

Genomes of a large number of cyanobacteria have been sequenced, making them the most extensively sampled group of phototrophic organisms (Shih et al., 2013). Remarkably, the genome sizes of cyanobacteria span almost the complete range of sequenced bacteria, from less than 2 to 9 Mb. These organisms have undergone both genome reduction and genome expansion to adapt them to a wide range of environments (Swingley et al., 2008).

In contrast to the clustering of photosynthesis-related genes in purple bacteria, the cyanobacterial genes that are identified as coding for photosynthesis components are scattered over the entire genome, with little clustering and few cotranscribed operons. Those operons that are present are often smaller than the analogous operons found in purple photosynthetic bacteria. For example, the pet genes that code for the cytochrome $b_6 f$ (bc_1 in purple bacteria) complex are a single operon in purple bacteria, but are split into two smaller operons plus some individual genes in Synechocystis PCC 6803. Most of the genes coding for cyanobacterial photosystem components are constitutively (continuously) expressed in cyanobacteria. The role of post-transcriptional regulation of gene expression and assembly of photosystem components is therefore much greater in cyanobacteria than in the metabolically more versatile purple bacteria (Wollman et al., 1999). Transcriptional analysis of the model cyanobacterium Synechocystis PCC 6803 has revealed extensive transcription of noncoding and opposite strand DNA, which points to an important role of antisense RNA and other types of regulatory RNAs (Mitschke et al., 2011).

Exceptions to the general pattern of a lack of operons and gene clusters are the genes for the phycobilisome components in those species that are capable of chromatic acclimation (Chapter 5) (Sanfilippo et al., 2019). These organisms are capable of adapting to rapidly changing environmental conditions, so a higher level of regulation and control of gene expression is apparently required.

10.4   Chloroplast genomes

Chloroplasts of most eukaryotic photosynthetic organisms contain a circular DNA genome reminiscent of those found in bacteria. This genome is unquestionably a remnant of the evolutionary origin of chloroplasts as endosymbiotic cyanobacteria.

This issue is discussed in detail in Chapter 12. Complete genomes have been sequenced for many chloroplasts from plants and major classes of algae. The most striking aspect of the chloroplast genome compared with the genome of cyanobacteria is its much smaller size and therefore reduced genetic content. This reduced size results from the fact that most of the information needed to carry out photosynthesis has been transferred to the nucleus.

The size of chloroplast genomes that have been sequenced to date ranges from ~70 to 200 kb (Martin and Herrmann, 1998; Moore *et al.*, 2010). Chloroplast genomes from higher plants typically possess 50–80 coding sequences, which contain the information for many of the core proteins of the photosystems and the cytochrome $b_6 f$ complex, plus a complete complement of tRNA synthetases and most of the elements of the chloroplast protein synthesis machinery. Almost all other protein components of the chloroplast are coded by nuclear DNA and imported into the chloroplast and subsequently assembled into complexes (Section 10.5). The imported proteins greatly outnumber the chloroplast-encoded ones. Estimates of the total number of chloroplast proteins range up to 5000 (Martin and Herrmann, 1998), so the vast majority of the proteins in chloroplasts are coded by nuclear DNA, synthesized on cytoplasmic ribosomes, and imported into the chloroplast.

Several groups of organisms have highly atypical chloroplast genomes. One is the dinoflagellates, in which a large number of very small circular segments of DNA are thought to comprise the chloroplast genome (Zhang *et al.*, 1999). These "minicircles" are similar in some ways to plasmids, although they each contain only a single gene. How the chloroplast genes are replicated and regulated in these organisms is not yet understood. The second type of atypical plastids are found in parasitic organisms that have lost the ability to carry out photosynthesis (Wolfe *et al.*, 1992). Their chloroplasts have lost all the genes that code for photosynthetic proteins. Many parasites, including the organism that causes malaria, are not now photosynthetic, but their ancestors almost certainly were. These organisms, known as apicomplexans, have retained a small vestigial chloroplast genome

that functions primarily in fatty acid synthesis (McFadden *et al.*, 1996). Photosynthetic members of this group, *Chromera velia*, have been discovered (Moore *et al.*, 2008). Finally, a remarkable amoeboid organism, *Paulinella chromatophora*, has only recently acquired a cyanobacterium and much of the genome content has not yet transferred to the nucleus, with the result that the chloroplast genome is over 1 Mb in size (Nowack and Grossman, 2012; Nowack and Weber, 2018).

The majority of chloroplast-encoded proteins are involved in translation, photosynthesis, or cofactor biosynthesis (Martin and Herrmann, 1998). A typical feature of chloroplast genomes is the large inverted repeat region. The functional importance of this structure is not entirely clear, as its size varies greatly in many plants, and it is completely missing in some. It is thought to assist in promoting the stability of the genome, as the genes in the inverted repeat exhibit slower rates of nucleotide substitution than the rest of the chloroplast genome (Maier *et al.*, 1995). The chloroplast contains a single origin of DNA replication. The majority of chloroplast genes are monocistronic, and the transcription rate is generally thought not to be the main mechanism of chloroplast gene regulation (Section 10.6).

The complete nuclear genome sequences of many algae and plants have been determined and have transformed research into the mechanisms of photosynthesis in eukaryotic photosynthetic organisms. Many important crop plants have had their genomes sequenced, and this information will be critical for future crop improvement (Morrell *et al.*, 2012).

10.5 Pathways and mechanisms of protein import and targeting in chloroplasts

The most important consequence of the split genome in photosynthetic eukaryotes is the requirement that the vast majority of the proteins forming the chloroplast be coded for in the nucleus, synthesized on cytosolic ribosomes, and then imported into the chloroplast. Once inside the chloroplast, they must be targeted to the proper place and then, in many cases, assembled into large oligomeric complexes. There is a remarkably complex orchestration of the assembly and regulation of the chloroplast and the interaction between the chloroplast and the nucleus (Eberhard *et al.*, 2008; Nickelsen and Rengstl, 2013). First, we will consider the mechanisms of the import of proteins into chloroplasts and their targeting to the proper place (Inaba and Schnell, 2008; Li and Chiu, 2010; Schleiff and Becker, 2011; Celedon and Cline, 2013; Richardson and Schnell, 2020). There are at least four distinct targeting pathways of proteins to chloroplasts. Here we focus on the Toc–Tic pathway, which accounts for import of ~90% of proteins targeted to the chloroplast (Inaba and Schnell, 2008). Toc and Tic are large protein complexes that are given the abbreviations Toc (translocon at the outer membrane of chloroplasts) or Tic (translocon at the inner membrane of chloroplasts).

Almost all nuclear-encoded proteins that are destined for the chloroplast have an N-terminal extension to the translation product, compared with the mature protein or with an analogous protein found in cyanobacteria. This N-terminal extension, or transit sequence, serves to target the protein to the chloroplast and is then cleaved off by a processing protease. Proteins that are targeted to the lumen of the chloroplast have to cross three membranes (in many types of algae they have to cross as many as five!) to reach their site of function. These proteins have a bipartite targeting domain, consisting of the normal chloroplast-targeting sequence plus an additional lumenal targeting domain. In these cases, the processing takes place in two steps, with the first step occurring on entry into the chloroplast, and the second step taking place upon transport into the thylakoid lumen.

Considerable evidence supports the view that the chloroplast-targeting sequence is both necessary and sufficient to direct a particular protein to the chloroplast. If this domain is removed, the resultant protein is not targeted to the chloroplast, and if the targeting domain is transferred to a protein not normally targeted to the chloroplast, the presence of the domain is sufficient to direct that protein to the chloroplast. Remarkably, there does not seem to be a consensus chloroplast-targeting sequence, but in most cases, these sequences are rich in the hydroxyl-containing amino acids serine and threonine and contain almost no acidic amino acids. The lumenal targeting domains are typical of signal sequences found in secreted proteins. In a few cases, proteins that are targeted to the outer chloroplast envelope membrane do not have a cleavable targeting sequence and are directly incorporated into the membrane.

Figure 10.2 illustrates the pathway and components involved in the chloroplast Toc–Tic import machinery (Li and Chiu, 2010). Import takes place in several stages, which can be manipulated experimentally to reveal the mechanisms involved. Some of these steps require energy in the form of either ATP or GTP, while others are energy-independent. The first step is the association of the protein that is to be transported with the outer chloroplast membrane. This is thought initially to involve energy-independent protein–lipid interactions of the transit peptide of the protein that is to be transported with the monogalactosyldiglyceride and digalactosyldiglyceride lipids that are the major lipid components of the chloroplast envelope. The next stage involves a series of energy-independent and energy-dependent protein–protein interactions. Large multiprotein complexes are present in both the outer and the inner chloroplast membranes. The various proteins that are involved in the Toc and Tic complexes are identified by their molecular masses in kDa. The Toc complex consists of three membrane proteins: Toc159, Toc75, and Toc34. The Toc75 protein forms a narrow protein-conducting channel of 8–9 Å diameter, through which the unfolded protein is threaded through the outer membrane. The early stages of protein association and import through the Toc75 channel require GTP, while the later stages require ATP. Both the Toc159 and Toc34 proteins bind GTP. The GTP requirement may be primarily a gating function, to ensure that only the correct proteins enter the pathway. The functions of

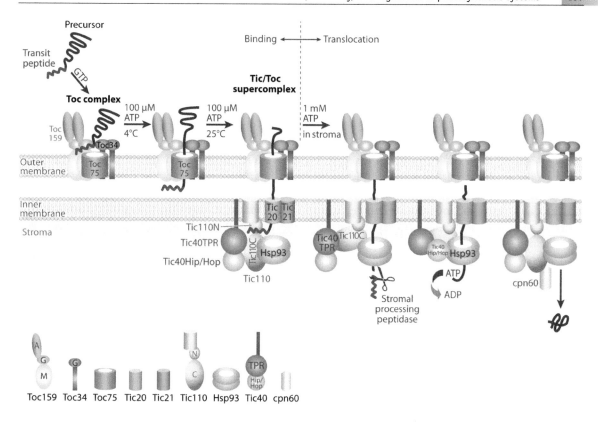

Figure 10.2   A working model for the import of proteins into chloroplasts. Proteins are identified by their molecular mass and whether they are part of the Toc or Tic complexes. Other abbreviations: Com70 and Hsp70-IAP are ATP-dependent molecular chaperones; ClpC and Cpn60 are chloroplast molecular chaperones; OM, IMS, and IM are outer membrane, intermembrane space, and inner membrane, respectively. *Source:* Li and Chiu (2010)./Annual Reviews.

the Toc159 protein were earlier attributed to a Toc89 protein, which is a proteolytic fragment of the Toc159 protein. The ATP is probably utilized by molecular chaperones, Com70 and Hsp70 (heat shock protein of 70 kDa mass). The Com70 protein is present on the cytoplasmic side of the chloroplast membrane and may help to stabilize the unfolded protein prior to translocation. In addition, Hsp70 is found in the intermembrane space between the inner and outer chloroplast membranes, where it may help draw the proteins across the outer membrane.

The Tic complex has a more transitory nature than does the stably assembled Toc complex and is assembled only when it associates with the Toc complex to form a supercomplex. The Tic/Toc supercomplex forms sites of adhesion between the outer and inner chloroplast membranes, so that the imported proteins can be directly translocated across both membranes. The Tic complex consists of three main proteins: Tic22, Tic110, and Tic20. Tic22 is localized at the outer side of the inner membrane and is the first component which the translocating protein associates with after crossing the outer membrane. Tic20 is an integral membrane protein that is thought to be intimately involved in the translocation event itself. Tic110 is a transmembrane protein that has most of its mass on the stromal side of the inner membrane. It associates with another molecular chaperone, ClpC, which may help to pull the protein through the membrane in an ATP-dependent manner. The next stop is the GroEL/GroES molecular chaperone, Cpn60, which helps the protein fold to its final conformation if it is a stromal protein. The stromal processing metalloprotease cleaves off the transit sequence. The protein can now have a number of possible fates. It can remain a stromal protein, be further translocated to the thylakoid lumen, or be assembled into a thylakoid membrane complex.

Lumenal and thylakoid membrane targeting is surprisingly complex, with a minimum of four distinct pathways identified, each of which works with certain components but not with others (Fig. 10.3) (Celedon and Cline, 2013). One pathway is similar to the secretory (Sec) system found in many bacteria.

It requires ATP hydrolysis and is responsible for translocating plastocyanin, the 33 kDa protein of the Photosystem II oxygen-evolving complex, and the Photosystem I subunit that binds plastocyanin, the PsaF protein. The second pathway is similar to the signal recognition particles (SRP) that are involved in secretion by the endoplasmic reticulum in eukaryotes and many bacteria. It is a GTP-dependent pathway and serves to insert the LHC antenna complexes into the thylakoid membrane. Both the Sec and the SRP pathways are stimulated by a pH difference across the thylakoid membrane, but the nucleotide triphosphates (NTP) are absolutely required. The third pathway operates only on ΔpH

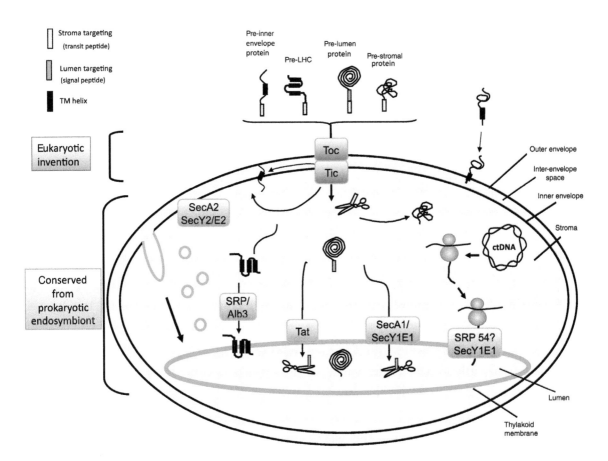

Figure 10.3   A model for the routing of lumenal targeted proteins and thylakoid integral membrane proteins. The four targeting pathways are shown, from left to right in the figure: the Sec pathway, which is ATP-dependent and ΔpH-stimulated; the signal recognition particle (SRP) pathway, which is GTP-dependent and ΔpH-stimulated; the ΔpH-dependent pathway; and the spontaneous pathway. *Source:* Celedon and Cline (2013)/Elsevier.

and has no NTP requirement. It is related to a bacterial pathway that has recently been discovered. An unusual aspect of the ΔpH pathway is that it can apparently translocate folded proteins without unfolding them. The 17 and 23 kDa oxygen-evolving proteins utilize this pathway. The last pathway for protein import into thylakoids has no known energy or ΔpH requirement and appears to consist of spontaneous integration into the thylakoid membrane. For lumenal proteins with a second targeting sequence, it is cleaved off by a thylakoid protease in the lumen.

It is not clear why there is only a single dominant import pathway into the chloroplast – the Toc/Tic system – yet so many distinct protein import pathways into thylakoids. Some of the reasons for this pattern may lie in the specific aspects of particular proteins that utilize each pathway. For example, an integral membrane protein needs to avoid aggregation. This is accomplished by the SRP pathway. Similarly, the need to transport an already folded protein is compatible only with the ΔpH pathway (Celedon and Cline, 2013).

10.6   Gene regulation and the assembly of photosynthetic complexes in cyanobacteria and chloroplasts

Many proteins that function in photosynthesis are part of multiprotein complexes, such as the photosystems, the cytochrome $b_6 f$ complex, or the rubisco complex. These complexes always have a fixed stoichiometry of the protein constituents, such as the $L_8 S_8$ rubisco complex. The fixed stoichiometry introduces the need to regulate the amount of each constituent protein. If this regulation did not occur, there would often be significant amounts of unassembled photosystem components, which could be deleterious to the organism. This negative effect would be particularly severe if the excess components included

chlorophylls, because uncomplexed chlorophyll rapidly leads to production of singlet oxygen, which is highly damaging to all biological components. In addition, the overall amounts of each of the complexes must be adjusted for the most efficient functioning of the photosynthetic apparatus. Finally, short-term regulation of the complexes by effects such as phosphorylation or carotenoid levels provides a fine-tuning of the distribution of energy. These latter regulatory mechanisms were discussed in Chapters 5–7 and are not considered further here.

For the reasons given above, the synthesis and assembly of photosynthetic constituents is very tightly regulated. Although the details of this complex regulatory system are not yet all understood, considerable progress has been made, and some general principles are becoming evident (Eberhard et al., 2008). In contrast to the situation with anoxygenic photosynthetic bacteria described above (Section 10.2), which exhibit typical bacterial mechanisms of transcriptional control over gene expression, the situation in cyanobacteria, and especially in chloroplasts, is quite different. Chloroplast mRNA transcripts are typically very long-lived, with half-lives of 6–40 hours (Kim et al., 1993). This is to be compared with mRNA half-lives of minutes in typical bacteria and less than an hour in cyanobacteria. The factors that give chloroplast mRNAs such a long lifetime appear to lie mainly in the 5′ untranslated regions just upstream from the beginning of the coding region of the gene. Nuclear-encoded translation factors bind to these regions and modulate the ability of the message to interact with the chloroplast ribosome, thereby controlling the rate of production of chloroplast-encoded proteins. Thus, the control of synthesis is largely at the translational level and is primarily controlled by signals from the nucleus. As dramatic evidence of this fact, decreasing the copy number of chloroplast DNA by a factor of ten using inhibitors of DNA replication lowers the levels of transcripts dramatically but has almost no effect on the rates of synthesis of chloroplast-encoded proteins (Hosler et al., 1989).

In addition to the translational control, post-translational degradation of excess proteins or of proteins lacking essential cofactors is observed in many cases. For example, apo-plastocyanin, apo-cytochromes, and chlorophyll proteins without pigment are rapidly degraded by proteolysis. A second level of translational control of the rates of some subunits in oligomeric proteins by the amounts of other subunits is described in Section 10.7.

The chloroplast also sends out signals that affect the expression of nuclear genes. There is clear evidence for signals originating in the chloroplast that are sent to the nucleus, where they affect nuclear gene expression and possibly the translation of mRNAs on cytoplasmic ribosomes (Goldschmidt-Clermont, 1998; Nott *et al.*, 2006; Koussevitzky *et al.*, 2007; Chi *et al.*, 2013). This is especially true for the *Lhcb* genes for the LHC antenna proteins (also called *Cab* for chlorophyll *a/b*). These retrograde signals include Mg Protoporphyrin IX, an intermediate in chlorophyll biosynthesis (Koussevitzky *et al.*, 2007; Eberhard *et al.*, 2008). This signal is at least in part mediated by the products of several *gun* (genomes uncoupled) genes. Additional retrograde signals include reactive oxygen species and redox effects mediated by the oxidation–reduction state of the plastoquinone pool (Pfannschmidt *et al.*, 1999; Puthiyaveetil *et al.*, 2012).

How is the translational control of chloroplast gene expression accomplished at the molecular level? Although many of the details of this regulatory process are not understood, much progress has been made (Bruick and Mayfield, 1999). The translational control of chloroplast gene expression is mediated by a large number of nucleus-encoded protein factors. These nuclear proteins are synthesized on cytoplasmic ribosomes and are imported into the chloroplast via the import systems discussed above in Section 10.5. They then bind to the upstream 5′ untranslated region and affect the ability of the transcript to interact with the ribosome. The best-studied case is the *psbA* gene that codes for the D1 protein of the Photosystem II reaction center. Proteins of 60, 55, 47, and 38 kDa mass have been identified as specifically binding the 5′ untranslated region of the *psbA* gene. The 47 kDa protein is homologous to a poly A binding protein. This protein binds to the 5′ untranslated region of the *psbA* gene in a method that is sensitive to both

light and redox potential. The 60 kDa protein is a protein disulfide isomerase, whose cystine disulfide bond is reduced to SH groups in the light when conditions are more reducing in the chloroplast. The 60 kDa protein then reduces a disulfide on the 47 kDa poly A binding protein. When in the reduced state, the 47 kDa protein binds to the 5′ untranslated region of the *psbA* gene and enhances its association with the ribosome and, therefore, the translation of the D1 protein. In the dark state, the redox potential is higher, and the disulfide is reformed, inhibiting translation. In addition, an ADP-dependent protein kinase phosphorylates the 60 kDa protein disulfide isomerase in the dark when ADP levels are high. This inactivates the isomerase, providing an additional level of regulation of the transcription of the *psbA* gene.

10.7   The regulation of oligomeric protein stoichiometry

In all the major protein complexes of the photosynthetic apparatus of eukaryotic chloroplasts, some subunits are encoded by chloroplast DNA and synthesized in the chloroplast on chloroplast ribosomes. Other subunits are coded for by nuclear genes and synthesized on cytoplasmic ribosomes, imported into the chloroplast, and then the oligomeric complex assembled. This requires an extraordinary level of communication and regulation between the chloroplast and the rest of the cell, some of which is described above in Section 10.6. The result is a remarkable coordination of the two genomes to produce a smoothly functioning system that is able to respond rapidly to changes in conditions or stress. Figure 10.4 shows which subunits of the photosynthetic apparatus are coded for by the nucleus and chloroplast genomes (Allen *et al.*, 2011).

In many cases, the level of accumulation of the subunits of a multiprotein complex is controlled by one master component, which is termed the dominant subunit. The other subunit(s) respond to its

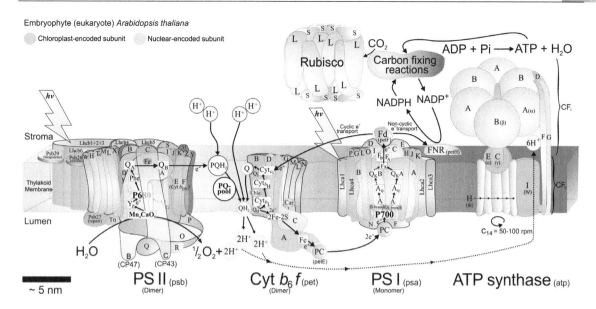

Figure 10.4   Origin of the major proteins of the photosynthetic apparatus in chloroplasts. Subunits coded for and synthesized in the chloroplast are colored green, while subunits coded for in the nucleus, synthesized in the cytoplasm, and then imported into the chloroplast are colored yellow. *Source:* Allen *et al.* (2011)/Elsevier.

level of accumulation; in particular, their rate of synthesis is reduced if the dominant subunit is not present or is present in reduced amounts. The dependent subunit is said to be controlled by eptistatic synthesis (CES) and is therefore designated the CES subunit. In most of the multiprotein oligomeric complexes of photosynthetic membranes, dominant and CES subunits are found (Wollman *et al.*, 1999; Eberhard *et al.*, 2008). For example, the D2 protein of the Photosystem II reaction center complex is the dominant subunit, and the D1 protein is the CES subunit. In turn, the D1 protein is dominant with respect to the CP47 antenna complex.

In one case, the mechanism of the CES effect is understood relatively well. Cytochrome f is the CES subunit in the assembly of the cytochrome $b_6 f$ complex. Cytochrome b and subunit IV are the dominant subunits (see Chapter 7 for a discussion of the cytochrome $b_6 f$ complex). If cytochrome f is not assembled into the complex, then a C-terminal portion of the protein binds to the 5′ untranslated region of its own mRNA, inhibiting its interaction with the ribosome. This binding does not take place if the

cytochrome is assembled into the complex, because the C-terminal region is sequestered internally in the oligomeric protein. This is an autoregulation effect, in which the CES subunit inhibits its own synthesis. However, it is dependent on the levels of the dominant subunit, and therefore fits the CES pattern. Much of the evidence for the CES effect has been obtained from the green alga *Chlamydomonas*, which is an excellent model system for study of eukaryotic photosynthetic gene expression.

10.8 Assembly, photodamage, and repair of Photosystem II

Photosynthetic systems face a special challenge. They absorb large amounts of light energy and convert it into chemical energy. At the molecular level, the energy in a photon is a huge perturbation, which the system processes efficiently under normal conditions. However, under some conditions, they may not be able to process all the incoming energy. The excess energy can lead to production of toxic species and damage to the system if it is not dissipated safely. Photosynthetic organisms therefore contain a complex set of regulatory and repair mechanisms. Some of these mechanisms regulate energy flow in the antenna system to avoid excess excitation of the reaction centers and to ensure that the two photosystems are equally driven. These antenna processes were discussed in Chapter 5. While these processes are effective, they are not perfect and sometimes toxic species are produced. Additional

mechanisms are needed to dissipate these compounds, in particular, active oxygen species. Even with all of these protective and scavenging mechanisms, damage to the photosynthetic apparatus still takes place and additional mechanisms are present to repair the system.

The most dramatic of the damaging processes in photosynthetic organisms is the phenomenon of photodamage of Photosystem II. When Photosystem II is activated, there is approximately a one in a million chance that it will suffer a significant damaging event (Anderson et al., 1998). The molecular nature of the photodamage

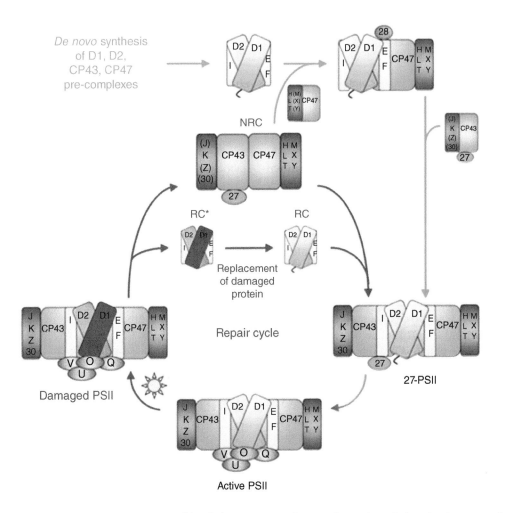

Figure 10.5 Scheme describing the assembly of Photosystem II in cyanobacteria and also the damage and repair cycle that maintains the complex in a functional form. *Source:* Weisz *et al.* (2019)/National Academy of Sciences.

suffered is not yet clear, although it seems to be largely confined to the D1 protein that is part of the core Photosystem II reaction center complex. If this photodamage is not repaired, the system will very rapidly lose all activity. Instead, a remarkable **damage-repair cycle** is present, in which the damaged Photosystem II reaction center is disassembled, a new copy of the D1 protein is synthesized, and the complex is reassembled and returned to service (Melis, 1999; Vass, 2012; Nickelsen and Rengstl, 2013; Weisz et al., 2019). Figure 10.5 illustrates the Photosystem II damage and repair cycle in cyanobacteria (Weisz et al., 2019).

There are indications that photodamage largely originates from over-reduction of the quinone acceptor complexes and also paradoxically that the major focus is on the oxidizing side of Photosystem II (Zavafer et al., 2015). Both these statements may be true, but are operative only under certain situations. Some data suggest that there is a threshold intensity for photodamage, while other work suggests that there is a low but constant probability for photodamage with every photon absorbed. If the rate of photodamage is relatively low, the repair cycle can keep up with it and the system operates at nearly maximum efficiency. However, if the rate of photodamage is higher than the rate of repair, then the damaged Photosystem II reaction centers accumulate and the system exhibits the condition of **photoinhibition**. Photoinhibition is a complex state that may have different molecular origins under different types of conditions (Pinnola and Bassi, 2018).

Photosystem 1 seems to be generally much less susceptible than Photosystem II to photodamage, especially under optimal conditions. The major threat to Photosystem I comes from active oxygen species that can be generated by auto-oxidation of the ferredoxin acceptors (Asada, 1999; Lima-Melo et al., 2019). The active oxygen scavenging systems effectively prevent photodamage to Photosystem 1, except under stress conditions, where it may be much more vulnerable. Photosystem I is especially sensitive to damage from chilling (Sonoike, 1999).

References

Allen, J. F., de Paula, W. B. M., Puthiyaveetil, S., and Nield, J. (2011) A structural phylogenetic map for chloroplast photosynthesis. *Trends in Plant Science* 16: 645–655.

Anderson, J. M., Park, Y. I., and Soon, W. S. (1998) Unifying model for the photoinactivation of Photosystem II in vivo; under steady-state photosynthesis. *Photosynthesis Research* 56: 1–13.

Asada, K. (1999) The water–water cycle in chloroplasts: Scavenging of active oxygens and dissipation of excess photons. *Annual Review of Plant Physiology and Plant Molecular Biology* 50: 601–639.

Bauer, C., Setterdahl, A., Wu, J., and Robinson, B. (2009) Regulation of gene expression in response to oxygen tension. In: N. Hunter, F. Daldal, M. Thurnauer and T. Beatty (eds.) *The Purple Phototrophic Bacteria*, Vol. 28. Dordrecht: Springer, pp. 707–725.

Bruick, R. K. and Mayfield, S. P. (1999) Light-activated translation of chloroplast mRNAs. *Trends in Plant Science* 4: 190–195.

Celedon, J. M. and Cline, K. (2013) Intra-plastid protein trafficking: How plant cells adapted prokaryotic mechanisms to the eukaryotic condition. *Biochimica et Biophysica Acta* 1833: 341–351.

Chi, W., Sun, X., and Zhang, L. (2013) Intracellular signaling from plastid to nucleus. *Annual Review of Plant Biology* 64: 559–582.

Choudhary M., Mackenzie C., Donohue T., and Kaplan S. (2009) Purple bacterial genomics. In: N. Hunter, F. Daldal, M. Thurnauer, and T. Beatty (eds.) *The Purple Phototrophic Bacteria*, Vol. 28. Dordrecht: Springer, pp. 691–706.

Eberhard, S., Finazzi, G., and Wollman, F. A. (2008) The dynamics of photosynthesis. *Annual Review of Genetics* 42: 463–515.

Goldschmidt-Clermont, M. (1998) Coordination of nuclear and chloroplast gene expression in plant cells. *International Review of Cytology* 177: 115–180.

Gregor, J. and Klug, G. (1999) Regulation of bacterial photosynthesis genes by oxygen and light. *FEMS Microbiology Letters* 57: 725–749.

Hosler, J. P., Wurtz, E. A., Harris, E. H., Gillham, N. W., and Boynton, J. E. (1989) Relationship between gene dosage and gene-expression in the chloroplast of *Chlamydomonas reinhardtii*. *Plant Physiology* 91: 648–655.

Imam, S., Noguera, D. R., and Donohue, T. J. (2014) Global analysis of photosynthesis transcriptional regulatory networks. *PLoS Genetics* 10: e1004837.

Inaba, T. and Schnell, D. J. (2008) Protein trafficking to plastids: One theme, many variations. *Biochemical Journal* 413: 15–28.

Kim, M. Y., Christopher, D. A., and Mullet, J. E. (1993) Direct evidence for selective modulation of *psbA*, *rpoA*, *rbcL* and 16s-RNA stability during barley chloroplast development. *Plant Molecular Biology* 22: 447–463.

Klug, G. and Masuda, S. (2009) Regulation of genes by light. In: N. Hunter, F. Daldal, M. Thurnauer, and T. Beatty (eds.) *The Purple Phototrophic Bacteria*, Vol. 28. Dordrecht: Springer, pp. 727–741.

Kolber, Z. S., Plumley, F. G., Lang, A. S., *et al.* (2001) Contribution of aerobic photoheterotrophic bacteria to the carbon cycle in the ocean. *Science* 292: 2492–2495.

Koussevitzky, S., Nott, A., Mockler, T. C., Hong, F., Sachetto-Martins, G., Surpin, M., Lim, I. J., Mittler, R., and Chory, J. (2007) Signals from chloroplasts converge to regulate nuclear gene expression. *Science* 316: 715–719.

Kumka, J. E., Schindel, H., Fang, M., Zappa, S., and Bauer, C. E. (2017) Transcriptomic analysis of aerobic respiratory and anaerobic photosynthetic states in *Rhodobacter capsulatus* and their modulation by global redox regulators RegA, FnrL and CrtJ. *Microbial Genomics* 3: 000125.

Li, H.-M. and Chiu, C.-C. (2010) Protein transport into chloroplasts. *Annual Review of Plant Biology* 61: 157–180.

Lima-Melo, Y., Alencar, V. T. C. B., Lobo A. K. M., Sousa, R. H. V., Tikkanen, M., Aro, E.-M., Silveira, J. A. G., and Gollan, P. J. (2019) Photoinhibition of Photosystem I provides oxidative protection during imbalanced photosynthetic electron transport in *Arabidopsis thaliana*. *Frontiers in Plant Science* 10: 916.

McFadden, G. I., Reith, M. E., Munholland, J., and Lang-Unnasch, N. (1996) Plastid in human parasites. *Nature* 381: 482.

Maier, R. M., Neckermann, K., Igloi, G. L., and Kössel, H. (1995) Complete sequence of the maize chloroplast genome: Gene content, hotspots of divergence and fine tuning of genetic information by transcript editing. *Journal of Molecular Biology* 251: 614–628.

Martin, W. and Herrmann, R. G. (1998) Gene transfer from organelles to the nucleus: How much, what happens, and why? *Plant Physiology* 118: 9–17.

Melis, A. (1999) Photosystem-II damage and repair cycle in chloroplasts: What modulates the rate of photodamage in vivo;. *Trends in Plant Science* 4: 130–135.

Mitschke, J., Georg, J., Scholz, I., Sharma, C. M., Dienst, D., Bantscheff, J., Voss, B., Steglich, C., Wilde, A., Vogel, J., and Hess, W. R. (2011) An experimentally anchored map of transcriptional start sites in the model cyanobacterium Synechocystis sp PCC6803. *Proceedings of the National Academy of Sciences USA* 108: 2124–2129.

Morrell, P. L., Buckler, E. S., and Ross-Ibarra, J. (2012) Crop genomics: Advances and applications. *Nature Reviews Genetics* 13: 85–96.

Moore, R. B., Oborník, M., Janouskovec, J., Chrudimsky, T., Vancova, M., Green, D. H., Wright, S. W., Davies, N. W., Bolch, C. J. S., Heimann, K., Slapeta, J., Hoegh-Guldberg, O., Logsdon, J. M., Jr., and Carter, D. A. (2008). A photosynthetic alveolate closely related to apicomplexan parasites. *Nature* 45: 959–963.

Moore, M. J., Soltis, P. S., Bell, C. D., Burleigh, J. G., and Soltis, D. E. (2010) Phylogenetic analysis of 83 plastid genes further resolves the early diversification of eudicots. *Proceedings of the National Academy of Sciences USA* 107: 4623–4628.

Naylor, G. W., Addlesee, H. A., Gibson, L. C. D., and Hunter, C. N. (1999) The photosynthesis gene cluster of *Rhodobacter sphaeroides*. *Photosynthsis Research* 62: 121–139.

Nickelsen, J., and Rengstl, B. (2013) Photosystem II assembly: From cyanobacteria to plants. *Annual Review of Plant Biology* 64: 609–635.

Nott, A., Jung, H. S., Koussevitzky, S., and Chory, J. (2006) Plastid-to nucleus retrograde signaling. *Annual Review of Plant Biology* 57: 739–759.

Nowack E. C. M., and Grossman A. R. (2012) Trafficking of protein into the recently established photosynthetic organelles of *Paulinella chromatophora*. *Proceedings of the National Academy of Sciences USA* 109: 5340–5345.

Nowack E. C. M., and Weber, A. P. M. (2018) Genomics-informed insights into endosymbiotic organelle evolution in photosynthetic eukaryotes. *Annual Review of Plant Biology* 69: 51–84.

Pfannschmidt, T., Nilsson, A., and Allen, J. F. (1999) Photosynthetic control of chloroplast gene expression. *Nature* 397: 625–628.

Pinnola, A. and Bassi, R. (2018) Molecular mechanisms involved in plant photoprotection. *Biochemical Society Transactions* 46: 467–482.

Puthiyaveetil, S., Ibrahim, I. M., and Allen, J. F. (2012) Oxidation–reduction signalling components in regulatory pathways of state transitions and photosystem stoichiometry adjustment in chloroplasts. *Plant, Cell and Environment* 35: 347–359.

Richardson, L. G. L. and Schnell, D. J. (2020) Origins, function, and regulation of the TOC–TIC general protein import machinery of plastids. *Journal of Experimental Botany* 71: 1226–1238.

Sanfilippo, J. E., Garczarek, L., Partensky, F., and Kehoe, D. M. (2019) Chromatic acclimation in cyanobacteria: A diverse and widespread process for optimizing photosynthesis. *Annual Review of Microbiology* 73: 407–433.

Sattley, W. M., Madigan, M. T., Swingley, W. D., Cheung, P. C., Clocksin, K. M., Conrad, A. L., Dejesa, L. C., Honchak, B. M., Jung, D. O., Karbach, L. E., Kurdoglu, A., Lahiri, S., Mastrian, S. D., Page, L. E., Taylor, H. L., Wang, Z. T., Raymond, J., Chen, M., Blankenship, R. E., and Touchman, J. W. (2008) The genome of *Heliobacterium modesticaldum,* a phototrophic representative of the *Firmicutes* containing the simplest photosynthetic apparatus. *Journal of Bacteriology* 190: 4687–4696.

Schleiff, E. and Becker, T (2011) Common ground for protein translocation: Access control for mitochondria and chloroplasts. *Nature Reviews Molecular and Cellular Biology* 12: 48–59.

Shih, P. M., Wu, D., Latifi, A, Axen, S. D., Fewer, D. P., Talla, E., Calteau, A., Cai, F., de Marsac, N. T., Rippka, R., Herdman, M.. Sivonen, K., Coursin, T., Laurent, T., Goodwin, L., Nolan, M., Davenport, K. W., Han, C. S., Rubin, E. M, Eisen, J. A., Woyke, T., Gugger, M., and Kerfeld, C. A. (2013) Improving the coverage of the cyanobacterial phylum using diversity-driven genome sequencing. *Proceedings of the National Academy of Sciences USA* 110: 1053–1058.

Sonoike, K. (1999) The different roles of chilling temperatures in the photoinhibition of Photosystem I and Photosystem II. *Journal of Photochemistry and Photobiology B* 48: 136–141.

Swingley, W. D., Blankenship, R. E., and Raymond, J. (2008) Integrating Markov clustering and molecular phylogenetics to reconstruct the cyanobacterial species tree from conserved protein families. *Molecular Biology and Evolution* 25: 1–12.

Swingley, W., Blankenship, R., and Raymond, J. (2009) Evolutionary relationships among purple photosynthetic bacteria and the origin of proteobacterial photosynthetic systems. In: N. Hunter, F. Daldal, M. Thurnauer, and T. Beatty (eds.) *The Purple Phototrophic Bacteria*, Vol. 28. Dordrecht: Springer, pp. 17–29.

Vass, I. (2012). Molecular mechanisms of photodamage in the Photosystem II complex. *Biochimica et Biophysica Acta* 1817: 209–217.

Weisz, D. A., Johnson, V. M., Niedzwiedzki, D. M., Shinn, M. K., Liu, H., Klitzke, C. F., Gross, M. L., Blankenship, R. E., Lohman, T. M., and Pakrasi, H. B. (2019). A novel chlorophyll protein complex in the repair cycle of Photosystem II. *Proceedings of the National Academy of Sciences USA* 116: 21907–21913.

Wolfe, K. H., Morden, C. W., and Palmer, J. D. (1992) Function and evolution of a minimal plastid genome from a nonphotosynthetic parasitic plant. *Proceedings of the National Academy of Sciences USA* 89: 10648–10652.

Wollman, F.-A., Minai, L., and Nechushtai, R. (1999) The biogenesis and assembly of photosynthetic proteins in thylakoid membranes. *Biochimica et Biophysica Acta* 1411: 21–85.

Xiong, J., Inoue, K., and Bauer, C. E. (1998) Tracking molecular evolution of photosynthesis by characterization of a major photosynthesis gene cluster from *Heliobacillus mobilis*. *Proceedings of the National Academy of Sciences USA* 95: 14851–14856.

Yurkov, V. V. and Beatty, J. T. (1998) Aerobic anoxygenic phototrophic bacteria. *Microbiology and Molecular Biology Reviews* 62: 695–724.

Yurkov, V. V. and Csotonyl, J. (2009) New light on aerobic anoxygenic phototrophs. In: N. Hunter, F. Daldal, M. Thurnauer, and T. Beatty (eds.) *The Purple Phototrophic Bacteria*, Vol. 28. Dordrecht: Springer, pp. 31–55.

Zavafer, A., Cheah, M. H., Hillier, W., Chow, W. S., and Takahashi, S. (2015) Photodamage to the oxygen evolving complex of Photosystem II by visible light. *Scientific Reports* 5: 16363.

Zhang, Z., Green, B. R., and Cavalier-Smith, T. (1999) Single gene circles in dinoflagellate chloroplast genomes. *Nature* 400: 155–159.

Chapter 11

The use of chlorophyll fluorescence to probe photosynthesis

Fluorescence is a remarkably informative technique in photosynthesis research. Fluorescence measurements are invaluable as probes of the inner workings of photosynthetic systems. We have already discussed their application in studying the properties of antenna complexes in Chapter 5, and the basic physical principles that underlie fluorescence are covered in the Appendix. However, there are specialized applications of fluorescence methods that permit exceptional insights, especially into Photosystem II. These measurements are now routinely used in physiological and stress measurements, and instruments are available to take into the field to assess the physiological state of crops during the growing season. In this chapter, we will explore the basis of these measurements and what they can tell us about photosynthesis.

Fluorescence measurements are perhaps the single most widely utilized spectroscopic technique in photosynthesis research. This versatile technique has had the biggest impact in analysis of processes associated with Photosystem II. Photosystem II exhibits a phenomenon known as variable fluorescence, in which the intensity of fluorescence reports on the properties of the reaction center complex and associated electron transport chain in a remarkably detailed way. There are a number of detailed reviews of various aspects of the use of fluorescence in understanding Photosystem II (Krause and Weiss, 1991; Dau, 1994; Schreiber et al., 1995, 1998; Horton et al., 1996; Lazar, 1999; Maxwell and Johnson, 2000; Stirbet and Govindjee, 2011, 2012; Ruban, 2013; Banks, 2017; Kalaji et al., 2017; Laisk and Oja, 2018; Stirbet et al., 2018).

Kautsky and Hirsch (1931) first pointed out the variable nature of chlorophyll fluorescence on the basis of observations by eye. This was before the 1932 Emerson and Arnold experiments (Chapter 3) and well before the concepts of photosynthetic energy transfer or the photosynthetic unit had been formulated. Govindjee (1995) has provided an interesting history of the early work on chlorophyll fluorescence. The first really quantitative analysis of the dependence of chlorophyll fluorescence was given by Duysens and Sweers (1963), in which they related the increase in fluorescence in Photosystem II upon illumination to the redox state of an early electron acceptor. This acceptor was known as Q (for quencher). Fortunately, the chemical species that was much later identified with Q is a quinone (the Q_A acceptor in Photosystem II), so the Q name was retained. Later analysis by Warren Butler and coworkers (Butler, 1978) provided the basis for

Molecular Mechanisms of Photosynthesis, Third Edition. Robert E. Blankenship.
© 2021 Robert E. Blankenship 2021 by John Wiley & Sons Ltd.
Companion website: https://www.wiley.com/go/blankenship/molecularphotosynthesis3e

much of our current understanding of this phenomenon.

The vast majority of studies of variable chlorophyll fluorescence have been carried out on higher plant systems. Similar measurements can be carried out on cyanobacteria and anoxygenic photosynthetic organisms, in some cases using the same instruments used on plants (Hohmann-Marriott and Blankenship, 2007; Ritchie and Mekjinda, 2015; Bernát *et al.*, 2018; Sipka *et al.*, 2018). However, the interpretation of these experiments is significantly different from in plants and instrumental factors that are optimized for higher plants may give misleading results.

11.1   The time course of chlorophyll fluorescence

If a sample of photosynthetic tissue, such as a leaf or an algal suspension, is first dark-adapted and then illuminated by a very weak measuring beam that is itself too weak to cause photochemistry in a significant number of reaction centers, the intensity of fluorescence that is observed is very low. The level of fluorescence in this situation is known as F_0. The fluorescence is low because most of the energy is trapped by photochemistry, leaving little to be emitted as fluorescence.

If a more intense actinic light is then turned on, the level of fluorescence that is induced by the weak measuring beam goes through a characteristic time course, which is shown in Figs. 11.1 and 11.2. The fluorescence first rises, with two inflections, to a maximum (F_M). At later times, the fluorescence goes through a series of complex changes. Under some conditions, these changes include oscillations that can last for many seconds. Eventually, a steady-state level is reached. This characteristic fluorescence intensity time course is usually called fluorescence induction or sometimes the Kautsky effect. The interpretation of the early changes is well understood and is described below. The slower steps reflect a number of distinct processes, including regulation of both energy transfer and carbon

(a)

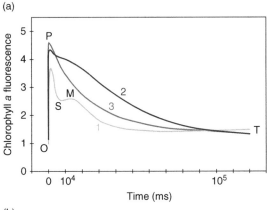

(b)

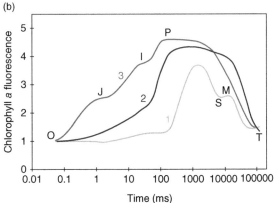

Figure 11.1 Chlorophyll *a* fluorescence induction transients of a pea leaf (kept in darkness for ~20 minutes before the measurement). (a) On a linear time scale; (b) on a logarithmic time scale. Wavelength of excitation: 650 nm. Excitation light intensity for curves labeled 1, 2, and 3 was, respectively, 32, 320, and 3200 μmol photons $m^{-2} s^{-1}$ at the leaf surface. For definition of OJIPSMT symbols, see Stirbet and Govindjee (2011). Fluorescence is given in arbitrary units. Source of the original figure: Strasser *et al.* (1995). *Source:* Stirbet and Govindjee (2011)/Elsevier.

metabolism. Their interpretation is often very difficult because more than one cause can give rise to similar observable effects.

The initial rise of fluorescence is correlated with the redox state of the Q_A electron acceptor. If an inhibitor such as DCMU is added to the sample, the fluorescence rises rapidly to F_M and then does not change further. DCMU blocks electron flow from Q_A to Q_B by displacing Q_B from its binding

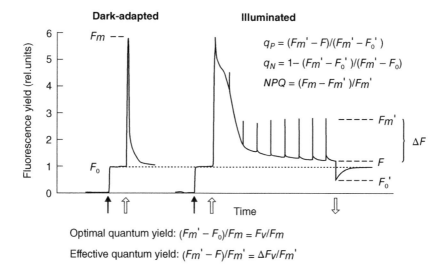

$$q_P = (Fm' - F)/(Fm' - F_0')$$

$$q_N = 1 - (Fm' - F_0')/(Fm' - F_0)$$

$$NPQ = (Fm - Fm')/Fm'$$

Optimal quantum yield: $(Fm' - F_0)/Fm = Fv/Fm$

Effective quantum yield: $(Fm' - F)/Fm' = \Delta Fv/Fm'$

Figure 11.2   Fluorescence measurements illustrate the induction of nonphotochemical quenching in Photosystem II. *Source:* Schreiber *et al.* (1998)/Cambridge University Press.

site. The slower rise without the inhibitor is due to the fact that until the pool of quinones in the membrane becomes reduced, Q_A is rapidly reoxidized and the fluorescence yield remains low. As the quinone pool gradually becomes reduced, more and more of Q_A becomes reduced and the fluorescence yield rises gradually to F_M, although the relation between fluorescence intensity and the amount of Q_A^- is not strictly linear.

The area under the fluorescence induction curve is proportional to the number of electrons that it takes to reduce the quinone pool. This sort of fluorescence induction analysis is widely used to investigate treatments or mutations that may affect the rate of linear electron flow and therefore the state of reduction of Q_A.

11.2   The use of fluorescence to determine the quantum yield of Photosystem II

Analysis of thylakoid fluorescence can be used to obtain quantitative estimates of the quantum yield of photochemistry in Photosystem II. This can be shown by simple manipulation of the equations that describe the yield of products from parallel first-order reactions. The basic equation for the yield of a given decay pathway is given by Eq. (A.40). It is given in specific form for Photosystem II in Eq. (11.1), where Φ_i is the quantum yield of the *i*th process and the rate constants for the various processes that can deactivate the excited state are summarized by k_p for photochemistry, k_f for fluorescence; and k_O for all other processes.

$$\Phi_i = \frac{k_i}{k_p + k_f + k_O} \qquad (11.1)$$

The quantum yield of one such process, photochemistry, with symbol Φ_p, is given by Eq. (11.2).

$$\Phi_p = \frac{k_p}{k_p + k_f + k_O} \qquad (11.2)$$

The quantum yield of fluorescence when all traps are open in the F_0 state is similarly given by Eq. (11.3):

$$F_0 = \frac{k_f}{k_p + k_f + k_O} \qquad (11.3)$$

When the traps are closed, at the F_M state, the effective rate constant for photochemistry is zero, and Eq. (11.4) results:

$$F_M = \frac{k_f}{k_f + k_O} \qquad (11.4)$$

If the difference between F_M and F_0 is taken and then divided by F_M, Eq. (11.5) is found, which simplifies by simple algebra to Eq. (11.6).

$$\frac{F_M - F_0}{F_M} = \frac{F_V}{F_M} = \frac{\dfrac{k_f}{k_f + k_O} - \dfrac{k_f}{k_p + k_f + k_O}}{\dfrac{k_f}{k_f + k_O}} \qquad (11.5)$$

$$\Phi_p = \frac{F_V}{F_M} = \frac{k_p}{k_p + k_f + k_O} \qquad (11.6)$$

In Eq. (11.6), F_V is called the variable fluorescence and is defined as $F_M - F_0$. Note that in this analysis that the symbols F_0, F_M, and F_V all refer to the quantum yields of fluorescence. The same symbols are often used for the intensities of fluorescence in these states. The intensities are the quantities that are measured in an actual experiment. The intensity of fluorescence is proportional to the quantum yield but also includes a term for the intensity of light absorbed and instrumental factors. However, because the ratio of the intensities is always used in the kinetic analysis, these terms cancel out.

Using typical values of F_M and F_0 measured in functioning systems, values of Φ_p of ~0.85 are found (Björkman and Demmig, 1987). There are, however, a number of assumptions that have been made in the derivation of Eq. (11.6), so that it is not strictly valid, especially under some conditions. The implicit assumption is made that all the fluorescence comes from Photosystem II, that none of the other rate constants change as the traps go from open to closed, and that all the fluorescence that is observed in both the F_0 and F_M states come from a homogeneous system in which all chlorophyll excited states are equivalent. This last assumption is clearly not valid, as all the complexity of the structure and function of the photosynthetic apparatus suggests and many experiments confirm. The variable fluorescence is relatively easy to measure experimentally and is a useful indicator of the maximum quantum yield of photochemistry in Photosystem II. However, it should not be taken as a rigorous way to measure this quantity because of these approximations.

A slightly different formulation of the equation relating fluorescence and the quantum yield of Photosystem II has enjoyed very wide use and is even easier to measure than the quantities F_V and F_M. The relationship is given by Eq. (11.7), where F_M' is the fluorescence maximum after a saturating flash, and F_t is the fluorescence intensity at a time t just prior to the application of the flash (Genty et al., 1989). The experiment is done in the presence of actinic light, so that F_t represents the yield of fluorescence under steady-state conditions.

$$\Phi_2 = \frac{F_M' - F_t}{F_M'} \qquad (11.7)$$

The advantage of Eq. (11.7) is that F_0 need not be determined, so that measurements can be made in field conditions in full sunlight without the need to fully dark-adapt the sample.

This method has become a standard technique in measurements of photosynthetic efficiency, especially ecophysiological studies, where the simplicity of the measurement permits analysis under a remarkably wide range of conditions and treatments. These studies have been made possible by the development of instruments that use the pulse amplitude modulation (PAM) technique, in which a series of weak pulses are given to the sample, and the modulated fluorescence recovered by selective pulse amplification (Schreiber et al., 1995). This technique removes any direct contribution from nonmodulated actinic light, and only records the yield of fluorescence stimulated by the modulated measuring beam. It is also relatively insensitive to perturbations induced by saturating flashes, and so provides a simple yet robust method for measurement of fluorescence parameters and photosynthetic efficiency.

11.3   Fluorescence detection of nonphotochemical quenching

The process of nonphotochemical quenching (NPQ) as a regulatory mechanism in Photosystem II was introduced in Chapter 5. This effect was first discovered by the use of fluorescence methods (Krause and Weiss, 1991). The clearest manifestation of the induction of nonphotochemical quenching is the change in the F_M value during illumination, as shown in Fig. 11.2. The F_M found in a dark-adapted sample is typically higher than F_M measured under intense illumination. This does not reflect any difference in the amount of Q_A^- that can be formed, because in both cases a saturating flash is used that fully reduces all Q_A in the sample. Instead, it tells us that some other process that is competing for the fluorescence has decreased. A quantitative relationship for the amount of this quenching effect is given by Eq. (11.8), where F_M is the value of F_M found just after dark adaptation, and F_M' is the value of F_M found after illumination has induced the quenching state.

$$NPQ = \frac{F_M - F_M'}{F_M'} \qquad (11.8)$$

This relation can be rearranged to give an equation that is similar to the Stern–Volmer relation Eq. (A.45), where $[Q]$ is the effective concentration of the quencher, and K_{SV} is the Stern–Volmer constant that reflects the efficiency of the quencher.

$$\frac{F_M}{F_M'} = 1 + K_{SV}[Q] \qquad (11.9)$$

There are multiple sources of quenchers that can contribute to the observed result of nonphotochemical quenching. This remains an active and controversial area of research.

11.4   The physical basis of variable fluorescence

Why does Photosystem II exhibit such a large change in fluorescence yield, while Photosystem I does not? The answer to this question has been difficult to establish, and even now some aspects of the reasons for the differences between the two photosystems are not clear. The major effect that gives rise to the fluorescence increase in Photosystem II upon accumulation of Q_A^- is a slowing of the rate of the primary electron transfer process from P680* to form the radical pair state P680$^+$Pheo$^-$. This is largely due to the electrostatic repulsive effect of the charge on Q_A^-. This negative charge raises the energy of the state P680$^+$Pheo$^-$Q$_A^-$ compared to P680$^+$Pheo$^-$Q$_A$. This electrostatic repulsion inhibits the charge separation process and promotes the back transfer of the excitation to the bulk antenna system where fluorescence takes place. Some of the increase is also due to recombination luminescence, in which the charge-separated state P680$^+$Pheo$^-$ recombines to form P680*, followed by back transfer of the excitation to the antenna system. Under most circumstances, P680$^+$ is never found, because it is such a strongly oxidizing species that it finds some species to reduce it (*see* Chapter 7). Therefore, it does not enter into the normal phenomenon of fluorescence induction in Photosystem II.

Photosystem I has several important differences compared to Photosystem II that all contribute to the fact that it does not exhibit variable fluorescence. First, the electron transfer process is largely irreversible, due to the steeper redox gradient on the acceptor side of Photosystem I. This rapidly removes the electron that is transferred farther from P700, compared to the situation in Photosystem II. This effect greatly reduces the probability of recombination luminescence due to charge recombination. In addition, the larger distance of the electron acceptor diminishes the electrostatic effect of the reduced acceptor on the primary electron transfer process. Finally, a significant amount of $P700^+$ is present under conditions of steady-state electron flow through Photosystem I. $P700^+$ quenches antenna excited states just as effectively as does P700, although the mechanism is quite different. $P700^+$ probably quenches the antenna by forming an excited state of the $P700^+$ complex, $(P700^{+*})$ which then rapidly decays by nonradiative pathways, although this is not definitely established. All these effects combined lead to the result that quenching of excited states of the Photosystem I antenna pigments takes place at the same rate regardless of the redox state of both the acceptor complex and the donor chlorophyll. Therefore, no variable fluorescence is observed in Photosystem I.

References

Banks, J. M. (2017) Continuous excitation chlorophyll fluorescence parameters: A review for practitioners. *Tree Physiology* 37: 1128–1136.

Bernát, G., Steinbach, G., Kaňa, R., Govindjee, Misra, A. N., and Prašil, O. (2018) On the origin of the slow M–T chlorophyll *a* fluorescence decline in cyanobacteria: interplay of short-term light-responses. *Photosynthesis Research* 136: 183–198.

Björkman, O. and Demmig, B. (1987) Photon yield of O_2 evolution and chlorophyll fluorescence characteristics at 77 K among vascular plants of diverse origins. *Planta* 170: 489–504.

Butler, W. L. (1978) Energy distribution in the photochemical apparatus of photosynthesis. *Annual Review of Plant Physiology* 29: 345–378.

Dau, H. (1994) Molecular mechanisms and quantitative models of variable photosystem II fluorescence. *Photochemistry and Photobiology* 60: 1–23.

Duysens, L. N. M. and Sweers, H. E. (1963) Mechanism of two photochemical reactions in algae as studied by means of fluorescence. In: Japanese Society of Plant Physiologists (ed.) *Studies on Microalgae and Photosynthetic Bacteria*. Tokyo: University of Tokyo Press, pp. 353–372.

Genty, B., Briantais, J.-M., and Baker, N. R. (1989) The relationship between the quantum yield of photosynthetic electron transport and quenching of chlorophyll fluorescence. *Biochimica et Biophysica Acta* 990: 87–92.

Govindjee (1995) Sixty-three years since Kautsky: Chlorophyll *a* fluorescence. *Australian Journal of Plant Physiology* 22: 131–160.

Hohmann-Marriott, M. and Blankenship, R. E. (2007) Variable fluorescence in green sulfur bacteria. *Biochimica et Biophysica Acta* 1767: 106–113.

Horton, P., Ruban, A. V., and Walters, R. G. (1996) Regulation of light harvesting in green plants. *Annual Review of Plant Physiology and Plant Molecular Biology* 47: 655–684.

Kalaji, H. M., Schansker, G., Brestic, M., *et al.* (2017) Frequently asked questions about chlorophyll fluorescence, the sequel. *Photosynthesis Research* 132: 13–66.

Kautsky, H. and Hirsch, A. (1931) Neue Versuche zur Kohlensäureassimilation. *Naturwissenschaften* 19: 964.

Krause, G. H. and Weiss, E. (1991) Chlorophyll fluorescence and photosynthesis: The basics. *Annual Review of Plant Physiology and Plant Molecular Biology* 42: 313–349.

Laisk, A. and Oja, V. (2018) Kinetics of photosystem II electron transport: a mathematical analysis based on chlorophyll fluorescence induction. *Photosynthesis Research* 136: 63–82.

Lazár, D. (1999) Chlorophyll *a* fluorescence induction. *Biochimica et Biophysica Acta* 1412: 1–28.

Maxwell, K. and Johnson, G. N. (2000) Chlorophyll fluorescence – A practical guide. *Journal of Experimental Botany* 51: 659–668.

Ritchie, R. J. and Mekjinda, N. (2015) Measurement of photosynthesis using PAM technology in a purple sulfur bacterium *Thermochromatium tepidum* (Chromatiaceae). *Photochemistry and Photobiology* 91: 350–358.

Ruban, A. (2013) *The Photosynthetic Membrane: Molecular Mechanisms and Biophysics of Light Harvesting*. Chichester: Wiley.

Schreiber, U., Bilger, W., Hormann, H., and Neubauer, C. (1998) Chlorophyll fluorescence as a diagnostic tool: Basics and some aspects of practical relevance. In: A. S. Rahavendra (ed.) *Photosynthesis: A Comprehensive Treatise*. Cambridge: Cambridge University Press.

Schreiber, U., Hormann, H., Neubauer, C., and Klughammer, C. (1995) Assessment of photosystem II photochemical quantum yield by chlorophyll fluorescence quenching analysis. *Australian Journal of Plant Physiology* 22: 209–220.

Sipka, G., Kis, M., Smart, J. L., and Maroti, P. (2018) Fluorescence induction of photosynthetic bacteria. *Photosynthetica* 56: 125–131.

Stirbet, A. and Govindjee (2011) On the relation between the Kautsky effect (chlorophyll *a* fluorescence induction) and photosystem II: Basics and applications of the OJIP fluorescence transient. *Journal of Photochemistry and Photobiology B* 104: 236–257.

Stirbet, A. and Govindjee (2012) Chlorophyll *a* fluorescence induction: a personal perspective of the thermal phase, the J–I–P rise. *Photosynthesis Research* 113: 15–61.

Stirbet, A., Lazar, D., Kromdijk, J., and Govindjee (2018) Chlorophyll *a* fluorescence induction: Can just a one-second measurement be used to quantify abiotic stress responses? *Photosynthetica* 56: 86–104.

Strasser, R. J., Srivastava, A., and Govindjee (1995) Polyphasic chlorophyll *a* fluorescent transient in plants and cyanobacteria. *Photochemistry and Photobiology* 61: 32–42.

Chapter 12

Origin and evolution of photosynthesis

12.1   Introduction

In this chapter, we will examine the origin and evolution of photosynthesis. This issue is inextricably tied to the history of the Earth and the origin and evolution of life. As discussed in more detail below, there is compelling evidence that photosynthesis is an ancient metabolic process and arose relatively early in the history of life. The evolutionary history of photosynthesis is remarkably complex, with different classes of organisms and different parts of the photosynthetic apparatus having very different evolutionary paths. Understanding this process is complicated by the fact that the only organisms that we have to examine in detail are the ones that survive today. We may try to infer their relationship to ancient organisms that were their evolutionary precursors using the techniques of molecular evolution analysis, but this is often a very difficult process, especially when the length of time separating these organisms is measured in billions of years. Important information about the history of life and photosynthesis is provided by geological analysis including fossils of various sorts, geochemical studies, and biomarkers.

12.2   Early history of the Earth

The Earth is 4.55 billion years old. During the early period of the Earth's history, it was subject to intense bombardments of material left over from the creation of the Solar System (Sleep *et al.*, 1989; Arndt and Nisbet, 2012). This bombardment was so intense, and some of the objects were so large, that their impact was of sufficient energy to vaporize the oceans and raise the temperature of the Earth high enough that any incipient life would surely have been destroyed, effectively autoclaving the entire Earth. These bombardments finally tapered off at approximately 4 billion years ago, although the time of the last Earth-sterilizing impact is not known with certainty. The best evidence for these events comes from the cratering record from the Moon, which preserves impact craters essentially permanently, because, unlike the Earth, it has no active geological processes that reshape the surface. This intense early bombardment provided an "impact frustration" of the origin of life in that, even if life had begun, it would

Molecular Mechanisms of Photosynthesis, Third Edition. Robert E. Blankenship.
© 2021 Robert E. Blankenship 2021 by John Wiley & Sons Ltd.
Companion website: https://www.wiley.com/go/blankenship/molecularphotosynthesis3e

probably have been destroyed by these events (Maher and Stevenson, 1988; Sleep, 2018). The possibility that life began elsewhere and was then transported to Earth cannot be completely ruled out, a process known as panspermia. However, while there is no evidence yet of this, and also because it merely pushes the origin of life questions to some other location, we will not treat this scenario as useful to consider here.

The nature of the physical environment on the early Earth has been the subject of considerable discussion (Nisbet and Sleep, 2001; Arndt and Nisbet, 2012; Sleep et al., 2012). The solar radiation intensity was ~30% less than the Earth receives today, because the Sun had not yet reached its full intensity. However, evidence suggests that the temperature on Earth may not have been much different than it is today. This situation is called the "faint young sun paradox." The higher-than-expected temperatures may have been due to a very high CO_2 content of the atmosphere, which created a greenhouse effect that retained heat more efficiently, and also because heat from radioactive decay within the Earth also contributed to the environment. The early atmosphere was probably almost completely devoid of O_2 (Kasting, 1993; Goldblatt and Zahnle, 2011; Catling and Zahnle, 2020). Current thinking is that the early atmosphere was not strongly reducing, but rather more neutral in redox balance. The atmosphere probably originated largely from volcanic outgassing.

12.3   Origin and early evolution of life

Despite the inhospitable conditions of the early Earth, life did begin and persist (Zubay, 2000; Deamer and Szostak, 2010; Muchowska et al., 2020). Exactly, how this took place is one of the great unsolved questions of science. Important early contributions to thinking in this area were made by the Russian scientist A. I. Oparin and British scientist J. B. S. Haldane (Miller et al., 1997).

They were among the first to address this question in a scientific manner, proposing what is now often called the "chicken soup" idea for the prebiotic environment, in which an aqueous solution of a mixture of prebiotically formed organic chemicals served as the building blocks for the earliest life forms. Where did the prebiotic organic matter come from? The American chemists Stanley Miller and Harold Urey performed a classic experiment in which electrical discharges were passed through a simulated reducing early Earth environment consisting of H_2, NH_3, and CH_4, all in a sealed container with water (Miller, 1953). After several days of reaction, they analyzed the solution and found that a remarkably rich collection of organic molecules had been formed, including a number of amino acids and even some nucleotide bases. When this same experiment is performed under the less reducing conditions now thought to have been present on the early Earth, less organic matter is produced, so it is not clear whether this mechanism could have provided sufficient prebiotic organic material. An interesting re-evaluation of some of Miller's archived samples using modern techniques revealed an even wider range of organic material than the original analysis found (Bada, 2013).

Other possible sources for organics include delivery from meteorites, especially from a class of carbon-rich meteorites known as carbonaceous chondrites, or from the infall of interstellar dust particles (Oro et al., 1990). Another potential source of prebiotic organic matter is deep-sea hydrothermal vents, where the energy-rich hydrothermal fluid can synthesize some organic molecules from CO_2 and H_2 (Wächterhäuser, 1990; Martin et al., 2008). In addition, a variety of other possible prebiotic environments can be envisaged, including shallow ponds in which periodic wet and dry conditions can drive certain types of synthesis (Damer and Deamer, 2020).

Regardless of the source of the prebiotic organic matter, life probably began when a chance assemblage of these molecules acquired the ability to self-replicate. Eventually, a membrane enclosed the newly formed cell, and the beginnings of metabolism and information storage developed.

There are huge gaps in the story of the origin and early evolution of life. It is not clear that a sufficient variety of organic molecules could be made by prebiotic chemistry to provide all the essential building blocks of life. The earliest life forms have almost certainly completely vanished from the Earth, devoured by later, more sophisticated, organisms, so we may never know the details of the origin of life. This problem is further complicated by the fact that, unlike many branches of science, the origin of life has a historical component that precludes most types of controlled experiments. There were probably many random events, or "frozen accidents," which were perpetuated in life, but are not necessarily the way that life would develop a second time. These include developments such as the selection of the identities of the essential biological molecules such as the amino acids, nucleotide bases, sugars, and lipids that make up cells, as well as the chiralities of the asymmetric molecules such as sugars and amino acids (Davankov, 2018). However, we cannot judge how much of the developmental progression of life is historical as opposed to deterministic, because we only have one example to study: life on Earth. The unity of life at the molecular level very strongly suggests that life on Earth has a single origin, or, more precisely, that only a single form of life survived to the present day.

An important concept in thinking about the origin and early development of life is the idea of the **RNA world**. In modern organisms, RNA is intermediate between the information storage molecule DNA and the catalytic proteins. RNA is itself capable of catalyzing a number of chemical transformations and is a temporary information storage molecule that results from the transcription of DNA prior to its translation into proteins. RNA can carry out both information storage and catalysis, although it is not as efficient at either as DNA and proteins, respectively. The idea of an RNA world, in which early life consisted of self-replicating RNA molecules that were assembled from prebiotic chemicals, was first proposed by Wallace Gilbert shortly after the discovery of the catalytic properties of RNA by Thomas Cech and Sidney Altman (Gilbert, 1986). This idea has been embraced by most researchers in the origin of life studies, despite significant problems with the stability of the constituents of RNA and the lack of plausible prebiotic synthetic pathways for many of these components (Orgel, 1994, 1998; Vázquez-Salazar and Lazcano, 2018). However, there has been significant progress in achieving the synthesis of ribonucleotides under plausible prebiotic conditions and polymerization to form RNA molecules (Powner et al., 2009; Benner et al., 2012; Sutherland, 2017).

Additional critical developments that had to take place for life to resemble its current form include the conversion to the DNA and protein world, which requires the development of the genetic code and the origin of translation from RNA to protein. Finally, cellular membranes encapsulated the protocell, providing a distinction between the inside and the outside of the cell and permitting the development of concentration gradients across the membrane, which is essential for modern metabolism (Griffiths, 2007; Szostak, 2011; Deamer, 2017). The exact order in which developments took place is not known, and several steps may have taken place in parallel rather than in a distinct order, but might possibly have followed the progression shown in Fig. 12.1. This eventually produced a cell that has been termed the **last universal common ancestor** (LUCA) of all life on Earth. This cell had a DNA/RNA/protein information transfer system, an extensive set of enzyme-catalyzed reactions that constituted a complex metabolism, and most of the core capabilities that we find in modern-day bacteria. In short, LUCA was a moderately complex, sophisticated cell, well beyond the early stages represented by the RNA world (Woese, 1987; Penny and Poole, 1999; Doolittle, 2000; Becerra et al., 2007; Koonin et al., 2020).

LUCA subsequently diverged into three lines of cells that survived to the present day: bacteria, archaea, and eukaryotes (Chapter 2). Although this scenario is appealing in its simplicity, it is a gross oversimplification, possibly to the point of being largely wrong. Early cells undoubtedly exchanged genetic information very readily, and this has continued to a greater or lesser degree until the present

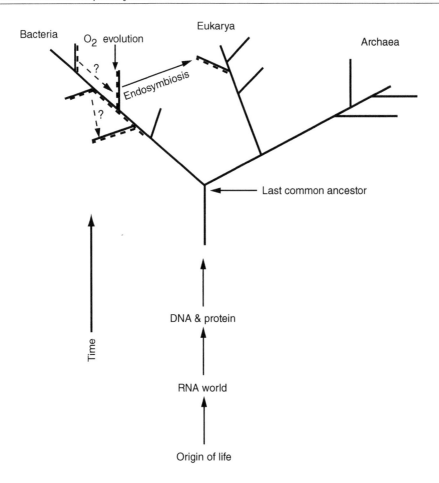

Figure 12.1   Schematic picture of the origin and early evolution of life. The solid line is the evolutionary tree based on small subunit RNA. Dashed lines indicate photosynthetic cells. Dashed arrows indicate possible horizontal gene transfer events.

day in the bacteria and archaea, although probably not so much in the eukaryotes. This extensive **lateral gene transfer** (LGT) (also called **horizontal gene transfer** (HGT)) transforms the tree of life into a multi-stalked bush, as illustrated in Fig. 12.2 (Doolittle, 1999, 2000; Martin, 1999). So, LUCA may not have really existed as a single type of organism but rather may have been a population of diverse cells that freely exchanged genetic information, in essence, a meta-organism. There is considerable evidence for substantial amounts of HGT during the evolution of photosynthesis (Raymond *et al.*, 2002).

12.4   Geological evidence for life and photosynthesis

The only way to probe the early Earth directly is by using techniques of geology, including paleontology, geochemistry, and mineralogy. Collectively, the application of geological methods to understand early life is called geobiology (Knoll *et al.*, 2012).

The oldest presumptive experimental evidence for life on Earth is in the form of an isotope fractionation of carbon in 3.95-billion-year-old rocks

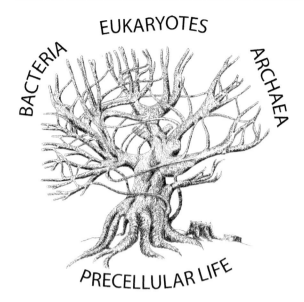

EUKARYOTES

BACTERIA

ARCHAEA

PRECELLULAR LIFE

Figure 12.2 An alternative picture of the tree of life, incorporating massive lateral gene transfer, especially in the early period of evolution. *Source:* Courtesy of W. Ford Doolittle.

from Canada, which are the oldest known rocks to survive on Earth (Tashiro *et al.*, 2017). These rocks contain inclusions in which the carbon is isotopically light compared with inorganic carbon. Isotopes are variants of the same element with different numbers of neutrons. Almost all carbon on Earth is ^{12}C, with approximately 1% being ^{13}C. Chemical reactions discriminate slightly between the two isotopes, a phenomenon called a **kinetic isotope effect**. The kinetic isotope effect originates because of the mass difference between isotopes, with the result that the vibrational frequency of the molecule with the heavier isotope will be lower (Eq. (A.61)). This lowers the zero-point vibrational energy, which increases the activation energy and slows the reaction. All biological carbon fixation mechanisms have a preference for the lighter isotope, so biogenic carbon is isotopically lighter than inorganic carbon. The effect is subtle, so the selectivity (δ^{13}C) is reported in mil percent ‰ or parts per thousand. Typical isotope fractionation in photosynthetic carbon fixation in modern organisms is about −25‰, meaning that the organic

matter produced during photosynthesis is 25 parts per thousand enriched in the lighter isotope than the inorganic carbon pool. These ancient rocks contain carbon with δ^{13}C = −30 to −37‰, which is considered to be evidence for some sort of autotrophic life 3.95 billion years ago, although purely chemical explanations cannot be rigorously ruled out. These putative autotrophic cells assimilated CO_2 to form organic matter. There is no reliable way to know if this life was photosynthetic, although most of the evidence argues that the earliest life forms were probably nonphotosynthetic. Note that the presumed earliest life forms described above in the origin of life scenario are heterotrophs, not autotrophs, in that they did not fix CO_2 but, rather, utilized the organic molecules already present in the prebiotic broth (see Chapter 2 for discussion of types of metabolisms). The 3.95-billion-year-old organisms would then correspond to a somewhat later form of life that developed after the prebiotic stores of chemicals had been exhausted and the switch to an autotrophic metabolism had taken place. It has also been proposed that the earliest life forms were autotrophic nonphotosynthetic (Wächterhäuser, 1990) or photosynthetic cells (Granick, 1965; Hartman, 1975, 1998; Mauzerall, 1992).

The overall structure of the tree of life shown in Fig. 12.1 argues against the idea that the earliest life forms were photosynthetic. All cells that are capable of chlorophyll-based photosynthesis are found in the bacterial domain, with the exception of eukaryotic phototrophs, which were unquestionably formed by the lateral transfer event of endosymbiosis (see Section 12.13). Based on this fact, it is reasonable to propose that photosynthesis arose somewhere in the radiation of bacterial species, and that the last common ancestor was not a photosynthetic cell. Of course, we cannot rule out the possibility that photosynthesis was invented at an earlier stage and that, for unknown reasons, the capability was lost in all but the bacterial lineages. However, this seems to be an *ad hoc* assumption that introduces unneeded complexity.

The first direct evidence of life on Earth is from 3.5-billion-year-old stromatolites and microfossils found in Australia (Schopf, 1992, 1993, 2011). Stromatolites are layered dome-shaped structures that are formed when there is a cycle of deposition of sediment and biogenic activity at the surface. Most evidence suggests that the biological activity present when the stromatolites were formed was photosynthetic. A few environments on Earth today, such as hypersaline lagoons, exhibit the formation of stromatolite-like structures, and the dominant life form in those settings is always photosynthetic organisms such as cyanobacteria. The migration of organisms toward the light pitted against the influx of sediments gives rise to the characteristic layered, domed shape. These ancient stromatolite-forming organisms were almost certainly photosynthetic, as phototrophs are the only modern-day organisms that form similar structures. Laminated formations in South Africa and Australia from 3.4 billion years ago (BYA) have been interpreted in terms of reflecting anoxygenic phototrophs that probably used H_2 as an electron donor (Tice and Lowe, 2004; Schopf et al., 2017).

Microfossils that originated approximately 3.5 BYA have also been found in other rocks of the same age from the same area in Australia. These are fossilized casts of bacteria that have preserved the shape of the organism, although the organic matter has been replaced almost entirely by minerals. While it is thus far impossible to be sure of the metabolic capabilities of these ancient organisms, Schopf (1993) has argued that they are morphologically remarkably similar to certain groups of modern cyanobacteria, raising the possibility that these organisms were capable of oxygen evolution. The nature of the organism that gave rise to these microfossils has been controversial and questions even about whether they are biogenic in origin have been raised (Brasier et al., 2002), although current evidence suggests that they are indeed biogenic (Schopf, 2011; Schopf and Kudryavtsev, 2012).

The first evidence that indicates almost conclusively that oxygen-evolving organisms, presumably cyanobacteria, were present comes from the analysis of sulfur isotope fractionation in 2.4-billion-year-old rocks, which indicates that by that time free molecular oxygen was present in the atmosphere (Farquhar et al., 2000). While this time is generally (although not universally) accepted as the latest date by which oxygenic photosynthesis appeared, since no other known biotic or abiotic process is capable of producing so much free oxygen, it is usually argued that oxygenic photosynthesis appeared significantly earlier and it took a substantial period of time for enough oxygen to be produced to begin to accumulate in the atmosphere (Buick, 2008; Catling, 2014; Lyons et al., 2014; Catling and Zahnle, 2020).

The cyanobacteria, along with their oxygenic form of photosynthesis, had therefore almost certainly evolved prior to 2.4 BYA, although it is possible that they were already present at 3.5 BYA, as discussed above. This represents the latest date for the advent of oxygenic photosynthesis. However, considerable evidence suggests that O_2 evolution may have started significantly earlier, but that it took some time for free O_2 to accumulate to significant levels in the atmosphere. A date of 2.7 BYA for the appearance of cyanobacteria is often suggested (Buick, 2008). The buildup of atmospheric O_2 depended both on the production of O_2 and on the burial of reduced carbon compounds produced at the same time (Des Marais et al., 1992). The burial may depend on geological forces quite independent of life, such as plate tectonics and continent building. In addition, the early ocean contained large amounts of Fe^{2+}, which would have been oxidized immediately to Fe^{3+} by oxygen produced in the water column before it could diffuse to the atmosphere. This effect is thought to be the source of at least some of the banded-iron rock formations that were laid down during the geological period from 3.5 to 1.8 billion years ago. It is also possible that anoxygenic phototrophic bacteria were involved in the large-scale oxidation of Fe^{3+}. Some purple bacteria now can carry out this metabolism (see Section 12.5).

The subsequent history of the atmosphere has seen a gradual increase in the amount of O_2 and a decrease in the amount of CO_2 so that the latter is now a trace gas, as illustrated in Fig. 12.3 (Catling and Zahnle, 2020). This drawdown of CO_2 is due to its fixation by photosynthetic organisms and

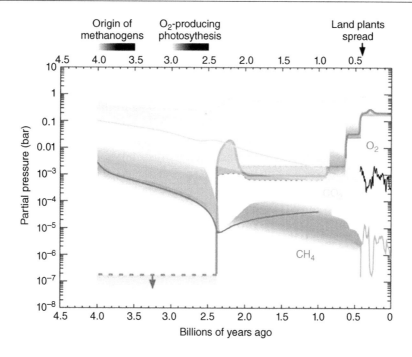

Figure 12.3   The composition of the Earth's atmosphere as a function of geological time. Curves are shown for O_2 (green), CO_2 (yellow), CH_4 (orange) and N_2 (blue). Uncertainties are indicated by shading. *Source:* Catling and Zahnle (2020)/American Association for the Advancement of Science. Licensed under CC BY 4.0.

subsequent burial of the reduced carbon organic matter. The increase of O_2 content in the atmosphere ultimately led to the present level of 21%, although there have been some notable fluctuations from that value, in particular, a spike in O_2 level to about 35% during the Carboniferous era, about 300 million years ago (Berner *et al.*, 2000). This spike is thought to have resulted from the evolution of land plants at about that time, which may have produced new types of biomass that was not readily degraded by existing organisms. This is also the period during which massive amounts of coal were being deposited, consistent with it being a time of high plant productivity. One problem that the plants of that era must have had to cope with was the probable significant losses due to photorespiration, because the high oxygen level would have enhanced oxygenation compared with carboxylation by rubisco (Chapter 9). This was well before C4 photosynthesis appeared in the fossil record, so the oxygenation reaction of rubisco could not be avoided in that manner. The CO_2 content of the atmosphere was also very low during this period, so this would not have reduced the oxygenation effect. The specificity factor for rubisco at this time is not known, but was probably lower than that found in modern plants. This further compounds the problem, because with a lower specificity factor, more oxygenation and less carboxylation take place. Therefore, plants almost certainly operated at a substantially lower efficiency during this period.

There is something of a paradox concerning the early fossil evidence for photosynthesis. As discussed above, the earliest known direct evidence for life on Earth is almost certainly photosynthetic, and such organisms are often interpreted as closely related to cyanobacteria. Cyanobacteria are remarkably complex cells in terms of photosynthetic capability, and they certainly do not occupy an early-branching position on the tree of life. It is inconceivable that the cyanobacteria were the first photosynthetic cells, as they contain most of the innovations that characterize the most advanced forms of photosynthesis. Anoxygenic photosynthetic bacteria are by any measure much more primitive as regards the structure of their photosynthetic apparatus and their lack of the pinnacle of photosynthetic capability, the oxygen-evolving complex. The earliest forms of photosynthesis must have been even more primitive than these cells, and the development of the major aspects of photosynthetic metabolism almost certainly culminated in the cyanobacteria; it did not begin with them.

If we accept this argument, then we are forced to conclude that the earliest forms of photosynthetic cells, and all even more primitive forms of life, are not at all well represented in the known fossil record. The origin of life and its progression through all the primitive stages to the last common ancestor, the divergence of the three domains, the development and evolution of the bacterial domain, including the advent of photosynthesis, must all have taken place between 4.0 and 3.5 BYA. The refinement of the process of photosynthesis to the advanced oxygenic form found in cyanobacteria must have taken place somewhere between 3.5 and 2.4 BYA, with the time of 2.7 BYA as a likely time for that transition, which is often called a "singular event," since it appears to have happened only once and had such a profound impact on the subsequent history of both life and the physical Earth.

The evolutionary history of the cyanobacteria has been discussed extensively (Fischer *et al.*, 2016; Shih *et al.*, 2017; Martin *et al.*, 2018; Sanchez-Barracaldo and Cardona, 2019). Outside of some of the photosynthesis-related genes, they do not exhibit any obvious evolutionary relationship to any other group of photosynthetic prokaryotes. The cyanobacteria are most closely related to another group of bacteria, the melainabacteria (Blankenship, 2017; Soo *et al.*, 2017; Grettenberger *et al.*, 2020). However, the melainabacteria do not exhibit any evidence of ever having been phototrophic or of containing any genetic material that relates to phototrophy. It therefore seems clear that the cyanobacteria invented, or more likely acquired by horizontal gene transfer, the ability to carry out

photosynthesis after their divergence from the melainabacteria. How the acquisition of photosynthesis by cyanobacteria relates to the question of linked photosystems is discussed below in Section 12.10.

Another apparent paradox is the discovery that enzyme systems that either use O_2, such as various oxidases, or protect against its reactive byproducts, such as superoxide or hydrogen peroxide, are found widely distributed throughout the tree of life. This suggests that these enzyme systems were present in the last common ancestor, which we earlier argued was not even photosynthetic, let alone oxygenic. So, the ability to use or protect against oxygen appears on the surface to have been present prior to the ability to make oxygen (Castresana and Moreira, 1999; Brochier-Armanet *et al.*, 2009; Ducluzeau *et al.*, 2014). There are two possible explanations for this apparent paradox. First, low levels of O_2 and other reactive oxygen species were almost certainly produced on the early Earth by nonbiological processes such as UV photolysis of water, so even the earliest cells may have needed protection from these toxic species. Second, the ability to use and cope with oxygen species is such a huge advantage for any life form that early enzyme systems that developed in response to the advent of oxygenic photosynthesis may have been very widely disseminated by horizontal gene transfer.

12.5   The nature of the earliest photosynthetic systems

What was the nature of the earliest form of photosynthesis and what might have been its evolutionary antecedents? Unfortunately, there is little definitive information to constrain our thinking on this question. All known existing photosynthetic organisms are highly sophisticated cells, far removed from the first forms. We will base this discussion on extrapolations from what seem to be the constant features of all photosynthetic organisms, but this will carry us only to relatively primitive photosynthetic systems, not to the very earliest forms. Therefore, what will be presented here is more speculative than the rest of our discussion and should be considered to be a plausible scheme, but only one of many such possible schemes.

Early ideas regarding the origin of photosynthesis were proposed by Granick (1965) and refined by Olson and Pierson (1987), Mauzerall (1992), and Olson (1999, 2006). Membranes are central to all modern photosynthetic organisms, so it seems likely that photosynthesis began as a membrane phenomenon. One proposed early primitive reaction center is shown in Fig. 12.4a. The proposed reaction center consists of a porphyrin, probably a simple symmetric molecule, which, upon excitation, transfers an electron from a soluble Fe^{2+} donor ($FeOH^+$ in mildly alkaline aqueous solution) to a membrane-bound Fe–S acceptor and, finally, to a soluble low-potential acceptor similar to Photosystem I-type reaction centers. An alternative proposal, also shown in Fig. 12.4a, is a cyclic system in which the final acceptor is a quinone. Both protons and electrons are driven in a cyclic manner, as occurs in contemporary purple bacteria. Probably only one of these two possibilities was actually the case in the early system. Figure 12.4b shows a proposed more advanced reaction center, in which chlorophyll-type pigments have been incorporated,

as well as a cytochrome, which serves as a soluble cyclic electron carrier.

Fe^{2+} as an early electron donor is appealing, based on the fact that the early ocean is thought to have contained a large concentration of reduced Fe, so it would presumably have been available on the early Earth, first to be replaced by other reduced compounds such as H_2S and eventually to be replaced as a donor by the ubiquitous, but difficult to oxidize, H_2O. Purple photosynthetic bacteria that have the capability to oxidize Fe^{2+} have been identified (Widdel et al., 1993; Bird et al., 2011).

Other proposals for the possible evolutionary precursors of photosynthesis are phototactic systems (Nisbet et al., 1995) or ultraviolet protection systems (Mulkidjanian and Junge, 1997). However, there is no direct evidence to support these proposals. There is, nevertheless, an evidently deep connection between photosynthesis and nitrogen fixation, in that some of the enzymes that carry out key steps of chlorophyll biosynthesis are clearly homologous to nitrogenase. This is discussed in more detail below in Section 12.7.

A very primitive reaction center probably did not have an associated antenna, which were later added multiple times to improve the efficiency of the light collection in various environments. Whether the primitive organism that contained

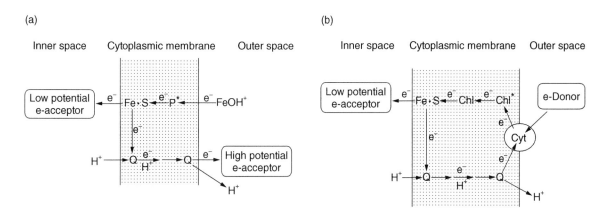

Figure 12.4   Proposed primitive reaction centers. (a) A possible very primitive reaction center. (b) An improved reaction center. Fe•S, iron–sulfur electron acceptor; Q, quinone; Cyt, cytochrome. *Source:* Olson and Pierson (1987)/Springer Nature.

this complex was capable of autotrophic carbon fixation is unclear, but a photoheterotrophic metabolism, in which light is used merely as a supplemental energy source, seems to be simpler. Autotrophic metabolism may have come later, as the available organic matter became scarce.

12.6 The origin and evolution of metabolic pathways with special reference to chlorophyll biosynthesis

How do metabolic pathways originate and evolve? This is an issue that goes well beyond photosynthesis, but certainly includes photosynthesis as an example of a metabolic pathway, albeit an extraordinarily complex one. Several conceptual frameworks have been proposed for the origin and evolution of metabolic pathways (Lazcano and Miller, 1999; Muchowska et al., 2020). The first of these is known as the retrograde hypothesis (Fig. 12.5a) (Horowitz, 1945). The basic idea is that early organisms lived off prebiotic organic matter and originally had no need for biosynthetic pathways. As they gradually depleted the more complex prebiotic compounds, they evolved a mechanism to synthesize them from simpler components that were still available. When these also began to run out, the synthesis was pushed back to even simpler compounds; so, the pathway evolved backward from the most complex to the simplest compounds, hence the term "retrograde." This concept works only if the desired end product of the pathway was available in the prebiotic soup.

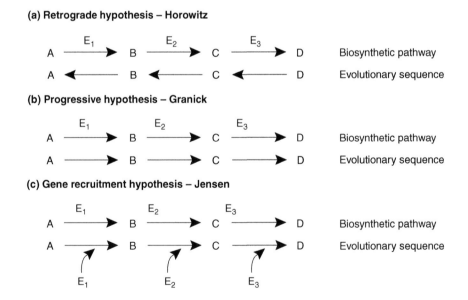

Figure 12.5 Evolution of metabolic pathways. (a) Retrograde hypothesis of Horowitz. As prebiotic resources are depleted, the metabolic pathway evolves backward to produce simpler precursors. This mechanism requires that the end products of the pathway are initially available in the environment. (b) Progressive hypothesis of Granick. More complex compounds are elaborated from simpler ones, so the pathway evolves forward. This mechanism requires that each intermediate in the pathway be useful to the organism as an end product. (c) Gene recruitment hypothesis of Jensen. New pathways originate by a series of recruitments of previously existing enzymes, with modified function. The efficiency is refined over time.

Just the opposite idea was proposed by Granick; this is known as the Granick, or progressive, hypothesis (Fig. 12.5b) (Granick, 1965). He argued that it is not conceivable that highly complex molecules such as chlorophylls were synthesized by prebiotic chemistry, given their very specific functional groups and multiple chiral centers. Instead, they are the end product of progressive evolutionary development, in which simple molecules are the start of the biosynthesis chain and are then progressively elaborated in later steps. In this view, each intermediate in the modern pathway was at some point the endpoint in the pathway. This requires that each intermediate in the modern pathway be usable in the past as an end product. In the case of chlorophyll biosynthesis, Granick proposed that simple porphyrins or porphyrin precursors were the starting points, and that successive steps were added to improve the efficiency of the pigments or to extend light absorption into new spectral ranges. This is an appealing idea and is probably at least partially true. The Granick hypothesis in the context of photosynthesis has been championed by Mauzerall (1992), as well as embraced by others (Olson and Pierson, 1987; Olson, 1999, 2006; Martin *et al.*, 2018).

A third possible mechanism for the origin and evolution of metabolic pathways is the patchwork, or gene recruitment, hypothesis, as shown in Fig. 12.5c (Jensen, 1976). The basic idea is that new pathways are formed by recruiting genes that evolved earlier for another pathway. The newly formed pathway is therefore a patchwork, or mosaic, of steps borrowed from other sources, which were then adapted to their new function by mutation of the genes that code for the proteins. This requires that the gene first be duplicated and that one copy be used for its original function, while the other can then be modified and optimized for the new pathway. This proposal does not address how the original pathways evolved, but instead focuses on how an organism that already has an extensive metabolism might extend it into new regimes.

These three conceptual ideas for the evolution of metabolic pathways are not necessarily mutually exclusive, and it may well be that the retrograde hypothesis applies for some of the early steps, the progressive hypothesis for some of the later steps, and the gene recruitment hypothesis at selected places all along the pathway. In fact, there is now some evidence that this compromise view has validity for the chlorophyll biosynthetic pathway, as discussed next.

12.7 Origin and evolution of photosynthetic pigments

One specific area where the issue of metabolic pathway evolution has been widely discussed is the question of whether chlorophyll or bacteriochlorophyll is the more primitive pigment. This might be revealed by examining the pathway of biosynthesis of these pigments, which might then be a proxy for the overall development of photosynthesis, according to the Granick hypothesis. The modern-day biosynthetic pathway for chlorophylls was described in Chapter 4. It consists of the buildup of tetrapyrroles from smaller substituents, the conversion to fully conjugated porphyrins, and then the elaboration of the porphyrins into the more asymmetric chlorophylls, with a variety of substituents added along the way. The order of the biosynthetic steps and the identities of the enzymes that carry out the transformations have been well established for both chlorophyll *a* and bacteriochlorophyll *a* synthesis by a combination of biochemical and genetic experiments (Beale, 1999; Chew and Bryant, 2007; Bryant *et al.*, 2020).

It is difficult to imagine that the earliest precursors in the chlorophyll biosynthetic pathway were ever useful as photopigments, as they are colorless compounds. So, the early steps could not have evolved by the strict application of the progressive hypothesis, but could possibly have evolved backward to replace prebiotically available porphyrins and their precursors (Taniguchi *et al.*, 2012). The early steps of the biosynthetic pathway are almost certainly much older than the later steps and have clearly been recruited from heme biosynthesis. The later steps are more likely to have evolved by the progressive mechanism, as the concept that the earliest photosynthetic organisms utilized simpler pigments that were later improved by evolution seems logical.

In the biosynthetic pathway of these pigments (Fig. 4.6) chlorophyll is a precursor of bacteriochlorophyll. According to the strict interpretation of the Granick hypothesis, this implies that the bacteriochlorophyll-containing organisms are more recently evolved than are the chlorophyll-containing organisms. This seems inconsistent with the clearly simpler structure and mechanism of the bacteriochlorophyll-containing organisms. Many lines of evidence strongly suggest that these bacteriochlorophyll-containing organisms are more closely related to the early primitive phototrophs than are the chlorophyll-containing organisms, all of which contain two photosystems and the capability of oxygen evolution. This is discussed in more detail in the next section. However, a simple solution to this problem has been suggested that also provides strong evidence for the gene recruitment hypothesis.

The later steps of the chlorophyll/bacteriochlorophyll biosynthetic pathway are shown in Fig. 12.6. The first reductive step in all cases is the reduction of ring D. Biosynthesis of bacteriochlorophyll a is nearly identical to chlorophyll synthesis through the chlorophyllide a intermediate, but contains two additional steps: the reduction of ring B and the conversion of the vinyl group at C3 to an acetyl moiety.

In all anoxygenic photosynthetic organisms, ring D reduction is carried out in a reaction by an enzyme complex known as light-independent protochlorophyllide oxidoreductase (LI-POR) consisting of the BchL, BchN, and BchB proteins (Chapter 4). In higher plants, an entirely unrelated light-driven enzyme known as light-dependent protochlorophyllide oxidoreductase (LD-POR) carries out the same step, and the light-independent pathway is absent.

Lower plants, algae, and oxygenic photosynthetic prokaryotes contain both enzyme systems. The light-independent complex in these oxygenic organisms consists of the ChlL, ChlN, and ChlB proteins. This distribution pattern clearly indicates that the light-independent enzyme complex is the more primitive system. Therefore, in this discussion, we will focus our attention primarily on it.

In the anoxygenic bacteria that contain bacteriochlorophyll a, a second enzyme complex carries out the reduction of ring B (Fig. 12.6). The proteins that comprise this complex are the BchX, BchY, and BchZ proteins. The complex is known as the light-independent chlorophyllide reductase (LI-COR). They exhibit a distant, but unambiguous, sequence similarity to the BchL, BchN, and BchB proteins and almost certainly result from an ancient gene duplication and divergence (Armstrong, 1998). In addition, all these proteins also show a similarity to three proteins involved in nitrogen fixation: the NifH, NifD, and NifK proteins. The NifH protein, also called the Fe-protein, exhibits a significant (~40%) degree of identity to the ChlL, BchL, and BchX proteins, with the most highly conserved regions corresponding to the functionally important regions of the Fe-protein. The NifD and NifK proteins, which together make up the nitrogenase Mo–Fe protein, have a weaker, but still clearly apparent, similarity to the ChlN, ChlB, BchN, BchB, BchY, and BchZ proteins. The NifD and NifK proteins are also similar to each other, suggesting that they originated from an ancient gene duplication and divergence. The same is true for the various pairs of the chlorophyll reductases. The exact number and timing of the gene duplications that gave rise to the current pattern of relationships among all these proteins are not clear (Raymond et al., 2004; Boyd et al., 2011). The LI-POR chlorophyll biosynthesis enzyme system has been found to have a complex evolutionary path including significant episodes of HGT (Gupta, 2012; Sousa et al., 2013).

The crystal structures of the LI-POR complex from the purple bacterium *Rhodobacter capsulatus* (Muraki et al., 2010) and the cyanobacterium *Prochlorococcus marinus* (Moser et al., 2013) have been solved and clearly support the idea that these enzymes are evolutionarily related to nitrogenase. Structural features of the *P. marinus* enzyme and its relationship to nitrogenase are shown in Fig. 12.7.

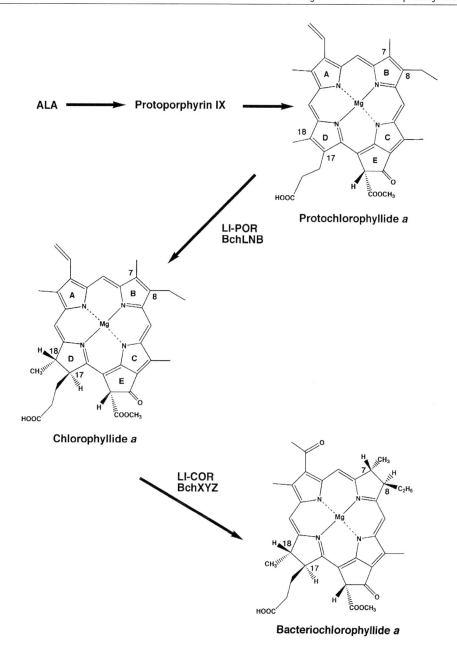

Figure 12.6 Ring reduction steps of the biosynthetic pathway for chlorophyll and bacteriochlorophyll. The light-independent protochlorophyllide reductase (LI-POR) reduces the double bond between C-17 and C-18 in protochlorophyllide to form chlorophyllide. This reaction is catalyzed by the BchLNB enzyme complex. For bacteriochlorophyll biosynthesis, an additional reduction of the double bond between C-7 and C-8 is catalyzed by the BchXYZ enzyme complex.

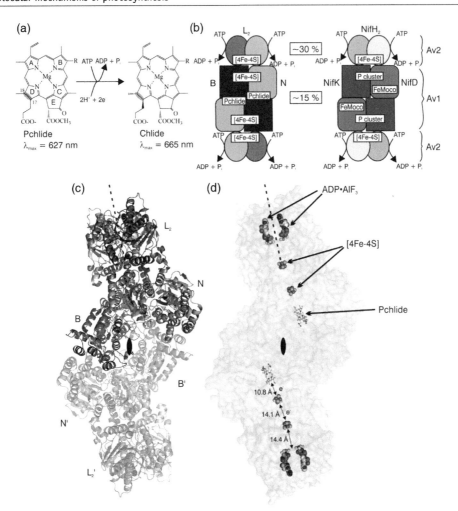

Figure 12.7 Chemistry and structure of the light-independent protochlorophyllide reductase enzyme from *Prochtorococcus marinus*. (a) Chemical reaction catalyzed by LI-POR. (b) Subunit structure of the LI-POR enzyme (left) compared to nitrogenase (right). (c) Structure of the LI-POR complex, consisting of the ChlL, ChlN, and ChlB proteins. (d) Cofactors of the LI-POR enzyme, showing the pathway of electron transfer. *Source:* Moser *et al.* (2013)/National Academy of Sciences.

Sequence analysis of the reductase enzymes discussed above concluded that the data supported the interpretation that a series of ancient duplication events gave rise to the line of descent that led to the chlorophyll-like pigments, and that then a second duplication gave rise to the two distinct classes of reductases: those that reduce ring B (BchX) and those that reduce ring D (BchL and ChlL) (Burke *et al.*, 1993). The ancient reductase was proposed to be nonspecific and to reduce protochlorophyllide all the way to bacteriochlorophyllide in two successive turnovers of the same enzyme. The second duplication and divergence permitted a more specific, and therefore probably more efficient, set of enzymes to develop, which were specialized to reduce just one of the two rings. This set of enzymes was subsequently lost, leading to organisms that make chlorophyll and not bacteriochlorophyll.

This scenario, while not verified by reconstruction of the ancient enzymes, explains the apparent paradox of the Granick hypothesis and the pigment content of different groups of photosynthetic organisms (Lockhart *et al.*, 1996a).

The origin and evolution of the light-independent protochlorophyllide reductase appear to constitute a clear case of gene recruitment, in that the enzymes for an existing metabolic pathway, nitrogen fixation, or possibly a precursor reductase enzyme (Raymond *et al.*, 2004), have been modified to carry out one of the key steps of chlorophyll biosynthesis.

In addition to the evolutionary connection between photosynthesis and nitrogen fixation exemplified by the similarity of the Nif and LI-POR enzymes, both also have a distant similarity to an enzyme important in methanogenesis, the reductase enzyme that is involved in the biosynthesis of the Ni-containing cofactor F430 (Staples *et al.*, 2007; Zheng *et al.*, 2016; Ghebreamlak and Mansoorabadi, 2020). Thus, these three critical biochemical pathways share common ancestors, although the precise relationship has not yet been established.

A second likely example of gene recruitment is the step in which Mg is inserted into the porphyrin ring of the partially assembled chlorophyll. In this case, the enzyme that carries out this step is clearly related to presumably more ancient enzymes that insert Co or Ni into porphyrins (Walker and Willows, 1997). In addition, the first portion of the chlorophyll biosynthetic chain, up to the metal insertion step, is certainly adapted from the essentially identical pathway of heme biosynthesis. Heme biosynthesis is almost certainly a more ancient pathway, as it is found in all three domains of life, so presumably was already present in LUCA (Bryant *et al.*, 2020).

Some discussion of the position of chlorophyll *c* in the evolutionary development of chlorophylls is necessary (Larkum, 2006). Chlorophyll *c* (Fig. 4.3) is clearly the simplest chlorophyll-type pigment, because it is a porphyrin rather than a chlorin (ring D is not reduced), and it has no tail. One might imagine that it is the most primitive chlorophyll,

and indeed it, or a closely related pigment, may have been present in ancient photosynthetic organisms. However, the distribution and function of chlorophyll *c* do not suggest that it is a primitive pigment. Chlorophyll *c* is found only in oxygenic photosynthetic organisms, and almost exclusively in eukaryotic algae such as kelp and dinoflagellates. In all cases, the chlorophyll *c* is exclusively an antenna pigment; it has never been found associated with a reaction center complex. For these reasons, it seems most likely that chlorophyll *c* is an adaptation to a particular environment in which its light absorption properties are well suited to the collection of light in the water column (Chapter 1).

The biosynthetic pathway of chlorophylls tells us much about the evolution of photosynthesis. It clearly suggests that many, if not most, of the genes that code for the biosynthetic enzymes were recruited from other metabolic pathways and adapted for use in chlorophyll biosynthesis. This view is also consistent with the assertion made earlier that photosynthesis was not present in the earliest cells and arose only after the divergence from the last common ancestor. It seems likely that this mosaic pattern of ancestry is also true for many, if not most, of the other constituents of the photosynthetic apparatus (Xiong *et al.*, 2000; Hohmann-Marriott and Blankenship, 2011).

12.8 Evolutionary relationships among reaction centers and other electron transport components

Structural and mechanistic aspects of reaction centers and other electron transfer carriers from various classes of photosynthetic organisms are discussed in detail in Chapters 6 and 7. Figure 12.8 summarizes much of this information in a grand comparison of the energetics of electron transport in these systems. A pattern is immediately recognizable in this diagram: namely, that there are two broad classes of reaction centers, those that resemble Photosystem I and those that resemble Photosystem II. These two classes of reaction centers are therefore designated Type I and Type II. Alternatively, Type I reaction centers are called Fe–S-type centers, and Type II are called pheophytin–quinone reaction centers (also sometimes called Q-type). A wealth of sequence data, biophysical comparisons, and structural similarities all indicate clearly that reaction centers within a class have an evolutionary relationship to each other (Olson and Pierson, 1987; Mathis, 1990; Nitschke and Rutherford, 1991; Blankenship, 1992, 1 994, 2001, 2010; Nitschke *et al.*, 1996, 1998; Olson, 1999, 2006; Hohmann-Marriott and Blankenship, 2011; Fischer *et al.*, 2016; Khadka *et al.*, 2017; Martin *et al.*, 2018; Cardona *et al.*, 2019; Cardona and Rutherford, 2019). In this section, we will explore these relationships in more detail, and attempt to draw some more general conclusions.

In all cases, modern reaction centers are dimeric in structure. Two similar or identical integral membrane proteins form the core of the complex, along with additional subunits. Photosystem I, Photosystem II, and the reaction centers of the anoxygenic purple bacteria and the filamentous anoxygenic phototropic bacteria all have a heterodimeric arrangement, in which there are two distinct, yet related, core reaction center subunits. An evolutionary scenario that explains how this pattern came about is shown in Fig. 12.9. The protein that is the ancestor of photosynthetic reaction centers was probably monomeric, although it may or may not have had photosynthetic capabilities. It developed the ability to dimerize, creating a homodimeric reaction center, with two identical copies of the subunits coded by a single reaction center gene. The reaction center gene underwent gene duplication, followed by divergence, to give a heterodimeric complex. This created two potential electron transfer pathways, one in each half of the dimeric complex. In many heterodimers it is either certain or very likely that electron transfer takes place down only one of the two-electron transfer pathways, creating a functional asymmetry to the complex. The gene duplication and subsequent divergence that took place in Photosystem I are clearly distinct from similar events that took place

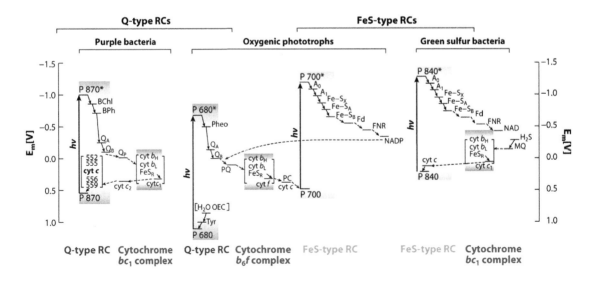

Figure 12.8 Electron transport diagrams for anoxygenic (left and right sides) and oxygenic (middle) photosynthetic organisms. The two classes of reaction centers are known as Q type (Type II) and Fe–S type (Type I), based on the nature of the early electron acceptors. Details of the pathways are given in Chapters 6 and 7. *Source:* Hohmann-Marriott and Blankenship (2011)/Reproduced with permission from the Annual Review of Plant Biology, Volume 62 © 2011 by Annual Reviews, http://www.annualreviews.org.

in Photosystem II, because in each case the two halves of a given heterodimer are much more similar to each other than they are to either half of the other heterodimer. It is clear that at least two distinct gene duplication events took place during the evolution of reaction centers.

There is some uncertainty as to whether or not a third duplication took place, involving Photosystem II and the purple/filamentous anoxygenic phototroph reaction centers from anoxygenic bacteria. The latter organisms contain L and M subunits that form the core of the heterodimer, whereas Photosystem II contains the D1 and D2 subunits as its core (Chapter 7). Sequence analysis clearly indicates a pattern in which the L and M peptides originated from one gene duplication, whereas D1 and D2 originated from a distinct gene duplication, as shown in Fig. 12.9. However, functional data, in particular, the Q_A/Q_B acceptor system found in both types of complexes, appear to support a single ancient gene duplication, with L and D1 as homologs, as well as M and D2. This creates a paradox, which cannot currently be resolved unambiguously. The sequence data, as well as structural comparisons, strongly suggest that the functional similarities must result from convergent evolution. However, the functional data indicate that the genes in a given organism have not evolved independently after duplication. This pattern may be a result of convergent evolution, in which the asymmetrical electron transfer and the two-electron gate of the quinones have arisen independently two separate times. This seems on the face of it an unlikely possibility, although the functional benefit of a system that avoids free radicals, efficiently pumps protons, and produces products with paired electrons such as quinones may be sufficiently strong that it has driven the evolutionary path in this way (see below). Another possibility is that there has been concerted evolution of the L/M or D1/D2 genes, in which changes to one gene must somehow be transferred to the other before they can be fixed. This process is called gene conversion and is well-known in some systems, although not in a situation like the case under discussion here. Alternatively, it may be a case in which the

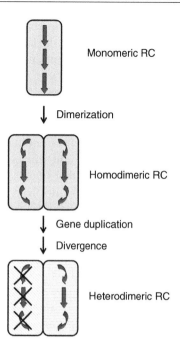

Figure 12.9   Scenario for the evolutionary development of reaction centers from monomeric through homodimer to heterodimer. *Source:* Blankenship (1992)/ Springer Nature Limited.

tree-building algorithms have been fooled by very different constraints on the two different sets of genes (Blankenship, 1994; Lockhart *et al.*, 1996b; Graur and Li, 2000). However, all tree-building algorithms, including direct structural comparisons, give trees with the same topology despite significantly different underlying assumptions, which argues against this possibility.

The heterodimeric arrangement of most modern reaction centers virtually demands that a homodimeric arrangement was present in an ancestral form (Nelson, 2013; Orf *et al.*, 2018). Remarkably, three groups of modern organisms have retained the homodimeric subunit structure for their reaction centers. These are the green sulfur bacteria, the heliobacteria, both of which are strictly anaerobic organisms that contain a Type I reaction center (Büttner *et al.*, 1992; Liebl *et al.*, 1993) and the chloracidobacteria, which are aerobic (Tsukatani *et al.*, 2012). Although it is not yet definitively established experimentally, all available evidence is consistent with the idea that there are two equivalent electron transfer pathways in these organisms.

What is the functional advantage of having a heterodimeric reaction center? In the Type II reaction centers, it probably relates to the functional distinctions between Q_A and Q_B, where one quinone (Q_A) is in a nonpolar environment and carries out only electron transfer, while the other (Q_B) is in a more polar environment, and is also involved in proton translocation. However, this argument cannot apply for Photosystem I, where the two-electron transfer chains reconverge at the $Fe-S_X$ acceptor. There seems to be no obvious advantage to having an asymmetric reaction center in Photosystem I-type complexes. Indeed, the two groups of organisms with homodimeric reaction centers discussed above clearly do not require this arrangement. In addition, the two halves of the Photosystem I core reaction center complex are more similar to each other than are the L and M or D1 and D2 proteins, suggesting a more recent duplication event for Photosystem I. Both these lines of evidence suggest that the functional advantage of a heterodimer in the Type I reaction center is less stringent, and that the explanation for it may be more subtle, possibly involving sensitivity of the organism to oxygen, that being one variable that usually seems to correlate with whether a homodimeric or heterodimeric complex is present. It is now clear that Photosystem I has electron transfer down both branches of the electron transfer chain, although the possible functional benefits of this pattern are not yet apparent (Joliot and Joliot, 1999; Guergova-Kuras et al., 2001).

12.9   Do all photosynthetic reaction centers derive from a common ancestor?

Are all photosynthetic reaction centers related to each other, or have there been multiple independent inventions of them? The two different classes of reaction centers have only minimal sequence similarity to each other, not significantly above what would be expected randomly. However, it is well known that very distantly related proteins can exhibit minimal sequence identity, yet still be homologous (descended from a common ancestor) (Doolittle, 1994). The homology can be revealed when detailed structures of the proteins are compared. This situation appears to be the case with the two classes of photosynthetic reaction centers. The central core of the Type I Photosystem I reaction center is remarkably similar in three-dimensional structure to the purple bacterial reaction center, which is a Type II complex (Schubert et al., 1998; Sadekar et al., 2006). In Photosystem I, each half of the core reaction center complex consists of 11 transmembrane helical stretches (Chapter 7). The last five of these helices form the electron transfer heart of the complex, while the first six form an antenna domain. A similar pattern of five transmembrane helices is also found in the purple bacterial, filamentous anoxygenic phototrophic bacterial, and Photosystem II reaction centers. Histidine residues in the same relative position in the protein sequence coordinate the special pair of chlorophyll or bacteriochlorophyll molecules that are the heart of the complex. In all cases, the five transmembrane helical segments are arranged in an interlocking C. Further evidence of the relationship between the two classes of photosystems comes from the structure of Photosystem II compared with Photosystem I (Chapter 7). In addition to the similar core electron transfer portion of the complex, the two integral membrane antenna complexes in Photosystem II, CP43, and CP47, each have six transmembrane helices arranged in three groups of two. This is remarkably similar to the structural arrangement of the antenna domain in Photosystem I. Taken as a whole, this evidence strongly suggests that the two types of reaction centers, while very distantly related, share a common ancestor. These relationships are illustrated in Figs 12.10 and 12.11 (Schubert et al., 1998).

Further evidence of the single origin of all photosynthetic reaction centers comes from comparative structural analysis, in which the structures of

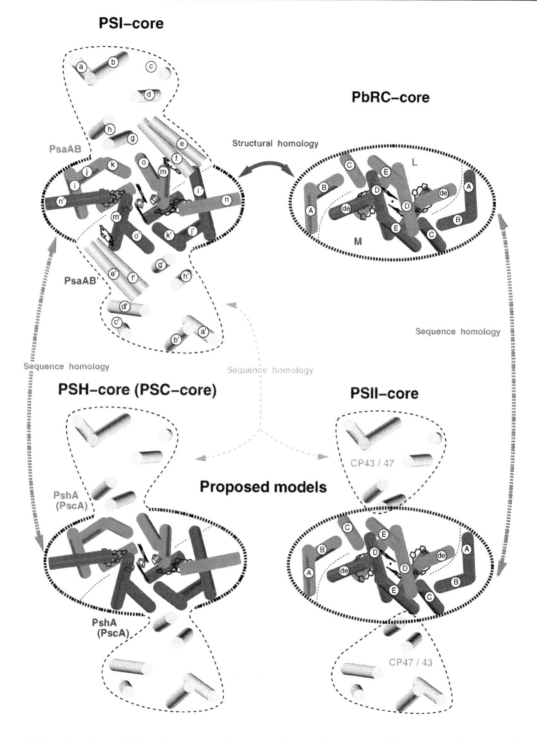

Figure 12.10   Structural relationships among photosynthetic reaction centers. All known reaction centers have an electron transfer core of two integral membrane proteins, each with five transmembrane helices, arranged in an interlocking C-shape. A core antenna complex is found in Photosystem I (where it is fused to the electron transfer core) and in Photosystem II (where it is composed of the CP43 and CP47 antenna complexes), and probably also in the green sulfur bacteria and heliobacteria. *Source:* Schubert *et al.* (1998)/Elsevier.

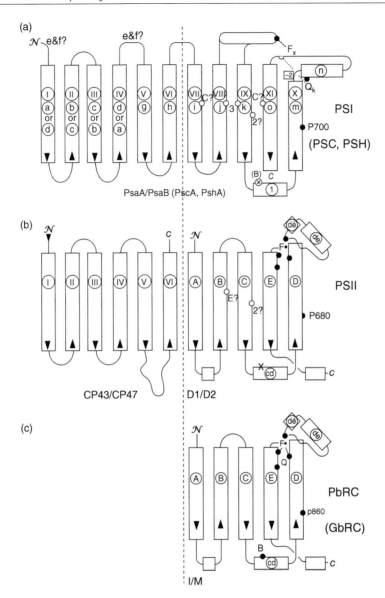

Figure 12.11   Membrane topology diagram describing the arrangement of the transmembrane helices in the 11-and 5-helix reaction center complexes. (a) Photosystem I, in which the N-terminal antenna and C-terminal electron transfer domains are fused into a single combined complex. (b) Photosystem II, in which the two domains are present, but on distinct proteins. (c) Purple bacterial reaction center, in which only the electron transfer domain is present. *Source:* Schubert *et al.* (1998)/Elsevier.

all the RC complexes that are known were compared using an algorithm that finds the best structural alignment of multiple protein structures
(Sadekar *et al.*, 2006; Cardona *et al.*, 2019). This

method was used directly to construct evolutionary
trees based on the structure of the complex instead
of the sequence of the proteins. Evolutionary trees
generated using this method have the same basic

topology as trees built using sequence comparisons, suggesting that this is the correct topology. This method also reveals the three putative gene duplication events discussed above, and strongly supports the idea that the primitive reaction centers were homodimers and that the organisms were anoxygenic phototrophs.

Is the Type I or Type II reaction center more similar to the primordial complex? The traditional thinking has been that the Type I reaction center, which is found in many strictly anaerobic organisms, is the more primitive. The longer length of the Type I reaction center polypeptides compared with the Type II polypeptides, 11 versus 5 transmembrane helices, can be interpreted either way. One view has it that the 11-helix form was the ancestral form and underwent gene fission into the 5-helix form plus an antenna domain. Alternatively, the 5-helix protein may have been the ancestral form, which underwent gene fusion with a separate antenna domain to give the 11-helix form (Schubert *et al.*, 1998; Baymann *et al.*, 2001; Mix *et al.*, 2005). The structure of the evolutionary tree that is derived from structural comparisons of reaction centers suggests that the primordial reaction center may have been intermediate between the Type I and Type II complexes, and is sometimes called Type 1.5 (Sadekar *et al.*, 2006).

If we accept that all reaction centers are related to an ancient primitive complex, is it possible to identify how this primordial form arose or what might have been the function of the precursor? The evolutionary precursor of reaction centers has been suggested to come from an ultraviolet (UV) protection system (Mulkidjanian and Junge, 1997) or from cytochrome *b* (Meyer, 1994; Xiong and Bauer, 2002). The latter possibility is interesting, in that cytochrome *b* is a widely distributed and, therefore, presumably ancient, integral membrane protein that is involved in energetic metabolism in all cells (Schütz *et al.*, 2000). However, there does not appear to be any significant structural similarity between the reaction center and cytochrome *b*, arguing against this possibility (Sadekar *et al.*, 2006).

12.10 The origin of linked photosystems and oxygen evolution

Perhaps the most important developments in the evolution of the photosynthetic process are the origin of linked photosystems and the oxygen evolution capability. All known organisms that are capable of oxygen evolution also contain linked photosystems. Whether these events are inextricably linked to evolutionary developments is not clear. It is entirely possible to imagine a scenario in which Photosystem II and the ability to split water developed in an organism that lacked Photosystem I. This organism might have oxidized water using Photosystem II, delivering electrons to the quinone pool. The reduced quinone may have been oxidized in a manner similar to the reverse electron flow that purple bacteria use to reduce NAD^+ for use in carbon fixation, thereby reoxidizing the quinone pool and permitting continued electron transport. This putative organism then later added or developed Photosystem I, which is more efficient at producing the reductant needed for carbon fixation. Although this scenario is logically consistent and appealing in its simplicity, there is no evidence in support of it. In every known oxygenic photosynthetic organism, oxygen evolution and the two photosystems are found together.

The ability to carry out anoxygenic photosynthesis with Photosystem I using H_2S as an electron donor can be expressed under conditions of high sulfide in some cyanobacteria (Padan, 1979; Cohen *et al.*, 1986; Liu *et al.*, 2020). However, these organisms retain the ability to carry out normal oxygen-evolving photosynthesis using both photosystems. Therefore, it seems likely that they have successfully adapted to high sulfide environments by developing the ability to oxidize sulfide, but are probably not evolutionary intermediates in the conversion from anoxygenic to oxygenic photosynthesis, although Hamilton (2019) has argued that they are more reflective of ancestors of oxygenic phototrophs.

Cyanobacteria that have lost Photosystem II entirely and are specialized to carry out nitrogen fixation in symbiosis with eukaryotic algae have been discovered (Thompson *et al.*, 2012), although this also seems likely to be a case of adaptation to a particular environment and gene loss from a precursor that had the ability to carry out oxygen evolution.

There are several unanswered questions concerning the transition from anoxygenic to oxygenic photosynthesis. First, how did two photosystems become present in the same organism, and how did they become connected in series? Second, how did the ability to make O_2 from H_2O originate and develop? Third, how does the pigment composition of the organisms relate to this issue? Various possibilities have been discussed in the literature to explain the distribution of types of reaction centers and metabolic capabilities of the anoxygenic and oxygenic photosynthetic organisms (Olson and Pierson, 1987; Nitschke and Rutherford, 1991; Blankenship, 1992, 1994, 2001, 2010; Nitschke *et al.*, 1996, 1998; Olson, 1999, 2006; Larkum, 2006; Hohmann-Marriott and Blankenship, 2011; Sousa *et al.*, 2013; Fischer *et al.*, 2016; Khadka *et al.*, 2017; Martin *et al.,* 2018; Cardona *et al.*, 2019; Cardona and Rutherford, 2019). All the proposals can be described as variants of two basic mechanisms known as selective loss or fusion (Fig. 12.12). The selective loss mechanism entails that the two photosystems developed early in a single organism and that the various anoxygenic forms arose by loss of one or other photosystem. The fusion mechanism entails that the two classes of photosystem developed independently (although ultimately they are probably derived from a common ancestor as discussed above) and became linked by genetic fusion or a significant lateral transfer of genetic information. This is similar to the endosymbiotic event that unquestionably took place to form chloroplasts, except that the resultant organism is still bacterial in nature. An important point to appreciate is that, because of these possible events, the evolutionary path of much of the cyanobacterial genome may have been very different from the evolutionary history of the photosynthetic apparatus.

The fusion mechanism predicts that different parts of the photosynthetic apparatus may have quite different evolutionary histories. This proposal is testable, in that it may be possible to identify that different parts of the photosynthetic apparatus came from different ancestors. Photosystem II and the carbon fixation pathways may have come from the purple bacterial ancestor, while Photosystem I and the cytochrome b part of the cytochrome b_6f complex that links the photosystems may have come from the green sulfur bacterial or heliobacterial ancestor (Nitschke *et al.*, 1998; Schütz *et al.*, 2000). The origin of other parts, such as cytochrome f and the oxygen-evolving complex, is still largely a mystery.

A different scenario has been proposed by Allen (2005), in which changes in gene regulation in an organism that contained both Type I and Type II reaction centers that did not cooperate led to an organism in which both photosystems worked together. This is a variant of the selective loss hypothesis.

12.11 Origin of the oxygen-evolving complex and the transition to oxygenic photosynthesis

Although the mechanisms discussed above address the question of how the two photosystems may have ended up in the same organism, they do not really address the more interesting question of how they became functionally linked and, most important of all, how the ability to oxidize water to form molecular oxygen originated and whether there were any transitional forms. This important development has had central importance in the evolution of photosynthesis and profound consequences for both the biosphere and the physical Earth.

In order for the transition from anoxygenic to oxygenic photosynthesis to take place, two essential developments had to happen. First, the redox span created when the principal photopigment absorbs a

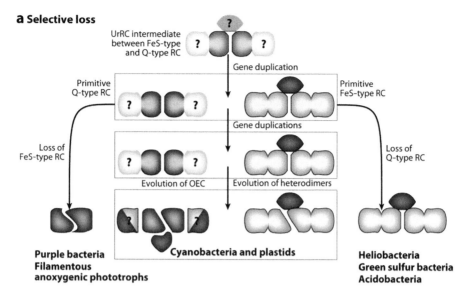

a Selective loss

UrRC intermediate
between FeS-type
and Q-type RC

Gene duplication

Primitive
Q-type RC

Primitive
FeS-type RC

Gene duplications

Loss of
FeS-type RC

Loss of
Q-type RC

Evolution of OEC   Evolution of heterodimers

**Purple bacteria
Filamentous
anoxygenic phototrophs**

Cyanobacteria and plastids

**Heliobacteria
Green sulfur bacteria
Acidobacteria**

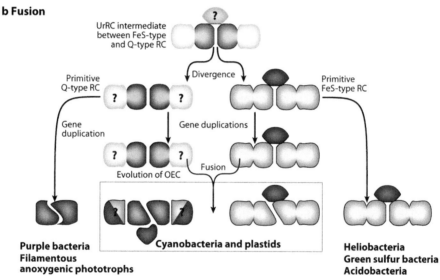

b Fusion

UrRC intermediate
between FeS-type
and Q-type RC

Primitive
Q-type RC

Divergence

Primitive
FeS-type RC

Gene
duplication

Gene duplications

Fusion

Evolution of OEC

**Purple bacteria
Filamentous
anoxygenic phototrophs**

Cyanobacteria and plastids

**Heliobacteria
Green sulfur bacteria
Acidobacteria**

Figure 12.12 Proposals regarding the evolutionary development of organisms with two linked photosystems. (a) Selective loss mechanism, in which both classes of reaction centers develop in a single organism, and the various groups of anoxygenic bacteria result from the loss of one or the other photosystem. (b) Fusion mechanism, in which the two classes of reaction centers develop separately and are joined in the same organism by a genetic fusion. *Source:* Hohmann-Marriott and Blankenship (2011)/Reproduced with permission from the Annual Review of Plant Biology, Volume 62 © 2011 by Annual Reviews, http://www.annualreviews.org.

photon had to increase. The vertical arrows in Fig. 12.8 illustrate this redox span. The photon absorption first creates a highly reducing excited state and then, after electron transfer, an oxidized reaction center that is capable of oxidizing electron donors (Blankenship and Prince, 1985). (See Appendix 1 for additional discussion of this point.) Photosystem II is unique among all photosynthetic reaction centers, in that it has a much more highly oxidizing redox potential, over $+1\,V$ versus NHE (Fig. 12.8).

The other critical development is the ability to store oxidizing equivalents. This is necessary because the redox chemistry of water is intrinsically a four-electron process, whereas photochemistry is a one-electron process. In modern-day oxygenic organisms this charge accumulation is carried out by a multinuclear Mn/Ca cluster in the oxygen-evolving complex that is part of the Photosystem II reaction center complex (Chapter 7).

The first of these critical evolutionary developments necessitated a change in pigments from the long-wavelength-absorbing bacteriochlorophylls found in anoxygenic organisms to the shorter-wavelength-absorbing chlorophyll pigments found in all oxygenic organisms (Blankenship and Hartman, 1998). This may have taken place through the loss of the *bchX*, *bchY*, and *bchZ* genes that carry out the reduction of ring B of the bacteriochlorophyll precursor (see Section 12.7). This would produce a pigment, 3-acetyl chlorophyll, that is both structurally and energetically intermediate between chlorophyll *a* and bacteriochlorophyll *a*. Further deletion of the enzymes that convert the C3 vinyl group into an acetyl leads to chlorophyll *a*. Structures of these pigments are shown in Fig. 12.13. Interestingly, a pigment that is chemically similar to the intermediate pigment 3-acetyl chlorophyll is known: chlorophyll *d*. It is the major pigment in a cyanobacterium, *Acaryochloris marina* (Miyashita *et al.*, 1996). This organism at one point was thought to be a "missing link" between anoxygenic and oxygenic photosynthesis. However, it now seems more likely to be a more recent adaptation to a particular photic environment (Swingley *et al.*, 2008).

The second critical evolutionary development, that along the way to oxygenic photosynthesis was proposed to have taken place, did so via an intermediate step in which a Mn catalase enzyme – which shows some structural similarities to the OEC (Dismukes, 1996; Whittaker *et al.*, 1999) – served as the evolutionary source for the OEC, and an intermediate stage took place in which hydrogen peroxide was an electron donor for the incipient Photosystem II (Blankenship and Hartman, 1998; Liang *et al.*, 2006). Structural comparisons of the oxygen evolution complex with known multinuclear Mn enzymes found the closest match of the OEC to Mn catalases and related enzymes, but the overall structural similarity is low (Raymond and Blankenship, 2008).

An evolutionary scenario that incorporates these two ideas is shown in Fig. 12.14 (Blankenship and Hartman, 1998). Alternative scenarios in which abiotically formed Mn minerals (Sauer and Yachandra, 2002) or Mn bicarbonate complexes (Baranov *et al.*, 2000; Dismukes *et al.*, 2001) served as the source for the Mn cluster of the OEC have been suggested.

12.12   Antenna systems have multiple evolutionary origins

The remarkable variety of distinct antenna complexes found in different classes of photosynthetic organisms was discussed in Chapter 5 (Green and Parson, 2003). These different classes of antennas, including the phycobiliproteins, the bacterial LH1 and LH2 complexes, the integral membrane three-helix LHC complexes, the chlorosome antenna complexes, and the peridinin-chlorophyll complexes, almost certainly are the results of multiple independent evolutionary inventions. These complexes utilize different types of pigments and proteins, which show no apparent similarities and have no apparent common structural or functional features other than the energy transfer process. It seems clear that the diversity of antenna types has resulted from the adaptation of organisms to a wide range of photic environments that differ in terms of light intensity or light quality (spectral distribution).

While the major classes of antennas were apparently derived independently, clear evolutionary relationships are evident within these classes and

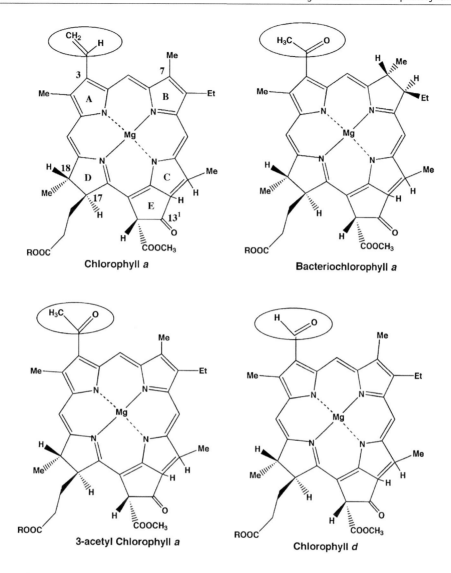

Figure 12.13   Chemical structures of chlorophyll *a*, bacteriochlorophyll *a*, 3-acetyl chlorophyll *a*, and chlorophyll *d*.

have been documented in numerous cases. Examples of these relationships are given below.

The evolutionary history of the LHC family of antennas is particularly interesting. All the chlorophyll *a/b*-containing LHCII complexes and the minor LHCs found in Photosystem II are integral membrane proteins and have a similar three-trans-membrane-helix structure. They are also related to the LHCI complexes associated with eukaryotic Photosystem I and to the chlorophyll *a/c* antennas

found in several classes of eukaryotic algae (Green and Durnford, 1996; Green, 2001, 2003).

Where did the LHC superfamily of antennas originate? They are found in all photosynthetic eukaryotes, but not in any prokaryotes. However, a cyanobacterial ancestry is thought to be likely for this family. The LHC proteins share a distant relationship to two classes of stress-induced antenna complexes: the early light-inducible proteins (ELIP) found in chloroplasts and the high light-inducible

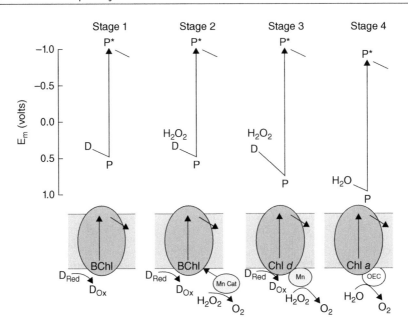

Figure 12.14 Evolutionary scenario for the conversion of the reaction center from bacteriochlorophyll to chlorophyll and the incorporation of the Mn complex. *Source:* Blankenship and Hartman (1998)/Elsevier.

protein (HLIP) found in cyanobacteria. Internal sequence similarities suggest that the HLIP is the distant ancestor for the entire family, and a series of gene duplications, fusions, and losses are proposed to have led to the three-helix family of LHC complexes (Dolganov *et al.*, 1995; Green and Kühlbrandt, 1995; Montané and Kloppstech, 2000; Green, 2003; Engelken *et al.*, 2010; Niyogi and Truong, 2013).

A plausible evolutionary scenario for these antenna complexes is shown in Fig. 12.15. The single-helix HLIP-type antenna is the ancestor, with a two-helix complex as the next stage. This complex developed into a four-helix form by gene duplication followed by fusion. The last helix may then have been lost, and the modern three-helix complex was formed (Engelken *et al.*, 2010). Some of the ancestral forms may have functioned primarily in photoprotection instead of in antenna processes. The PsbS antenna protein that is implicated in the xanthophyll cycle that regulates and protects Photosystem II antenna function (Chapter 5) is a four-helix complex that is very similar to the

predicted four-helix evolutionary intermediate (Li *et al.*, 2000).

The bacterial LH1 and LH2 complexes are evidently related to each other, given their similar sequences and structures. The α and β peptides have some sequence similarity and are also likely to be related to an ancient single-helix or homodimeric antenna complex. However, these antennas do not exhibit any apparent similarity to any of the other classes of antennas.

The phycobilisome antenna system found in cyanobacteria as well as eukaryotic red algae consists of a related pair of α and β proteins for each of the classes of biliproteins. Certain key amino acids are conserved in the biliproteins, and they have evolved by way of a multiple series of gene duplications and divergences from an ancestral form that could form short rods (Apt *et al.*, 1995; Grossman *et al.*, 1995). Phycoerythrin, which binds at the distal ends of the phycobilisome rods, is the most recently evolved member of the bilin family and also contains the most elaborate set of pigment binding sites. Remarkably, the biliproteins have

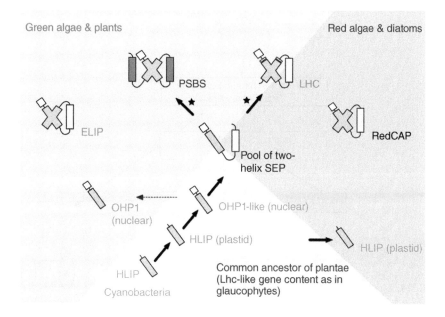

Green algae & plants

Red algae & diatoms

PSBS

LHC

ELIP

RedCAP

Pool of two-
helix SEP

OHP1
(nuclear)

OHP1-like (nuclear)

HLIP (plastid)

HLIP (plastid)

HLIP

Cyanobacteria

Common ancestor of plantae
(Lhc-like gene content as in
glaucophytes)

Figure 12.15 Evolutionary scenario for the development of the LHC class of antenna complexes. *Source:* Engelken *et al.* (2010)/Springer Nature.

structural similarities to the globin family of proteins, which includes hemoglobin, and they may indeed be distantly related proteins (Schirmer *et al.*, 1985; Kikuchi *et al.*, 2000).

Cryptophyte algae also have biliprotein antennas that are related to those in cyanobacteria and red algae, although the organization is very different. The phycobiliproteins are not organized into phycobilisomes and are found in the thylakoid lumen, which is the opposite side of the membrane in comparison with those organisms that have phycobilisomes. The subunit composition is also quite different, with a β protein that is related to the β proteins in organisms that contain phycobilisomes, and α proteins that are unrelated to either the α or the β proteins found in phycobilisomes (Wilk *et al.*, 1999). How the cryptophyte antennas relate evolutionarily to the phycobilisomes is not yet understood. However, they are probably closest to those of the red algae, as the cryptophytes are almost certainly derived from red algae by secondary endosymbiosis (see Section 12.14).

The phycobilisome antennas are widely distributed in the cyanobacteria and almost certainly

originated in this group. The primary endosymbiont that was the ancestor of the chloroplast in eukaryotic photosynthetic organisms probably also contained a phycobilisome antenna. This complex has been retained in the red algae, and portions of it are retained in the cryptophyte algae, but it has been completely lost in the green algae, which are the ancestor of plants (Green, 2019). This process is illustrated in Fig. 12.16.

12.13 Endosymbiosis and the origin of chloroplasts

What is the evolutionary origin of the chloroplasts found in eukaryotic photosynthetic organisms? As discussed in Chapters 2 and 10, chloroplasts in eukaryotic photosynthetic organisms are semi-autonomous cell organelles that contain DNA and ribosomes. The chloroplast DNA codes for, and the ribosomes make, many of the proteins that form the photosynthetic apparatus. Others are coded by nuclear DNA and are imported from the cytoplasm.

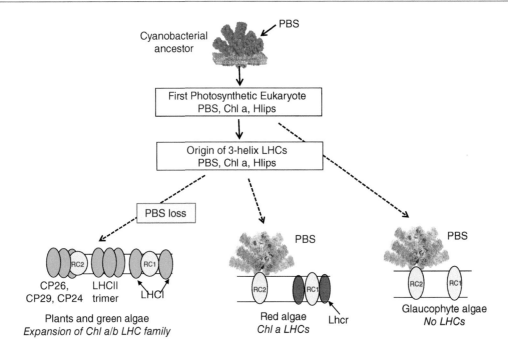

Figure 12.16   Evolutionary processes involving phycobilisome antennas and LHC antennas. *Source:* Green (2019)/ MDPI/CC BY 4.0.

This curious division of responsibility for the provision of the information needed to build the organelles gives us an important message about the evolutionary histories of chloroplasts as well as mitochondria.

At one time there were two main theories for the evolutionary origin of chloroplasts and mitochondria. One was the autogenous theory, according to which the chloroplast or mitochondrion was thought to be derived in some manner from cellular components, including nuclear DNA (Cavalier-Smith, 1992). The other was the endosymbiosis theory, according to which the organelles were considered to be the remnants of a symbiotic association between a protoeukaryote and a bacterium. This issue has now been settled definitively in favor of the endosymbiosis theory.

The endosymbiotic theory was proposed over a century ago by Schimper and was further developed early in the twentieth century by Mereschkowsky; but it was not finally accepted until the 1970s, largely due to the efforts of Lynn Margulis (1993). The issue

now is not *whether* but *how many times* chloroplasts arose from bacterial symbionts, and what were the natures of the symbiont and the host cell (Gould *et al.*, 2008; Howe *et al.*, 2008; Dorrell and Howe, 2012; Nowack and Weber, 2018; Sato, 2020).

What is the evidence that supports the endosymbiotic theory so overwhelmingly? Many aspects of chloroplast structure and function are typically bacterial and strongly support the endosymbiotic hypothesis (Gray, 1992, 1999). These features include a circular DNA genome without associated histone proteins, bacterial-like 70S ribosomes, the sensitivity pattern of protein synthesis inhibitors, the fact that translation begins with *N*-formylmethionine, the lack of polyadenylation of chloroplast mRNA, and the typical bacterial patterns for promoters and ribosome binding sites. In one group of algae, the glaucocystophytes, represented by *Cyanophora paradoxa* (Price *et al.*, 2012), the chloroplast envelope retains remnants of a cell wall typical of Gram-negative bacteria. However, the strongest evidence for the endosymbiotic theory comes from comparative genomics of chloroplasts and cyanobacteria. In essentially every case in which sequence comparisons have been made, chloroplast genes cluster in a phylogenetic tree with homologous genes from cyanobacteria. The only reasonable explanation for this finding is that chloroplasts share a much closer common ancestor with cyanobacteria than with the rest of the host cell in which they are contained. This effectively disproves the autogenous theory for chloroplast origins.

The nature of the cyanobacterial organism(s) that underwent endosymbiosis has long been uncertain, as no extant cyanobacterial lineage matches the expected properties of the endosymbiont very well. It has been suggested that the endosymbiont was a freshwater cyanobacterium, possibly capable of nitrogen fixation (Blank and Sanchez-Baracaldo 2010; Dagan *et al.*, 2013), but other possibilities have also been suggested (Moore *et al.*, 2019).

If we accept the endosymbiotic origin of chloroplasts as fact, a secondary question then becomes whether the primary endosymbiotic event happened only once, or whether it occurred many times. The former is described as a **monophyletic** origin of chloroplasts, and the latter is known as a **polyphyletic** origin. This has been a controversial issue, although most researchers now favor a monophyletic origin, with the symbiont being a cyanobacterium and the host a phagocytic cell that engulfed the cyanobacterium and truly failed to digest its dinner (Douglas, 1998).

One of the strongest pieces of evidence cited in support of the monophyletic origin of chloroplasts is the fact that certain genes are clustered together in a distinctive way in all chloroplast genomes, and not in any known cyanobacterium (Stoebe and Kowallik, 1999). This suggests that the common ancestor of all chloroplasts had the same clustering, and that it has been maintained in all chloroplast lineages. In addition, in several well-documented cases, nuclear gene trees are consistent with the various classes of eukaryotic photosynthetic organisms comprising sister groups. If these organisms were related only very distantly, it might suggest that they had acquired their plastids independently. However, this is apparently not the case, and the relationships are entirely consistent with a scenario in which the ancestor of all eukaryotic photosynthetic organisms first acquired plastids and then evolved into the existing groups of algae and plants. There is one important complication to this picture, in that many groups of eukaryotic algae have clearly arisen by a secondary endosymbiosis, in which a eukaryotic alga was incorporated into a second host. This issue is discussed below in Section 12.14. A more precise statement of the monophyletic hypothesis is therefore that a single primary endosymbiotic event took place, and that all chloroplasts are derived from this cell. A clear exception to the monophyletic rule is the case of *Paulinella,* discussed in Chapters 2 and 10, in which it is well established that the endosymbiosis is very recent and that gene transfer to the nucleus is still at an early stage.

Although most data support the monophyletic hypothesis for plastid origins (Douglas, 1998; Martin *et al.*, 1998; Stoebe and Kowallik, 1999; Tomitani *et al.*, 1999; Cavalier-Smith, 2000; Moreira *et al.*, 2000), there are some unanswered questions. If this view is correct, then the founder cyanobacterium must have been a very unusual cell, unlike any known existing organism. It must have contained both chlorophyll *a* and *b* as well as phycobilisomes. The only existing organisms that are close to this pattern are the prochlorophytes, because they contain chlorophylls *a* and *b* and, in some cases, certain phycobiliproteins. However, none of the known prochlorophytes contains the full complement of phycobilisome components that are found in the cyanobacteria and red algae. No known class of organisms is configured precisely as required by the monophyletic hypothesis. A compromise between the monophyletic and polyphyletic views has been advanced by Larkum (1999) and Howe *et al.* (2008), in which a single primary endosymbiont paved the way for subsequent distinct endosymbiotic events.

Another unanswered question concerns the evolutionary history of chlorophyll *b*. There are two major groups of oxygenic photosynthetic prokaryotes: the cyanobacteria and the prochloro-phytes. The former lack chlorophyll *b*, and the latter contain it. When the first chlorophyll *b*-containing prochlorophyte, *Prochloron*, was discovered in the 1970s, it was hailed as the likely evolutionary precursor to the chloroplasts of green algae and plants. This was because chlorophyll *b* is found in these lineages of photosynthetic eukaryotes and is not found in any cyanobacteria. But this idea became less attractive with the finding that the different groups of prochlorophytes did not appear to be a monophyletic group (Palenik and Haselkorn, 1992; Urbach *et al.*, 1992). Instead, they appeared to be related to different distinct groups of cyanobacteria. This prompted the idea that the ability to make chlorophyll *b* may have arisen independently several times during the course of evolution. However, the pendulum has now swung back the other way somewhat, in that the enzyme that makes chlorophyll *b* has been identified from several prochlorophytes and has been found to be clearly homologous to the enzyme that makes chlorophyll *b* in green algae and plants (Tomitani *et al.*, 1999). In addition, while the proteins that bind chlorophyll *b* in green algae and plants are distinct from the chlorophyll *b*-binding proteins in prochlorophytes (LaRoche *et al.*, 1996), the latter proteins appear to be homologous to each other, as would be expected if the prochlorophytes were sister taxa (Durnford *et al.*, 1999). These two findings reopen the question of whether the prochlorophytes are indeed a polyphyletic group.

When did endosymbiosis happen? No precise dates are available, although the geological record does provide significant constraints. The first evidence for eukaryotes is biomarker molecules in 2.7-billion-year-old rocks (Brocks *et al.*, 1999). Biomarkers are organic compounds found in rocks that are derived from particular compounds that occur only in a single group of organisms. They are thus a molecular "fingerprint" that tells us that organisms of a certain class were present at a given time in geological history. No microfossils of these organisms are known, so it is difficult to know very much about their nature and other methods, such as evolutionary trees derived using a relaxed molecular clock, give significantly later times of ~1800 MYA (Parfrey *et al.*, 2011). The early record of eukaryotic fossils is very incomplete, but examples that are clearly identified as being of photosynthetic origin have been found in 2.1-billion-year-old formations (Han and Runnegar, 1992) and were reasonably well represented by about 1.9–2.0 billion years ago (Knoll, 1992). Much of the early fossil record of eukaryotes consists of acritarchs. The word "acritarch" is derived from the Greek *akritos*, meaning confused. Acritarchs are microscopic fossils that are almost certainly derived from eukaryotic phytoplankton, although precisely which group of modern organisms they are most closely related to is not clear. The available evidence thus places the appearance of eukaryotic photosynthetic organisms at 2.0–2.1 billion years ago at the latest, and possibly several hundred million years prior to that. Land plants evolved from green algae and first made an appearance on land approximately 450–500 million years ago (Bhattacharya and Medlin, 1998).

12.14 Most types of algae are the result of secondary endosymbiosis

Evidence is now overwhelming that several groups of eukaryotic algae originated from a secondary endosymbiosis, in which a eukaryotic alga was incorporated into a second host (Palmer and Delwiche, 1996; Delwiche and Palmer, 1997; Delwiche, 1999; Keeling, 2010, 2013; Curtis *et al.*, 2012). In some cases, it is clear that these secondary endosymbionts were themselves incorporated into a third host, forming a tertiary plastid. The image of a small fish being eaten by a larger fish, which in turn is eaten by a larger fish, comes to mind. Most groups of algae appear to have arisen in this manner, with only three groups – the green algae, the red algae, and the glaucocystophytes – thought to represent primary endosymbiotic plastids. This is illustrated in Fig. 12.17.

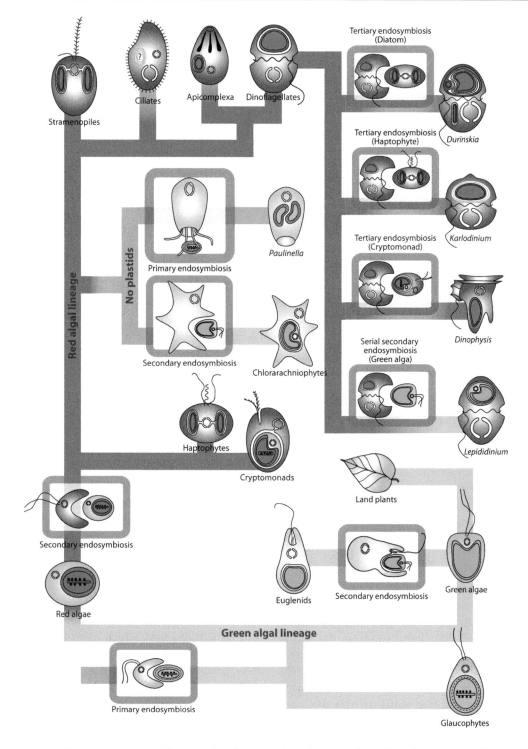

Figure 12.17   Secondary endosymbiotic origin of algae. The primary endosymbiont is the source for the green algae, red algae, and glaucocystophytes, while all other groups of algae result from secondary endosymbiosis of a eukaryotic alga. *Source:* Keeling (2010)/The Royal Society.

The first evidence for these secondary and tertiary endosymbiotic events came from ultrastructural studies, in which it was found that the chloroplasts from these algae (including, among others, the kelps, diatoms, dinoflagellates, and euglenoids) contain four, or in some cases three, limiting membranes surrounding the chloroplast. Recall that higher plant chloroplasts (which are almost certainly the products of primary, rather than secondary, endosymbiosis) are surrounded by a double membrane (Chapters 2 and 10). These membranes are thought to have derived from the inner cell membrane of the original symbiotic cyanobacterium plus the membrane that formed the food vacuole during phagocytosis (or possibly from the inner and outer cell membranes of the symbiotic cyanobacterium). Two additional membranes would arise in the same manner from secondary endosymbiosis, giving a total of four membranes. This leaves only the dinoflagellates and euglenoids with three membranes to be explained. Here, it is possible that one of the outer membranes was lost, or, more likely, that ingestion was by a pathway other than phagocytosis (Cavalier-Smith, 2000). The third and fourth membranes in those algal groups that contain them are derived from the cell membrane of the eukaryotic symbiont and/or membranes of the secondary host.

The strongest piece of evidence in favor of secondary endosymbiosis is the presence in two groups of algae (cryptophytes and chlorarachniophytes) of a vestigial nucleus, called a **nucleomorph**, from the eukaryotic photosynthetic organism that became the secondary endosymbiont. This nucleomorph is found between the second and third membranes of the chloroplast, and analysis of its DNA is consistent with this proposal (Curtis et al., 2012). Presumably, in the other groups of algae that arose via secondary endosymbiosis, the nucleomorph has been lost completely. In a few cases, it appears that a tertiary endosymbiosis has taken place, in which an alga that contained a secondary plastid was itself incorporated (Delwiche, 1999; Keeling, 2010).

In a few cases, it seems clear that the ability to carry out photosynthesis has been lost from some lines of cells, but the plastid has not entirely disappeared. This appears to be so for a number of parasitic organisms, including the pathogens that cause malaria and toxoplasmosis (Delwiche, 1999; Cavalier-Smith, 2000; Moore et al., 2008).

12.15 Following endosymbiosis, many genes were transferred to the nucleus, and proteins were reimported to the chloroplast

As discussed in Chapter 10, the number of genes found in chloroplasts is much reduced from what is found in a typical bacterium. Most of the genetic information needed to make a chloroplast is found in the nucleus. The majority of chloroplast genes code either for integral membrane proteins associated with the photosynthetic apparatus or for elements of the chloroplast's protein synthesis machinery. Typical green plant or green algal chloroplasts contain only about 60–80 protein-coding genes. The two other groups of primary endosymbionts, red algae, and glaucocystophytes, contain significantly larger numbers, 200 and 136, respectively (Martin and Herrmann, 1998; Green, 2011). This number is much less than the ~3000 protein-coding genes found in a typical cyanobacterium such as *Synechocystis*.

What has happened to all the missing genes? Presumably, the primary endosymbiont had a full complement of genes, which have, over the years, disappeared from the chloroplast genome. Some of these genes were obviously not needed for green algal or plant chloroplasts, such as those that code for cell wall biosynthesis or phycobilisome components. Some of these genes were already present in the host and were therefore redundant, so many of them have been lost. However, most genes have been transferred to the nucleus, and their protein gene products are reimported into the chloroplast. Estimates suggest that between 1000 and 5000 proteins are imported into the chloroplast, dwarfing the number coded for and synthesized inside the chloroplast (Martin and Herrmann, 1998).

How did this wholesale gene transfer take place? How were the products reimported? And what functional advantage does this give to the organism? Definitive answers to all these questions are not yet available, although a number of aspects of this complex story are beginning to become clear (Martin and Herrmann, 1998; McFadden, 1999). Several steps need to take place before a chloroplast gene is successfully transferred to the nucleus. First, and most easily, some DNA must be lost from a chloroplast and must find its way to the nucleus in which it is subsequently incorporated. This is facilitated by the fact that most photosynthetic eukaryotes contain multiple chloroplasts, and each chloroplast usually contains multiple copies of the chloroplast genome. So, it is easy to imagine how a single copy of a gene might find its way to the nucleus without disrupting the function of the gene in the chloroplast. Second, the transferred gene must be expressed and targeted to the chloroplast. This might happen all at once, or it might go through an intermediate stage in which the protein is expressed in the cytoplasm (Martin and Herrmann, 1998). Finally, the transferred gene is properly expressed and targeted from the nuclear copy of the gene, and the chloroplast copy becomes redundant and is lost, completing the process of genetic transfer.

In a number of cases, it is clear that a gene lost from the chloroplast is replaced by the gene product not from that gene but, rather, from a different gene that codes for a similar protein. In some cases, the nuclear genes for chloroplast proteins are thought to have originated in the bacterium that became mitochondria. These genes were transferred to the nucleus and acquired chloroplast targeting sequences, and the gene products were subsequently imported to the chloroplast. It is therefore clear that there need not be a strict one-to-one correspondence between genes transferred to the nucleus and proteins reimported.

The gene transfer has been so extensive in chloroplasts (and even more so in mitochondria) that there must clearly be strong selective pressures driving the process. There have been several suggestions as to what these pressures might be, probably all of which are at least partially valid; but a completely satisfying explanation is not yet at hand. First, an organellar gene is essentially a clonal, asexual entity and may accumulate deleterious mutations if it does not have the ability to eliminate these mutations by recombination during sexual reproduction. This phenomenon is called "Muller's ratchet" (Muller, 1964) and is known to operate in a number of cases, particularly in animal mitochondrial genomes. However, occasional recombination with other chloroplasts and the multiple genome copies of chloroplast DNA within a chloroplast may mitigate this process, so that, surprisingly, the observed rate of nucleotide substitution in chloroplast genomes is not very different from that in the nucleus. This suggests that Muller's ratchet may not be a dominant factor.

An additional factor that has been identified is the fact that reactive oxygen species are produced as byproducts during the functioning of chloroplasts and mitochondria, and these will inevitably increase the amount of free radical-induced mutations to nearby genes. Transfer of genes to the nucleus minimizes this problem (Allen and Raven, 1996). However, the hydrogenosome, a hydrogen-producing organelle found in some primitive eukaryotes that lack mitochondria, has no oxygen metabolism, yet has completely transferred its genetic information to the nucleus, so other factors must also be involved (Martin and Müller, 1998).

Finally, it may be largely an issue of control, in that regulation and coordination of the three genomes in eukaryotic photosynthetic organisms (the nucleus, chloroplast, and mitochondrion) may be more efficient when directed from a single source than when decentralized. In support of this view is the fact that the genes most likely to be transferred are regulatory genes, while the catalytic genes for chloroplast functions tend to be left behind. An example of this is found in rubisco, in that the large catalytic subunit is always found in the chloroplast, while the small regulatory subunit is found in the nucleus in all green algae and plants. This pattern is repeated many times over with other chloroplast systems.

Why, then, have any genes remained in the organelle? Some genes may be most efficiently regulated by the ambient redox potential and are therefore retained in the organelle (Allen and Raven, 1996). It may also be that some proteins are not capable of being efficiently imported, so their genes have remained in the organelles, and the proteins are made close to their site of function. In one case, this idea has been tested, by deleting the gene for the large subunit of rubisco from the tobacco chloroplast and introducing it into the nucleus, along with an appropriate chloroplast targeting sequence (Kanevski and Maliga, 1994). The resulting transgenic plant is capable of autotrophic growth, suggesting that, at least for this protein, there is no insurmountable barrier to the relocation of the gene to the nucleus.

12.16   Evolution of carbon metabolism pathways

The Calvin–Benson cycle is clearly an example of a metabolic pathway that has recruited bits and pieces from a number of other pathways. As discussed in Chapter 9, many of the Calvin–Benson cycle enzymes are related to enzymes that function in glycolysis (or gluconeogenesis) plus the pentose phosphate pathway. The Calvin–Benson cycle is found in purple photosynthetic bacteria, and this was evidently the source of the cyanobacterial cycle (Martin and Schnarrenberger, 1997), although rubisco has a complex evolutionary history (Nisbet et al., 2007; Tabita et al., 2007).

There are two enzymes in the Calvin–Benson' cycle that are not found in the presumably more ancient carbohydrate metabolism pathways mentioned above: rubisco and phosphoribulokinase (PRK), the enzyme that synthesizes the substrate for rubisco. For many years, rubisco was thought to be a marker for photosynthesis, although this view

has now changed significantly. A variant of rubisco, known as form IV, has been found in many non-photosynthetic bacteria and archaea, where it is involved in sulfur metabolism (Tabita et al., 2007). As discussed briefly in Chapter 9, the selection of rubisco as the carboxylation enzyme for oxygenic photosynthesis is something of an enigma. Of all the known carboxylation enzymes (e.g., phosphoenolpyruvate carboxylase used in C4 and CAM), Rubisco seems the least well adapted to cope with an oxygen-containing atmosphere. Evolution has obviously improved rubisco over the years, increasing its specificity factor. However, the losses due to oxygenation are significant in normal C3 photosynthesis, typically 30% or more under adverse conditions. It may be that intrinsic chemical limitations have precluded the additional improvement of rubisco efficiency in higher plants.

The second response to the problem of oxygenation has been that organisms have evolved mechanisms for concentrating CO_2, so that the oxygenation reaction is thereby minimized. These mechanisms include C4 photosynthesis and CAM (although both these mechanisms also have the benefit of conserving water), as well as the active CO_2-concentrating mechanisms found in cyanobacteria and algae. These mechanisms are discussed in detail in Chapter 9.

The explanation that seems to fit the available data best is that rubisco is an enzyme that worked very well under the primordial conditions of no or low O_2 pressure and high CO_2 pressure. But after the advent of oxygenic photosynthesis, this mechanism began to be a limitation, so it was improved, first by increasing the specificity factor of rubisco and then by evolving the various CO_2-concentration mechanisms. Under present conditions, the rubisco system is decidedly less than perfect, but apparently it is not possible to change to a completely different system. Perhaps surprisingly, no oxygenic photosynthetic organism has abandoned rubisco completely for another primary carboxylation mechanism. It must be that the number of mutations and innovations that would be required to accomplish this change is so large that the probability of it happening is negligible. The aerobic anoxygenic phototrophic purple bacteria discussed in Chapter 10 have apparently lost the ability to do autotrophic carbon fixation. Their obligate aerobic lifestyle may have made the Calvin–Benson cycle so inefficient that they dispensed with it entirely and instead import organic matter from the environment and grow photoheterotrophically (Swingley *et al.*, 2007).

Considerable evidence suggests that C4 and CAM photosynthesis are innovations in response to decreasing atmospheric levels of CO_2. CAM is the older of the two and first appeared more than 250 million years ago (MYA) (Raven and Spicer, 1995). It may have appeared first as a CO_2-concentrating mechanism in aquatic plants during the Carboniferous era, 300 MYA, when CO_2 levels were low (see above) (Keeley, 1998). C4 photosynthesis is thought to be much more recent, dating from 15–30 MYA, and is thought to have evolved independently numerous times (Sage and Monson, 1999; Christin and Osborne 2013; Denton *et al.*, 2013).

Some anoxygenic photosynthetic organisms use carbon reduction cycles that do not involve rubisco or the Calvin–Benson cycle (Chapter 9). The green sulfur bacteria use a reverse TCA cycle, which is driven backwards by reduced ferredoxin (Buchanan and Arnon, 1990). Some of the enzymes of this system are apparently highly sensitive to oxygen, so it is not a good candidate for an alternative to the Calvin–Benson cycle. In addition, the filamentous anoxygenic phototrophic bacteria use a still different pathway involving hydroxypropionate (Fuchs, 2011). The oxygen sensitivity of this pathway has not been investigated in detail, but the organism is fully capable of tolerating oxygen and is often subjected to hyperoxic conditions in its native environment, where it grows under a layer of cyanobacteria, so this carbon fixation mechanism is probably stable to oxygen.

References

Allen, J. F. and Raven, J. A. (1996) Free-radical-induced mutation vs redox regulation: Costs and benefits of genes in organelles. *Journal of Molecular Evolution* 42: 482–492.

Allen, J. F. (2005) A redox switch hypothesis for the origin of two light reactions in photosynthesis. *FEBS Letters* 579: 963–968.

Apt, K. E., Collier, J. L., and Grossman, A. R. (1995) Evolution of the phycobiliproteins. *Journal of Molecular Biology* 248: 79–96.

Armstrong, G. A. (1998) Greening in the dark: Light-independent chlorophyll biosynthesis from anoxygenic photosynthetic bacteria to gymnosperms. *Journal of Photochemistry and Photobiology B: Biology* 43: 87–100.

Arndt, N. T. and Nisbet, E. G. (2012) Processes on the young earth and the habitats of early life. *Annual Review of Earth and Planetary Science* 40: 521–549.

Bada, J. L. (2013) New insights into prebiotic chemistry from Stanley Miller's spark discharge experiments. *Chemical Society Reviews* 42: 2186–2196.

Baranov, S. V., Ananyev, G. M., Klimov, V. V., and Dismukes, G. C. (2000) Bicarbonate accelerates assembly of the inorganic core of the water-oxidizing complex in manganese-depleted photosystem II: A proposed biogeochemical role for atmospheric carbon dioxide in oxygenic photosynthesis. *Biochemistry* 39: 6060–6065.

Baymann, F., Brugna, M., Mühlenhoff, U., and Nitschke, W. (2001) Daddy, where did (PS)I come from? *Biochimica et Biophysica Acta* 1507: 291–310.

Beale, S. I. (1999) Enzymes of chlorophyll biosynthesis. *Photosynthesis Research* 60: 43–73.

Becerra, A., Delaye, L., Islas, S., and Lazcano, A. (2007) The very early stages of biological evolution and the nature of the last common ancestor of the three major cell domains. *Annual Review of Ecology, Evolution and Systematics* 38: 361–379.

Benner, S. A., Kim, H.-J., and Carrigan, M. A. (2012) Asphalt, water, and the prebiotic synthesis of ribose, ribonucleosides, and RNA. *Accounts of Chemical Research* 45: 2025–2034.

Berner, R. A., Petsch, S. T., Lake, J. A., Beerling, D. J., Popp, B.N., Lane, R.S., Laws, E.A., Westley, M.B., Cassar, N., Woodward, F.I., and Quick, W. P. (2000) Isotope fractionation and atmospheric oxygen: Implications for phanerozoic O_2 evolution. *Science* 287: 1630–1633.

Bhattacharya, D. and Medlin, L. (1998) Algal phylogeny and the origin of land plants. *Plant Physiology* 116: 9–15.

Bird, L. J., Bonnefoy, V., and Newman, D. K. (2011) Bioenergetic challenges of microbial iron metabolisms. *Trends in Microbiology* 19: 330–340.

Blank, C. E. and Sanchez-Baracaldo, P. (2010) Timing of morphological and ecological innovations in the cyanobacteria – a key to understanding the rise in atmospheric oxygen. *Geobiology* 8: 1–23.

Blankenship, R. E. (1992) Origin and early evolution of photosynthesis. *Photosynthesis Research* 33: 91–111.

Blankenship, R. E. (1994) Protein structure, electron transfer and evolution of prokaryotic photosynthetic reaction centers. *Antonie van Leeuwenhoek Journal of Microbiology* 65: 311–329.

Blankenship, R. E. (2001) Molecular evidence for the evolution of photosynthesis. *Trends in Plant Science* 6: 4–6.

Blankenship, R. E. (2010) Early evolution of photosynthesis. *Plant Physiology* 154: 434–438.

Blankenship, R. E. (2017) How cyanobacteria went green. *Science* 355: 1372–1373.

Blankenship, R. E. and Hartman, H. (1998) The origin and evolution of oxygenic photosynthesis. *Trends in Biochemical Sciences* 23: 94–97.

Blankenship, R. E. and Prince, R. C. (1985) Excited state redox potentials and the Z scheme of photosynthesis. *Trends in Biochemical Sciences* 10: 382–383.

Boyd, E. S., Anbar, A. D., Miller, S., Hamilton, T. L., Lavin, M., and Peters, J. W. (2011) A late methanogen origin for molybdenum-dependent nitrogenase. *Geobiology* 9: 221–232.

Brasier, M. D., Green, O. R., Jephcoat, A. P., Kleppe, A. K., Van Kranendonk, M. J., Lindsay, J. F., Steele, A., and Grassineau, N. V. (2002) Questioning the evidence of Earth's oldest fossils. *Nature* 416: 76–81.

Brochier-Armanet, C., Talla, E., and Gribaldo, S. (2009) The multiple evolutionary histories of dioxygen reductases: Implications for the origin and evolution of aerobic respiration. *Molecular Biology and Evolution* 26: 285–297.

Brocks, J. J., Logan, G. A., Buick, R., and Summons, R. E. (1999) Archean molecular fossils and the early rise of eukaryotes. *Science* 285: 1033–1036.

Bryant, D. A., Hunter, C. N., and Warren, M. J. (2020) Biosynthesis of the modified tetrapyrroles—the pigments of life. *Journal of Biological Chemistry* 295: 6888–6925.

Buchanan, B. B. and Arnon, D. I. (1990) A reverse Krebs cycle in photosynthesis – Consensus at last. *Photosynthesis Research* 24: 47–53.

Buick, R. (2008) Early evolution of oxygenic photosynthesis. *Philosophical Transactions of the Royal Society B* 363: 2731–2743.

Burke, D. H., Hearst, J. E., and Sidow, A. (1993) Early evolution of photosynthesis – Clues from nitrogenase and chlorophyll iron proteins. *Proceedings of the National Academy of Sciences USA* 90: 7134–7138.

Büttner, M, Xie, D.-L., Nelson, H., Pinther, W., Hauska, G., and Nelson, N. (1992) Photosynthetic reaction center genes in green sulfur bacteria and in photosystemi are related. *Proceedings of the National Academy of Sciences USA* 89: 8135–8139.

Cardona, T. and Rutherford A. W. (2019) Evolution of photochemical reaction centres: More twists? *Trends in Plant Science* 24: 1008–1021.

Cardona, T., Sánchez-Baracaldo, P., Rutherford, A. W., and Larkum, A. W. (2019) Early archean origin of photosystem II. *Geobiology* 17: 127–150.

Castresana, J. and Moreira, D. (1999) Respiratory chains in the last common ancestor of living organisms. *Journal of Molecular Evolution* 49: 453–460.

Catling, D. C. (2014) The great oxidation event transition. In: K. K. Turekian and H. D. Holland (eds.) *Treatise on Geochemistry*, 2nd Edn, Vol. 6. Oxford: Elsevier Science, pp. 177–194.

Catling, D. C. and Zahnle, K. J. (2020) The archean atmosphere. *Science Advances* 6: eaax1420.

Cavalier-Smith, T. (1992) The number of symbiotic origins of organelles. *Bio Systems* 28: 91–106.

Cavalier-Smith, T. (2000) Membrane heredity and early chloroplast evolution. *Trends in Plant Science* 5: 174–182.

Chew, A. G. M. and Bryant, D. A. (2007) Chlorophyll biosynthesis in bacteria: The origins of structural and functional diversity. *Annual Review of Microbiology* 61: 113–129.

Christin, P.-A. and Osborne, C. P. (2013) The recurrent assembly of C4 photosynthesis, an evolutionary tale. *Photosynthesis Research* 117: 163–175.

Cohen, Y., Jorgensen, B. B., Revsbech, N. P., and Poplowski, R. (1986) Adaptation to hydrogen sulfide of oxygenic and anoxygenic photosynthesis among cyanobacteria. *Applied and Environmental Microbiology* 51: 398–407.

Curtis, B. A., Tanifuji, G., Burki, F., *et al.* (2012) Algal genomes reveal evolutionary mosaicism and the fate of nucleomorphs. *Nature* 492: 59–65.

Dagan, T., Roettger, M., Stucken, K., *et al.* (2013) Genomes of stigonematalean cyanobacteria (subsection v) and the evolution of oxygenic photosynthesis from prokaryotes to plastids. *Genome Biology and Evolution* 5: 31–44.

Damer, B. and Deamer, D. W. (2020) The hot spring hypothesis for an origin of life. *Astrobiology* 20: 429–452.

Davankov, V. A. (2018) Biological homochirality on the Earth, or in the Universe? A selective review. *Symmetry* 10: 749.

Deamer, D. W. (2017) The role of lipid membranes in life's origin. *Life* 7: 5.

Deamer, D. W. and Szostak, J. W., (eds.) (2010) *The Origins of Life*. Cold Spring Harbor, NY: Cold Spring Harbor Press.

Delwiche, C. F. (1999) Tracing the thread of plastid diversity through the tapestry of life. *American Naturalist* 154: S164–S177.

Delwiche, C. F. and Palmer, J. D. (1997) The origin of plastids and their spread via secondary symbiosis. In: D. Bhattacharya (ed.) *Origins of Algae and Their Plastids*. Vienna: Springer, pp. 53–86.

Denton, A. K., Simon, R., and Weber, A. P. M. (2013) C4 photosynthesis: From evolutionary analyses to strategies for synthetic reconstruction of the trait. *Current Opinion in Plant Biology* 16: 315–321.

Des Marais, D. J., Strauss, H., Summons, R. E., and Hayes, J. M. (1992) Carbon isotope evidence for the stepwise oxidation of the proterozoic environment. *Nature* 359: 605–609.

Dismukes, G. C. (1996) Manganese enzymes with binuclear active sites. *Chemical Reviews* 96: 2909–2926.

Dismukes, G. C., Klimov, V. V., Baranov, S. V., Kozlov, Y. N., DasGupta, J., and Tyryshkin, A. (2001) The origin of atmospheric oxygen on Earth: The innovation of oxygenic photosynthesis. *Proceedings of the National Academy of Sciences USA* 98: 2170–2175.

Dolganov, N. A. M., Bhaya, D., and Grossman, A. R. (1995) Cyanobacterial protein with similarity to the chlorophyll *a/b* binding-proteins of higher-plants – Evolution and regulation. *Proceedings of the National Academy of Sciences USA* 92: 636–640.

Doolittle, R. F. (1994) Convergent evolution: The need to be explicit. *Trends in Biochemical Sciences* 19: 15–18.

Doolittle, W. F. (1999) Phylogenetic classification and the universal tree. *Science* 284: 2124–2128.

Doolittle, W. F. (2000) Uprooting the tree of life. *Scientific American* 282(2): 90–95.

Dorrell, R. G. and Howe, C. J. (2012) What makes a chloroplast? Reconstructing the establishment of photosynthetic symbioses. *Journal of Cell Science* 125: 1865–1875.

Douglas, S. (1998) Plastid evolution: Origins, diversity, trends. *Current Opinion of Genetics and Development* 8: 655–661.

Ducluzeau A.-L., Schoepp-Cothenet, B., van Lis, R., Baymann, F., Russell, M. J., and Nitschke W. (2014) The evolution of respiratory O_2/NO reductases: An out-of-the-phylogenetic-box perspective. *Journal of the Royal Society Interface* 11: 20140196.

Durnford, D. G., Deane, J. A., Tan, S., McFadden, G. I., Gantt, E., and Green, B. R. (1999) A phylogenetic assessment of the eukaryotic light-harvesting antenna proteins, with implications for plastid evolution. *Journal of Molecular Evolution* 48: 59–68.

Engelken, J., Brinkmann, H., and Adamska, I. (2010) Taxonomic distribution and origins of the extended LHC (light-harvesting complex) antenna protein superfamily. *BMC Evolutionary Biology* 10: 233.

Farquhar, J., Bao, H., and Thiemens, M. (2000) Atmospheric influence of Earth's earliest sulfur cycle. *Science* 289: 756–758.

Fischer, W. W., Hemp, J., and Johnson, J. J. (2016) Evolution of oxygenic photosynthesis. *Annual Review of Earth and Planetary Sciences* 44: 647–683.

Fuchs, G. (2011) Alternative pathways of carbon dioxide fixation: Insights into the early evolution of life? *Annual Review of Microbiology* 65: 631–658.

Ghebreamlak, S. M. and Mansoorabadi, S. O. (2020) Divergent members of the nitrogenase superfamily: Tetrapyrrole biosynthesis and beyond. *CHEMBIOCHEM* 21: 1723–1728.

Gilbert, W. (1986) The RNA world. *Nature* 319: 618.

Goldblatt, C. and Zahnle, K. J. (2011) The faint young sun paradox remains. *Nature* 474: E3–E4.

Gould, S. B., Waller, R. F., and McFadden, G. I. (2008) Plastid evolution. *Annual Review of Plant Biology* 59: 491–517.

Granick, S. (1965) Evolution of heme and chlorophyll. In: V. Bryson and H. J. Vogel (eds.) *Evolving Genes and Proteins*. New York: Academic Press, pp. 67–88.

Graur, D. and Li, W.-H. (2000) *Fundamentals of Molecular Evolution*. Sunderland, MA: Sinauer Associates.

Gray, M. W. (1992) The endosymbiont hypothesis revisited. *International Review of Cytology* 141: 233–357.

Gray, M. W. (1999) Evolution of organellar genomes. *Current Opinion of Genetics and Development* 9: 678–687.

Green, B. R. (2001) Was "molecular opportunism" a factor in the evolution of different photosynthetic light harvesting pigment systems? *Proceedings of the National Academy of Sciences USA* 98: 2119–2121.

Green, B. R. (2003) The evolution of light-harvesting antennas. In: B. R. Green and W. W. Parson (eds.) *Light-Harvesting Antennas in Photosynthesis*. Dordrecht, Springer, pp. 129–168.

Green, B. R. (2011) Chloroplast genomes of photosynthetic eukaryotes. *Plant Journal* 66: 34–44.

Green, B. R. (2019) What happened to the phycobilisome? *Biomolecules* 9: 748.

Green, B. R. and Durnford, D. G. (1996) The chlorophyll–carotenoid proteins of oxygenic photosynthesis. *Annual Review of Plant Physiology and Plant Molecular Biology* 47: 685–714.

Green, B. R. and Kühlbrandt, W. (1995) Sequence conservation of light-harvesting and stress-response proteins in relation to the three-dimensional molecular structure of LHCII. *Photosynthesis Research* 44: 139–148.

Green, B. R. and Parson, W. W., (eds.) (2003) *Light-Harvesting Antennas*. Dordrecht: Kluwer Academic Press.

Grettenberger, C. L., Sumner, D. Y., Wall, K., Brown. C. T., Eisen, J. A., Mackey, T. J., Hawes, I., Jospin, G., and Jungblut, A. D. (2020) A phylogenetically novel cyanobacterium most closely related to *Gloeobacter*. *The ISME Journal* 14: 2142–2152.

Griffiths, G. (2007) Cell evolution and the problem of membrane topology. *Nature Reviews Molecular Cell Biology* 8: 1018–1024.

Grossman, A. R., Bhaya, D., Apt, K. E., and Kehoe, D. M. (1995) Light-harvesting complexes in oxygenic photosynthesis: Diversity, control, and evolution. *Annual Review of Genetics* 29: 231–288.

Guergova-Kuras, M., Boudreaux, B., Joliot, A., Joliot, P., and Redding, *K.* (2001) Evidence for two active branches for electron transfer in photosystem I. *Proceedings of the National Academy of Sciences USA* 98: 4437–4442.

Gupta, R. S. (2012) Origin and spread of photosynthesis based upon conserved sequence features in key bacteriochlorophyll biosynthesis proteins. *Molecular Biology and Evolution* 29: 3397–3412.

Hamilton, T. L. (2019) The trouble with oxygen: The ecophysiology of extant phototrophs and implications for the evolution of oxygenic photosynthesis. *Free Radical Biology and Medicine* 140: 233–249.

Han, T.-M. and Runnegar, B. (1992) Megascopic eukaryotic algae from the 2.1 billion-year-old Negaunee iron formation, Michigan. *Science* 257: 232–235.

Hartman, H. (1975) Speculations on origin and evolution of metabolism. *Journal of Molecular Evolution* 4: 359–370.

Hartman, H. (1998) Photosynthesis and the origin of life. *Origins of Life and Evolution of the Biosphere* 28: 515–521.

Hohmann-Marriott, M. F. and Blankenship, R. E. (2011) Evolution of photosynthesis. *Annual Review of Plant Biology* 62: 515–548.

Horowitz, N. H. (1945) On the evolution of biochemical synthesis. *Proceedings of the National Academy of Sciences USA* 31: 153–157.

Howe, C. J., Barbrook, A. C., Nisbet, R. E. R., Lockhart P. J., and Larkum, A. W. D. (2008) The origin of plastids. *Philosophical Transactions of the Royal Society B* 363: 2675–2685.

Jensen, R. A. (1976) Enzyme recruitment in evolution of new function. *Annual Review of Microbiology* 30: 409–425.

Joliot, P. and Joliot, A. (1999) in vivo; analysis of the electron transfer within photosystem I: Are the two phylloquinones involved? *Biochemistry* 38: 11130–11136.

Khadka, B., Adeolu, M., Blankenship, R. E., and Gupta, R. S. (2017) Novel insights into the origin of photosynthetic reaction centers I and II based on conserved indels in the core proteins. *Photosynthesis Research* 131: 159–171.

Kanevski, I. and Maliga, P. (1994) Relocation of the plastid *rbcL* gene to the nucleus yields functional ribulose-1,5-bisphosphate carboxylase in tobacco chloroplasts. *Proceedings of the National Academy of Sciences USA* 91: 1969–1973.

Kasting, J. F. (1993) Earth's early atmosphere. *Science* 259: 920–926.

Keeley, J. E. (1998) CAM photosynthesis in submerged aquatic plants. *Botanical Reviews* 64: 121–175.

Keeling, P. J. (2010) The endosymbiotic origin, diversification and fate of plastids. *Philosophical Transactions of the Royal Society B* 365: 729–748.

Keeling, P. (2013) The number, speed and impact of plastid endosymbioses in eukaryotic evolution. *Annual Review of Plant Biology* 64: 583–607.

Kikuchi, H., Wako, H., Yura, K., *et al.* (2000) Significance of a two-domain structure in subunits of phycobiliproteins revealed by the normal mode analysis. *Biophysical Journal* 79: 1587–1600.

Knoll, A. (1992) The early evolution of eukaryotes: A geological perspective. *Science* 256: 622–627.

Knoll, A. H., Canfield, D. E., and Konhauser, K. O., (eds.) (2012) *Fundamentals of Geobiology*. Oxford: Wiley-Blackwell.

Koonin, E. V., Krupovic, M., Ishino, S., and Ishino, Y. (2020) The replication machinery of LUCA: Common origin of DNA replication and transcription. *BMC Biology* 18: 61.

Larkum, A. W. D. (1999) The evolution of algae. In: J. Seckbach (ed.) *Enigmatic Microorganisms and Life in Extreme Environments*. Dordrecht: Kluwer Academic Publishers, pp. 29–48.

Larkum, A. W. D. (2006) The evolution of chlorophylls and photosynthesis. In: B. Grimm, R. J. Porra, W. Rüdiger, and H. Scheer (eds.) *Chlorophylls and Bacteriochlorophylls: Biochemistry, Biophysics, Functions and Applications*. Dordrecht, Springer, pp. 261–282.

LaRoche, J., van der Staay, G. W. M., Partensky, F., Ducret, A., Aebersold, R., Li, R., Golden, S. S., Hiller, R. G., Wrench, P. M., Larkum, A. W. D., and Green, B. R. (1996) Independent evolution of the prochlorophyte and green plant chlorophyll *a/b* light-harvesting proteins. *Proceedings of the National Academy of Sciences USA* 93: 15244–15248.

Lazcano, A. and Miller, S. L. (1999) On the origin of metabolic pathways. *Journal of Molecular Evolution* 49: 424–431.

Li, X. P., Bjorkman, O., Shih, C., Grossman, A. R., Rosenquist, M,, Jansson, S,, and Niyogi, K. K. (2000) A pigment binding protein essential for regulation of photosynthetic light harvesting. *Nature* 403: 391–395.

Liang, M.-C., Hartman, H., Kopp, R. E., *et al.* (2006) Production of hydrogen peroxide in the atmosphere of a Snowball Earth and the origin of oxygenic photosynthesis. *Proceedings of the National Academy of Sciences USA* 103: 18896–18899.

Liebl, U., Mockensturm-Wilson, M., Trost, J. T., Brune, D. C., Blankenship, R. E., and Vermaas, W. F. J. (1993) Single core polypeptide in the reaction center of the photosynthetic bacterium *Heliobacillus mobilis*: Structural implications and relations to other photosystems. *Proceedings of the National Academy of Sciences USA* 90: 7124–7128.

Liu, D., Zhang, J., Lu, C., Xia, Y., Liu, H., Jiao, N. Xun, L. Liu, J. (2020) *Synechococcus* sp. strain PCC7002 uses sulfide: Quinone oxidoreductase to detoxify exogenous sulfide and to convert endogenous sulfide to cellular sulfane sulfur. *mBio* 11: e03420.

Lockhart, P. J., Larkum, A. W. D., Steel, M. A., Waddell, P. J., and Penny, D. (1996a) Evolution of chlorophyll and bacteriochlorophyll: The problem of invariant sites in sequence analysis. *Proceedings of the National Academy of Sciences USA* 93: 1930–1934.

Lockhart, P. J., Steel, M. A., and Larkum, A. W. D. (1996b) Gene duplication and the evolution of photosynthetic reaction center proteins. *FEBS Letters* 385: 193–196.

Lyons, T. W., Reinhard, C. T., and Planavsky, N. J. (2014) The rise of oxygen in Earth's early ocean and atmosphere. *Nature* 506: 307–315.

Maher, K. A. and Stevenson, D. J. (1988) Impact frustration of the origin of life. *Nature* 331: 612–614.

McFadden, G. I. (1999) Endosymbiosis and evolution of the plant cell. *Current Opinion in Plant Biology* 2: 513–519.

Margulis, L. (1993) *Symbiosis in Cell Evolution:Microbial Communities in the Archean and Proterozoic Eons*. San Francisco: W. H. Freeman.

Mathis, P. (1990) Compared structure of plant and bacterial photosynthetic reaction centers. Evolutionary implications. *Biochimica et Biophysica Acta* 1018: 163–167.

Martin, W. (1999) Mosaic bacterial chromosomes: A challenge en route to a tree of genomes. *Bio Essays* 21: 99–104.

Martin, W. and Herrmann, R. G. (1998) Gene transfer from organelles to the nucleus: How much, what happens, and why? *Plant Physiology* 118: 9–17.

Martin, W. and Müller, M. (1998) The hydrogen hypothesis for the first eukaryote. *Nature* 392: 37–41.

Martin, W. and Schnarrenberger, C. (1997) The evolution of the Calvin–Benson' cycle from prokaryotic to eukaryotic chromosomes: A case study of functional redundancy in ancient pathways through endosymbiosis. *Current Genetics* 32: 1–18.

Martin, W., Stoebe, B., Goremykin, V., Hansmann, S.,Hasagawa, M., and Kowalik, K. V. (1998) Gene transfer to the nucleus and the evolution of chloroplasts. *Nature* 393: 162–165.

Martin, W., Baross, J., Kelley, D., and Russell, M. J. (2008) Hydrothermal vents and the origin of life. *Nature Reviews Microbiology* 6: 805–814.

Martin, W., Bryant, D. A., and Beatty, J. T. (2018) A physiological perspective on the origin and evolution of photosynthesis. *FEMS Microbiology Reviews* 42: 205–231.

Mauzerall, D. (1992) Light, iron, Sam Granick and the origin of life. *Photosynthesis Research* 33: 163–170.

Meyer, T. E. (1994) Evolution of photosynthetic reaction centers and light-harvesting chlorophyll proteins. *BioSystems* 33: 167–175.

Miller, S. L. (1953) Production of amino acids under primitive Earth conditions. *Science* 117: 528–529.

Miller, S. L., Schopf, J. W., and Lazcano, A. (1997) Oparin's "origin of life": Sixty years later. *Journal of Molecular Evolution* 44: 351–353.

Mix, L. J., Haig, D., and Cavanaugh, C.M. (2005) Phylogenetic analyses of the core antenna domain: Investigating the origin of photosystem I. *Journal of Molecular Evolution* 60: 153–163.

Miyashita, H., Ikemoto, H., Kurano, N., *et al.* (1996) Chlorophyll *d* as a major pigment. *Nature* 383: 402.

Montané, M.-H. and Kloppstech, K. (2000) The family of light-harvesting proteins (LHCs, ELIPs, HLIPs): Was the harvesting of light their primary function? *Gene* 258: 1–8.

Moore, K. R., Magnabosco, C., Momper, L., *et al.* (2019) An expanded ribosomal phylogeny of cyanobacteria supports a deep placement of plastids. *Frontiers in Microbiology* 10: 1612.

Moore, R. B., Oborník M., Janouskovec J., *et al.* (2008). A photosynthetic alveolate closely related to apicomplexan parasites. *Nature* 45: 959–963.

Moreira, D., LeGuyader, H., and Phillippe, H. (2000) The origin of red algae and the evolution of chloroplasts. *Nature* 405: 69–72.

Moser, J., Lange, C., Krausze, J., Rebelein, J., Schubert, W.-D., Ribbe, M. W., Heinz, D. W., and Jahn, D. (2013) Structure of ADP-aluminium fluoride-stabilized protochlorophyllide oxidoreductase complex. *Proceedings of the National Academy of Sciences USA* 110: 2094–2098.

Muchowska, K. B., Varma, S. J., and Moran, J. (2020) Nonenzymatic metabolic reactions and life's origins. *Chemical Reviews* 120: 7708–7744.

Mulkidjanian, A. Y. and Junge, W. (1997) On the origin of photosynthesis as inferred from sequence analysis – A primordial UV-protector as common ancestor of reaction centers and antenna proteins. *Photosynthesis Research* 51: 27–42.

Muller, H. J. (1964) The relation of recombination to mutational advance. *Mutation Research* 1: 2–9.

Muraki, N., Nomata, J., Ebata, K., Mizoguchi, T., Shiba, T., Tamiaki, H., Kurisu, G., and Fujita, Y. (2010) X-ray crystal structure of the light-independent protochlorophyllide reductase. *Nature* 465: 110–114.

Nelson, N. (2013) Evolution of photosystem I and the control of global enthalpy in an oxidizing world. *Photosynthesis Research* 116: 145–151.

Nisbet, E. G. and Sleep, N. H. (2001) The habitat and nature of early life. *Nature* 409: 1083–1091.

Nisbet, E. G., Cann, J. R., and van Dover, C. L. (1995) Origins of photosynthesis. *Nature* 373: 479–480.

Nisbet, E. G., Grassineau, N. V., Howe, C. J., *et al.* (2007) The age of Rubisco: the evolution of oxygenic photosynthesis. *Geobiology* 5: 311–335.

Nitschke, W. and Rutherford, A. W. (1991) Photosynthetic reaction centres – Variations on a common structural theme. *Trends in Biochemical Sciences* 16: 241–245.

Nitschke, W., Mattioli, T., and Rutherford, A. W. (1996) The FeS-type photosystems and the evolution of photosynthetic reaction centers. In: H. Baltscheffsky (ed.) *Origin and Evolution of Biological Energy Conversion.* New York: VCH Publishers, pp. 177–203.

Nitschke, W., Muhlenhoff, U., and Liebl, U. (1998) Evolution. In: A. S. Raghavendra (ed.) *Photosynthesis: A Comprehensive Treatise.* Cambridge: Cambridge University Press, pp. 285–304.

Niyogi, K. K. and Truong, T. B. (2013) Evolution of flexible non-photochemical quenching mechanisms that regulate light harvesting in oxygenic photosynthesis. *Current Opinion in Plant Biology* 16: 307–314.

Nowack, E. C. M. and Weber, A. P. M. (2018) Genomics-informed insights into endosymbiotic organelle evolution in photosynthetic eukaryotes. *Annual Review of Plant Biology* 69: 51–84.

Olson, J. M. (1999) Early evolution of chlorophyll-based photosystems *Chemtracts – Biochemistry and Molecular Biology* 12: 468–482.

Olson, J. M. (2006) Photosynthesis in the archean era. *Photosynthesis Research* 88: 109–117.

Olson, J. M. and Pierson, B. K. (1987) Evolution of reaction centers in photosynthetic prokaryotes. *International Review of Cytology* 108: 209–248.

Orf, G. S., Gisriel, C., and Redding, K. E. (2018) Evolution of photosynthetic reaction centers: Insights from the structure of the heliobacterial reaction center. *Photosynthesis Research* 138: 11–37.

Orgel, L. E. (1994) The origin of life on the earth. *Scientific American* 271: 77–83.

Orgel, L. E. (1998) The origin of life – A review of facts and speculations. *Trends in Biochemical Sciences* 23: 491–495.

Oró, J., Miller, S. S., and Lazcano, A. (1990) The origin and early evolution of life on Earth. *Annual Review of Earth and Planetary Science* 18: 317–356.

Padan, E. (1979) Facultative anoxygenic photosynthesis in cyanobacteria. *Annual Review of Plant Physiology* 30: 27–40.

Palenik, B. and Haselkorn, R. (1992) Multiple evolutionary origins of prochlorophytes, the chlorophyll *b*-containing prokaryotes. *Nature* 355: 265–267.

Palmer, J. D. and Delwiche, C. F. (1996) Second-hand chloroplasts and the case of the disappearing nucleus. *Proceedings of the National Academy of Sciences USA* 93: 7432–7435.

Parfrey, L. W., Lahr, D. J. G., Knoll, A. H., and Katz, L. A. (2011) Estimating the timing of early eukaryotic diversification with multigene molecular clocks. *Proceedings of the National Academy of Sciences USA* 108: 13624–13629.

Penny, D. and Poole, A. (1999) The nature of the last universal common ancestor. *Current Opinion in Genetics and Development* 9: 672–677.

Powner, M. W., Gerland, B., and Sutherland, J. D. (2009) Synthesis of activated pyrimidine ribonucleotides in prebiotically plausible conditions. *Nature* 459: 239–242.

Price, D. C., Chan, C. X., Yoon, H. S., *et al.* (2012) *Cyanophora paradoxa* genome elucidates origin of photosynthesis in algae and plants. *Science* 335: 843–847.

Raven, J. A. and Spicer, R. A. (1995) The evolution of crassulacean acid metabolism. In: K. Winter and J. A. C. Smith (eds.) *Crassulacean Acid Metabolism: Biochemistry, Ecophysiology and Evolution.* Berlin: Springer, pp. 360–385.

Raymond, J. and Blankenship, R. E. (2008) The origin of the oxygen-evolving complex. *Coordination Chemistry Reviews* 252: 377–383.

Raymond, J., Zhaxybayeva, O., Gerdes, S., Gogarten, J. P., and Blankenship, R. E. (2002) Whole genome analysis of photosynthetic prokaryotes. *Science* 298: 1616–1620.

Raymond, J., Siefert, J., Staples, C., and Blankenship, R. E. (2004) The natural history of nitrogen fixation. *Molecular Biology and Evolution* 21: 541–554.

Sadekar, S., Raymond, J., and Blankenship, R. E. (2006) Conservation of distantly related membrane proteins: Photosynthetic reaction centers share a common structural core. *Molecular Biology and Evolution* 23: 2001–2007.

Sage, R. F. and Monson, R. K., (eds.) (1999) *C4 Plant Biology*. San Diego: Academic Press.

Sanchez-Barracaldo, P. and Cardona, T. (2019) On the origin of oxygenic photosynthesis and cyanobacteria. *New Phytologist* 225: 1440–1446.

Sato, N. (2020) Complex origins of chloroplast membranes with photosynthetic machineries: Multiple transfers of genes from divergent organisms at different times or a single endosymbiotic event? *Journal of Plant Research* 133: 15–33.

Sauer, K. and Yachandra, V. K. (2002) A possible evolutionary origin for the Mn-4 cluster of the photosynthetic water oxidation complex from natural MnO_2 precipitates in the early ocean. *Proceedings of the National Academy of Sciences USA* 99: 8631–8636.

Schirmer, T., Bode, W., Huber, R., Sidler, W., and Zuber, H. (1985) X-ray crystallographic structure of the light-harvesting biliprotein C-phycocyanin from the thermophilic cyanobacterium *Mastigocladus laminosus* and its resemblance to globin structures. *Journal of Molecular Biology* 184: 257–277.

Schopf, J. W. (1992) *Major Events in the History of Life*. Boston: Jones and Bartlett.

Schopf, J. W. (1993) Microfossils of the early archean apex chert: New evidence of the antiquity of life. *Science* 260: 640–646.

Schopf, J. W. (2011) The paleobiological record of photosynthesis. *Photosynthesis Research* 107: 87–101.

Schopf, J. W. and Kudryavtsev, A. B. (2012) Biogenicity of Earth's earliest fossils: A resolution of the controversy. *Gondwana Research* 22: 761–771.

Schopf, J. W., Kudryavtsev, A. B., Osterhout, J. T., Williford, K. H., Kitajima, K., Valley, J. W., and Sugitani, K. (2017) An anaerobic ~3400 Ma shallow-water microbial consortium: Presumptive evidence of Earth's Paleoarchean anoxic atmosphere. *Precambrian Research* 299: 309–318.

Schubert, W. D., Klukas, O., Saenger, W., Witt, H. T., Fromme, P., and Krauss, N. (1998) A common ancestor for oxygenic and anoxygenic photosynthetic systems: A comparison based on the structural model of photosystem I. *Journal of Molecular Biology* 280: 297–314.

Schütz, M., Brugna, M., Lebrun, E., Baymann, F., Huber, R., Stetter, K. O., Hauska, G., Toci, R., Lemesle-Meunier, D., Tron, P., Schmidt, C., and Nitschke, W. (2000) Early evolution of cytochrome *bc* complexes. *Journal of Molecular Biology* 300: 663–675.

Shih, P. M., Hemp, J., Ward, L. M., Matzke, N. J. and Fischer W. W. (2017) Crown group oxyphotobacteria postdate the rise of oxygen. *Geobiology* 15: 19–29.

Sleep, N. H. (2018) Geological and geochemical constraints on the origin and evolution of life. *Astrobiology* 18: 1199–1219.

Sleep, N. H., Zahnle, K. J., Kasting, J. F., and Morowitz, H. J. (1989) Annihilation of ecosystems by large asteroid impacts on the early Earth. *Nature* 342: 139–142.

Sleep, N. H., Bird, D. K., and Pope, E. (2012) Paleontology of Earth's mantle. *Annual Review of Earth and Planetary Science* 40: 277–300.

Soo, R. M., Hemp, J., Parks, D. H., Fischer, W. W/, and Hugenholtz, P. (2017) On the origins of oxygenic photosynthesis and aerobic respiration in cyanobacteria. *Science* 355: 1436–1440.

Sousa, F. L., Shavit-Grievink, L., Allen, J. F., and Martin W. F. (2013) Chlorophyll biosynthesis gene evolution indicates photosystem gene duplication, not photosystem merger, at the origin of oxygenic photosynthesis. *Genome Biology and Evolution* 5: 200–216.

Staples, C. R., Lahiri, S., Raymond, J., Von Hubulis, L., Mukhophadhhyay, B., and Blankenship, R. E. (2007) The expression and association of group IV nitrogenase NifD And NifH homologs in the non-nitrogen fixing archaeon *Methanocaldococcus jannaschii*. *Journal of Bacteriology* 189: 7392–7398.

Stoebe, B. and Kowallik, K. V. (1999) Gene-cluster analysis in chloroplast genomics. *Trends in Genetics* 15: 344–347.

Sutherland, J. D. (2017) Studies on the origin of life — the end of the beginning *Nature Reviews Chemistry* 1: 12.

Swingley, W. D., Sadekar, S., Mastrian, S. D., Matthies, H. J., Hao, J., Ramos, H., Acharya, C. R., Conrad, A. L., Taylor, H. L., Dejesa, L. C., Shah, M. K., O'Huallachain, M. E., Lince, M. T., Blankenship, R. E., Beatty, J. T., and

Touchman, J. W. (2007) The complete genome sequence of *Roseobacter denitrificans* reveals a mixotrophic rather than photosynthetic metabolism. *Journal of Bacteriology* 189: 683–690.

Swingley, W. D., Chen, M., Cheung, P. C., Conrad, A. L., Dejesa, L. C., Hao, J., Honchak, B. M., Karbach, L. E., Kurdoglu, A., Lahiri, S., Mastrian, S. D., Miyashita, H., Page, L. E., Ramakrishna, P., Satoh, S., Sattley, W. M., Shimada, Y., Taylor, H. L., Tomo, T., Tsuchiya, T., Wang, Z. T., Raymond, J., Mimuro, M., Blankenship, R. E., and Touchman JW (2008) Niche adaptation and genome expansion in the chlorophyll *d*-producing cyanobacterium *Acaryochloris marina*. *Proceedings of the National Academy of Sciences USA* 105: 2005–2010.

Szostak, J. W. (2011) An optimal degree of physical and chemical heterogeneity for the origin of life? *Philosophical Transactions of the Royal Society B* 366: 2894–2901.

Tabita, F. R., Hanson, T. E., Li, H. Y., Satagopan, S., Singh, J., and Chan, S. (2007) Function, structure, and evolution of the rubisco-like proteins and their rubisco homologs. *Microbiology and Molecular Biology Reviews* 71: 576–599.

Taniguchi, M., Soares, A. R. M., Chandrashaker, V., and Lindsey, J. S. (2012) A tandem combinatorial model for the prebiogenesis of diverse tetrapyrrole macrocycles. *New Journal of Chemistry* 36: 1057–1069.

Tashiro, T., Ishida, A., Hori, M., Igisu, M., Koike, M., Méjean, P., Takahata, N., Sano, Y., and Komiya, T. (2017) Early trace of life from 3.95 Ga sedimentary rocks in Labrador, Canada. *Nature* 549: 516–518.

Thompson, A. W., Foster, R. A., Krupke, A., Carter, B. J., Musat, N., Vaulot, D., Kuypers, M. M. M., and Zehr, J. P. (2012) Unicellular cyanobacterium symbiotic with a singlecelled eukaryotic alga. *Science* 337: 1546–1550.

Tice, M. M. and Lowe, D. R. (2004) Photosynthetic microbial mats in the 3,416-Myr-old ocean. *Nature* 431: 549–552.

Tomitani, A., Okada, K., Miyashita, H., Matthijs, H. C. P., Ohno, T., and Tanaka, A. (1999) Chlorophyll *b* and phycobilins in the common ancestor of cyanobacteria and chloroplasts. *Nature* 400: 159–162.

Tsukatani, Y, Romberger, S. P., Golbeck, J. H. and Bryant, D. A. (2012) Isolation and characterization of homodimeric type-I reaction center complex from *Candidatus* Chloracidobacterium thermophilum, an aerobic chlorophototroph. *Journal of Biological Chemistry* 287: 5720–5732.

Urbach, E., Robertson, D. L., and Chisholm, S. W. (1992) Multiple evolutionary origins of prochlorophytes within the cyanobacterial radiation. *Nature* 355: 267–270.

Vázquez-Salazar, A. and Lazcano, A. (2018) Early life: Embracing the RNA world. *Current Biology* 28: R208–R231.

Wächterhäuser, G. (1990) Evolution of the first metabolic cycles. *Proceedings of the National Academy of Sciences USA* 87: 200–204.

Walker, C. J. and Willows, R. D. (1997) Mechanism and regulation of Mg-chelatase. *Biochemical Journal* 327: 321–333.

Whittaker, M. M., Barynin, V. V., Antonyuk, S. V., and Whittaker, J. W. (1999) The oxidized (3,3) state of manganese catalase. Comparison of enzymes from *Thermus thermophilus* and *Lactobacillus plantarum*. *Biochemistry* 38: 9126–9136.

Widdel, F., Schnell, S., Heising, S., Ehrenreich, A., Assmus, B., and Schink, B. (1993) Ferrous iron oxidation by anoxygenic phototrophic bacteria. *Nature* 362: 834–836.

Wilk, K. E., Harrop, S. J., Jankova, L., Edler, D., Keenan, G., Sharples, F., Hiller, R. G., and Curmi, P. M. G. (1999) Evolution of a light-harvesting protein by addition of new subunits and rearrangement of conserved elements: Crystal structure of a cryptophyte phycoerythrin at 1.63-Ångstrom resolution. *Proceedings of the National Academy of Sciences USA* 96: 8901–8906.

Woese, C. R. (1987) Bacterial evolution. *Microbiological Reviews* 51(2): 221–271.

Xiong, J., Inoue, K., and Bauer, C. E. (2000) Molecular evidence for the early evolution of photosynthesis. *Science* 289: 1724–1730.

Xiong, J. and Bauer, C. E. (2002) A cytochrome *b* origin of photosynthetic reaction centers: an evolutionary link between respiration and photosynthesis. *Journal of Molecular Biology* 322: 1025–1032.

Zheng, K., Ngo, P. D., Owens, V. L., Yang, X.-P., and Mansoorabadi, S. O. (2016) The biosynthetic pathway of coenzyme F430 in methanogenic and methanotrophic archaea. *Science* 354: 339–342.

Zubay, G. (2000) *Origins of Life on the Earth and in the Cosmos*, 2nd Edn. Burlington, MA: Harcourt Academic Press.

Chapter 13

Bioenergy applications and artificial photosynthesis

13.1 Introduction

In this final chapter, we will discuss two issues of significant contemporary interest: the bioenergy applications of photosynthesis and the long-sought goal of making an artificial system that mimics the energy storage processes of photosynthesis. Most of these systems seek to mimic the water-splitting activity of oxygenic photosynthesis, with the generation of hydrogen or some other reduced product instead of carbohydrate.

Abundant energy is essential to sustain advanced societies. The amount of energy usage of a society closely parallels the standard of living of the members of the society. The mixture of energy sources that is used to make up the total energy budget of a society has varied tremendously over time and by geography. Major sources of energy are all types of fossil fuels, including coal, natural gas, and products derived from crude oil; nuclear; various renewable sources such as wind, hydroelectric, solar, geothermal, and biomass. The mixture of energy resources currently in use on Earth is shown in Fig. 13.1 (BP, 2020). However, the different types of fuels are not used uniformly in all applications.

The largest source for all our energy needs is fossil fuels, including crude oil, natural gas, and coal, which together make up 84% of current use. Nearly all societies have a need for transportation fuels and for sources of electricity. Of these, coal is now used primarily for electricity generation, and fuels generated from crude oil are primarily used to power transportation. Certain types of transportation, for example, long-range airplanes, seem unlikely to be powered by electricity, as the amount of energy that can currently be stored by batteries per unit mass is very low. Carbon-based fuels, including gasoline, diesel, and natural gas, are unequaled in terms of the energy that can be extracted per unit mass and seem destined to be a significant part of the energy budget for the foreseeable future, especially for transportation needs.

13.2 Solar energy conversion

Solar energy is now a very small part of the total current energy budget, but it is destined to increase in the future. There are several features that make solar energy an attractive option, including the fact that it is free, widely distributed over the surface of

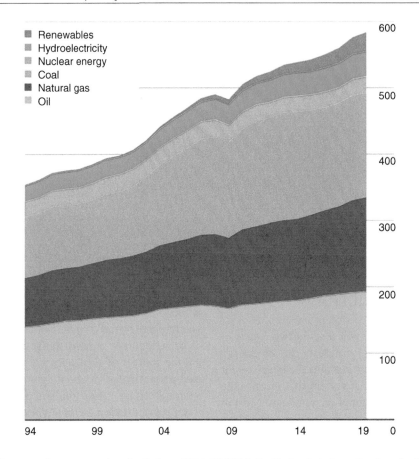

Figure 13.1   Sources of energy used on Earth from 1994 till 2019. Vertical axis is in units of exajoules. (1 **Exajoule** is equal to 10^{18} joules). *Source:* Data from BP (2020).

the Earth, reliable (albeit intermittent), and generally free of major environmental consequences such as carbon dioxide production or other heavy pollution. However, there are some disadvantages of solar; the principal one being that it is a relatively dilute energy source so that no matter what technology is used to convert solar energy to a usable form of energy, significant surface area is needed for collection of the energy. This is the result of the energy content of sunlight, which is distributed by wavelength as shown in Fig. 1.1. In turn, this is a function of the solar output determined by the type of star; our Sun is a G-type star and has an effective surface temperature of about 5600 K. Photosynthesis might look very different on a planet with a cooler M class star, whose output

spectrum will be significantly shifted to longer wavelengths (Kiang *et al.*, 2007a,b).

There are numerous ways to represent the amount of solar energy that falls on the Earth. The Sun continuously emits an enormous amount of energy: 3.8×10^{26} W. Only a small fraction of this energy strikes the Earth: 1.8×10^{17} W. Of this, much is absorbed and scattered in the atmosphere, so that the total that reaches the surface of the Earth is 1.1×10^{17} W. This is the total solar irradiance that reaches the surface of the Earth. Perhaps a more useful measure is to focus on the amount of solar energy that strikes a given area. The average power density of solar flux is 170 W m^{-2}, averaged over the diurnal cycle, the yearly fluctuation, and different geographical regions. Different regions receive significantly different amounts of solar energy, ranging from 300 W m^{-2} in the Red Sea to 200 W m^{-2} in Australia, 185 W m^{-2} in the United States, and 105 W m^{-2} in the United Kingdom. Of course, the intensity is not uniform throughout the course of a day, being effectively zero during the nighttime hours and reaching a peak of ~1000 W m^{-2} at noon. (World Energy Council, 2010).

There are a large number of technologies that are used to capture and store solar energy, including solar thermal, photovoltaic, and photochemical methods. The latter include dye-sensitized solar cells and both natural and artificial photosynthetic systems. Solar thermal installations, in which sunlight is used to heat a fluid that is then used either directly or indirectly as an energy source, include the simple rooftop hot water heater all the way to highly sophisticated installations that use mirrors to track the sun and focus the energy to a central tower where it is converted to heat and used to power an engine or electricity generator. Solar thermal systems utilize a surface that is effectively black, absorbing all incident photons from UV to IR and immediately converting the energy to heat. There is no threshold energy such as we will encounter in other types of solar technologies, but there is a limit imposed by the second law of thermodynamics on how much of the energy of the heat produced can be converted into useful work.

Photovoltaic systems produce electricity directly from absorbed photons (Dunlop, 2009). Photovoltaic devices utilize semiconductors and exploit the band structure of these materials, with an occupied valance band separated from a conduction band by a forbidden region called the bandgap. A photon that is absorbed promotes an electron from the valence band, leaving behind a hole or positive charge, to the conduction band, where it can be withdrawn to generate a photocurrent. A detailed comparison of the efficiency of photosynthetic and photovoltaic systems has been made, using as much as possible uniform definitions and considerations of the types of products (Blankenship et al., 2011). Traditional photovoltaic cells, such as Si cells, utilize a material with a bandgap energy that is in the near-infrared region.

Photons with energy higher than the bandgap are effectively absorbed. However, the energy available for storage is just that of the bandgap energy, so that any excess energy over this threshold is converted to heat and not available for storage. This concept of threshold energy applies to all photovoltaic cells, and also to photosynthetic organisms and to dye-sensitized solar cells. There is a fundamental limit to the amount of energy that can be extracted from sunlight using a photovoltaic cell. This limit is about 31% for a single-junction Si cell (Shockley and Queisser, 1961). The efficiency is in part a compromise between harvesting fewer photons at higher energy or more photons at lower energy, such that there is an optimum threshold energy that depends on the solar spectrum. The conversion efficiency can be higher if more sophisticated cells that contain multiple junctions are employed, with some that harvest longer wavelength photons and others that harvest shorter wavelength photons (Hanna and Nozik, 2006). In addition to the traditional Si photovoltaic cells, a wide range of other materials have been utilized, including organic compounds. The efficiency of photovoltaic cells varies widely, depending on the material utilized and its light-absorbing properties, as well as the construction of the cell. Typical commercial photovoltaic units that are utilized for rooftop installations are 10–20% efficient, considering the energy content of the entire solar spectrum as input. Inorganic materials called perovskites show great promise in inexpensive and high-efficiency photovoltaic cells (Tasleem and Tahir, 2020).

Dye-sensitized solar cells, often called Grätzel cells after their inventor, the Swiss chemist Michael Grätzel, are intermediate between photovoltaic cells and entirely photochemical systems (Oregan and Grätzel, 1991; Yella et al., 2011; Boschloo, 2019). They produce oxidized and reduced products, with the circuit completed by a mobile shuttle, often I_2/I_3^-, which acts as a shuttle between the anode and cathode. The photons are absorbed by dyes adsorbed on the surface of the substrate material, which is often nanocrystalline TiO_2. The excited state of the dye is sufficiently reducing (see Appendix 1) to reduce the TiO_2, and electrons are drawn off by an electrode and eventually re-reduce the dye through the action of the mobile shuttle. Dye-sensitized solar cells have efficiencies of up to 15% and can be very long lived.

13.3   What is the efficiency of natural photosynthesis?

In this section, we will investigate the efficiency of natural photosynthesis, with the goal of comparing it to some of the other systems described above. Before we make this detailed comparison, we need to discuss what we mean by efficiency. Efficiency is a simple concept, yet it can be very elusive to define precisely and to make meaningful comparisons of different types of solar energy conversion devices. There are many different types of efficiencies, all of which are correct, but which differ significantly in what they are trying to capture in terms of the performance of the system. This has led to considerable confusion, especially in the popular press, as to what really is the efficiency of natural photosynthesis, in particular as compared to photovoltaic devices.

13.3.1   Different types of efficiency of photosynthesis

Efficiency of a light-energy storage device is often defined loosely as the fraction (often expressed as a percentage) of the input energy that is converted to products. However, there are numerous subtleties that are essential to consider for a precise measure of efficiency, so that there are in reality several different measures of efficiency that are applicable to photosynthetic systems. Perhaps the most fundamental one is the **Photochemical Quantum Efficiency**, also often called the **Photochemical Quantum Yield**, which is defined as the percentage of photons absorbed that lead to photochemical products under optimum conditions. This is also discussed in the Appendix 1. Typical values for the photochemical quantum yield of isolated reaction

centers approach 100%, although fluorescence methods (which are somewhat indirect) suggest a lower maximum yield for Photosystem II.

In addition to the quantum efficiency, we can define the **Energy Transfer Efficiency**, which applies primarily to antenna systems. This measures the percentage of photons that are absorbed by pigments in the antenna system, and that are then transferred as excited states to other antenna pigments and ultimately to the reaction center. We discussed a simple way to measure energy transfer efficiency using fluorescence excitation spectra in Chapter 5. Measured values of energy transfer efficiency in intact antenna systems are also quite high, typically 80–95% under ideal conditions. So the earliest processes of photosynthesis are quite efficient when considering how absorbed photons are utilized.

Another measure of efficiency that is often quoted is here called the **Energy Storage Efficiency**. This is the percentage of absorbed photon energy that appears as photoproducts, using photons of optimum wavelengths and low intensities that avoid light saturation effects and losses during carbon fixation. As we will see below, this efficiency is ~27% for oxygenic photosynthetic organisms. Part of the loss of efficiency is due to the fact that oxygenic photosynthetic organisms utilize two photosystems operating in tandem, so two photons are needed to drive one electron from water to $NADP^+$. There is a loss of energy in the downhill reactions between the photosystems, although some of it is recovered in the form of ATP phosphate bond energy, and additional downhill losses on the oxidizing side of Photosystem II and especially on the reducing side of Photosystem I. There is an extensive literature describing the fundamental limits on energy storage in photosynthesis, including limits set by the second law of thermodynamics. An analysis of this question with references to earlier treatments is presented by Knox and Parson (2007).

The most conservative measure of the efficiency of photosynthesis is here called the **Solar Energy Storage Efficiency**. This considers the percentage of the incident energy of the entire solar spectrum that is converted into recoverable energy such as biomass. Not surprisingly, this efficiency is significantly lower than any of the other measures discussed above. Perhaps the most important reason for this lower efficiency is that approximately half of the energy in the solar spectrum is not absorbed by photosynthetic organisms and is therefore unavailable for energy storage, as only absorbed photons can be effective in photochemistry (see Appendix 1). Another major factor is the light saturation effect. In Chapter 3, we discussed the light saturation effect in flashing light that was discovered by Emerson and Arnold in 1932, although similar effects were known earlier using continuous light, which is more relevant to the situation in sunlight. Figure 13.2 shows a modeled curve of daily light intensity with the rate of photosynthesis superimposed. It is clear that at low light intensities the rate of photosynthesis is directly proportional to the light intensity, but that during most of the day, photosynthesis is light-saturated and much of the absorbed light must be dissipated. This light-saturation effect is a major loss factor in most photosynthetic organisms. Another major factor is the photorespiration effect that is present in C3 plants, which can represent a loss of 25–30% of the primary products to recover from the oxygenation

activity of rubisco (Chapter 9). When these and other known loss factors are included, the maximum efficiency of photosynthesis that is expected is 4.6% for C3 plants and 6.0% for C4 plants, which have dramatically lower rates of photorespiration (Zhu *et al.*, 2010). However, even this is higher than the highest rates of photosynthesis measured under field conditions, which are 2.4 and 3.7% for C3 and C4 plants, respectively. Estimates of the efficiency of photosynthesis under typical field conditions range from 0.2 to 1% (Bolton and Hall, 1979; Zhu *et al.*, 2010). Note that the values quoted here necessarily include the energy required to build and maintain the photosynthetic apparatus. The efficiencies quoted for photovoltaic cells typically do not include the construction and maintenance costs of the devices.

13.4 Calculation of the energy storage efficiency of oxygenic photosynthesis

As discussed above, there are many different measures of the efficiency of photosynthesis, each of which captures part of the overall picture of how photosynthesis works. Perhaps the most informative calculation concerns the percentage of the energy absorbed by a photosynthetic organism operating under ideal conditions that is stored as carbohydrate and oxygen, which we have called the energy storage efficiency. Here we will go through the details of how this is determined, using what we know of how the system works and key measured quantities.

To make this energetic calculation, it is necessary to know both the energy stored as chemical energy and the energy input as photon energy. To calculate the first quantity, we use Eq. (A20) and thermodynamic data, in particular the free energies of formation tabulated in Table A1. The balanced chemical equation of oxygenic photosynthesis must be used. Here we will write this equation in terms of the production of one O_2 molecule, as that

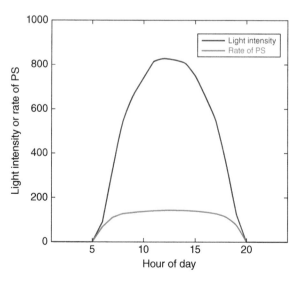

Figure 13.2 Diurnal cycle of light intensity (colored red) and modeled rate of photosynthesis (colored green). *Source:* Courtesy of Dr. Mark Hensen.

is how measurements of quantum requirement are traditionally reported. This process leads to the production of 1/6 of a glucose molecule, which as discussed in Chapter 9 is not the actual product but is close in energy.

$$H_2O + CO_2 \rightarrow \frac{1}{6}\left(Glucose\right) + O_2 \qquad (13.1)$$

So the standard state free energy stored is:

$$\Delta G° = \frac{1}{6}\Delta G_f°\left(Glucose\right) + \Delta G_f°\left(O_2\right) \\ - \Delta G_f°\left(H_2O\right) - \Delta G_f°\left(CO_2\right) \qquad (13.2)$$

$$\Delta G° = \frac{1}{6}\left(-914.54\right) + \left(0\right) \\ - \left(-237.19\right) - \left(-394.38\right) \qquad (13.3)$$

$$\Delta G° = +479 \text{ kJ mol}^{-1} \qquad (13.4)$$

This must be compared to the energy input from light. The energy content of the light energy can be calculated from a slightly modified version of Eq. (A3):

$$E = \left(QR\right)\frac{hc}{\lambda}N_A \qquad (13.5)$$

where QR is the quantum requirement and N_A is Avogadro's number, $6.022 \times 10^{23} \text{ mol}^{-1}$. We need to multiply by Avogadro's number because the thermodynamic calculation is made for a mole of molecules, while Eq. (A3) applies to a single photon of light, taken for this calculation to be at 680 nm.

The consensus modern values for the quantum requirement of O_2 production in oxygenic photosynthesis are 9–10 (Skillman, 2008). We will use 10 for our calculation. This gives an energy input of 1761 kJ mol^{-1}, which results in an efficiency of energy storage of 27% (479/1761). Recall that this value applies only to absorbed photons at the wavelength of maximum efficiency at low intensity, producing products at their thermodynamic standard state. So it represents a sort of "best-case scenario" for the overall process of photosynthesis, not including any of the loss processes discussed above, such as light saturation effects and photorespiration, and

also not including the losses due to absorbed photons with energy above the threshold value and due to the fact that much of the solar spectrum is not absorbed and therefore cannot be stored.

13.5   Why is the efficiency of photosynthesis so low?

Many people are surprised at the low overall efficiency of photosynthesis and ask why evolution hasn't done a better job of optimizing the process. The answer to this question is not entirely clear and has been the subject of much discussion and speculation. However, it is important to remember that evolution does not select by efficiency but rather by reproductive success. So a characteristic that actually lowers the efficiency of photosynthesis, such as the larger than needed antenna system found in most organisms, may improve the reproductive success of an organism by depriving its neighbor of light and thereby reducing the chance that it will be shaded.

Some other aspects of the low efficiency may result from evolutionary paths that were selected under one set of conditions, but are not optimal under current conditions. The clearest example of this is the oxygenation activity of rubisco, which leads to substantial losses due to photorespiration. The evolutionary choice of rubisco and the Calvin–Benson cycle for carbon fixation was almost certainly made at a time when the Earth was largely anoxic (Chapter 12). Under these conditions, the system operates at good efficiency with little loss (Chapter 9). When oxygen slowly began to increase in the atmosphere beginning about 2.4 BYA, organisms responded by improving the selectivity of rubisco for CO_2 vs O_2, and more recently by implementing carbon-concentrating strategies such as C4 photosynthesis. However, the evolutionary barrier to switch the basic carbon fixation mechanism to one that is better adapted to a world with 21% oxygen is apparently too great to overcome. The system is thus trapped using a carbon fixation mechanism that is not ideally suited to the modern world, but has been tweaked as much as possible to improve its performance.

The two-photosystem architecture and spectral utilization of oxygenic photosynthesis are other potentially limiting factors. The architecture of the system that we have now has resulted from a long series of evolutionary developments. The current system also certainly has limitations imposed by "legacy biochemistry" from earlier evolutionary developments that set the parameters for biological energy metabolism before photosynthesis was invented (Gust *et al.*, 2008). The system uses molecules as energy intermediates, such as quinones and ATP, which are well suited to other energetic systems but may not be able to capture a significant fraction of the larger quantities of energy that are present in photons, so the overall efficiency is lowered by the use of the "lowest common denominator" energetic intermediates.

13.6   How might the efficiency of photosynthesis be improved?

There is every reason to think that the low solar energy storage efficiency of natural photosynthesis can be improved in numerous ways, some of which are relatively minor changes to the system, while others are more sweeping and will require considerable effort. We will discuss some of these efforts that are already being tried or have been proposed, although there may be many other ways to improve the efficiency that have not yet been thought of. Many of these strategies have been devised in the context of algal bioenergy applications, although most of them should also apply to agricultural situations (Georgianna and Mayfield, 2012; Work *et al.*, 2012; Razeghifard, 2013; Ort *et al.*, 2015; Kromdijk *et al.*, 2016).

One of the most promising strategies to improve the efficiency of photosynthesis is to reduce the size of the antenna system that feeds energy into the reaction center. As discussed above, the light saturation behavior of the system shown in Fig. 13.2 indicates that the antenna system has more

pigments than are required to keep the reaction center functioning without wasting excitations at high light intensities. Various strategies have been adopted to reduce the antenna size and therefore spread the excitations throughout the culture or canopy. This is shown conceptually in Fig. 13.3 (Melis, 2009). If each cell in an algal culture has fewer pigments associated with a constant number of reaction centers, then light will penetrate more deeply into the culture and be more uniformly distributed. This will reduce the wasteful saturation effects in the upper portion of the culture and result in fewer absorbed photons that have to be quenched by nonphotochemical quenching processes. Strategies to reduce the antenna size are varied and include changes in regulatory processes, deletion or inactivation of genes that code for antenna proteins or accessory pigments such as chlorophyll *b* (Nakajima and Ueda, 1997; Beckman *et al.*, 2009; Melis, 2009; Ort *et al.*, 2010; Page *et al.*, 2012;

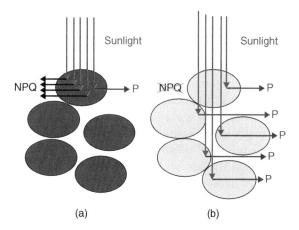

Figure 13.3   Concept of reduction of antenna size as a way to increase the efficiency of photosynthesis. Black arrows represent excess light absorption that must be dissipated using nonphotochemical quenching (NPQ) and blue arrows represent photosynthetic products. (a) shows wild-type algal cells in a dense culture, with the top layer of cells light saturated and the lower layers of cells light starved. (b) shows a culture with reduced antenna, represented by paler green cells, in which light penetrates more uniformly throughout the culture, minimizing NPQ and maximizing photosynthesis. *Source:* Melis (2009)/Elsevier.

Perrine *et al.*, 2012; Blankenship and Chen, 2013; Negi *et al.*, 2020). While these strategies are promising, it is possible that the balance of energy delivered to the two photosystems may be disrupted by the relatively crude antenna reduction methods that have been employed to date. More sophisticated efforts may need to include secondary changes to the natural regulatory processes. Similar antenna modulation effects may be possible in higher plants if leaves at the top of the canopy can be engineered to have smaller antennas, while those lower down have larger antennas. The goal of this "smart canopy" is to spread the light absorption as evenly as possible throughout the plant, while not permitting any photons to reach the soil where they would be dissipated as heat (Ort *et al.*, 2015).

A second relatively simple strategy that is being investigated is to expand the solar spectrum that is absorbed by the photosynthetic pigments. The key spectral region that is being focused on is the near infrared, from 700 to 750 nm. Most oxygenic photosynthetic organisms have no significant light absorption in this region of the solar spectrum. However, a few cyanobacteria are known that contain chlorophyll *d* or chlorophyll *f* as a photopigment, with absorption shifted about 50 nm to the red compared to organisms that contain chlorophyll *a* (Miyashita *et al.*, 1996; Chen *et al.*, 2010). Measurements indicate that the efficiency of energy storage in these organisms is equal to that in chlorophyll *a*-containing organisms (Mielke *et al.*, 2013). The expansion of the Photosynthetically Active Radiation (PAR) by 50 nm to include the 400–750 nm spectral region increases the photon flux by 19% over the standard 400–700 nm range of PAR, so this seems a useful strategy to pursue (Chen and Blankenship, 2011).

It is possible to conceive of systems that utilize only a single photosystem to carry out oxygenic photosynthesis, or systems that expand the spectral coverage using alternate pigments such as bacteriochlorophyll to capture a much larger fraction of the solar spectrum (Blankenship *et al.*, 2011). Other possible strategies are to replace the Calvin–Benson cycle with an alternative carbon fixation pathway that is not inhibited by O_2 (Bar-Even *et al.*, 2010; Peterhansel and Offermann, 2012; Schwander

et al., 2016; Liang *et al.*, 2018). These are more ambitious strategies and will require the tools of synthetic biology to accomplish (Zhao, 2013). Using these techniques, it may ultimately be possible to increase the efficiency of solar energy storage in photosynthetic organisms by a factor of two to four over the current limits. This could have a profound impact on bioenergy applications of photosynthesis as well as agriculture, with the promise of a second green revolution if they can be successfully implemented.

13.7   Artificial photosynthesis

An even more ambitious goal than the strategies outlined above to improve the efficiency of natural photosynthesis is the prospect of devising a completely synthetic method for accomplishing similar chemistry to that done by natural photosynthetic systems. This has been a dream of mankind for centuries. Perhaps the most colorful description of the prospect of artificial photosynthesis was made by the Italian chemist Giacomo Ciamician, considered to be the father of photochemistry (Ciamician, 1912).

> On the arid lands there will spring up industrial colonies without smoke and without smokestacks; forests of glass tubes will extend over the plants and glass buildings will rise everywhere; inside of these will take place the photochemical processes that hitherto have been the guarded secret of the plants, but that will have been mastered by human industry which will know how to make them bear even more abundant fruit than nature, for nature is not in a hurry and mankind is.

Considerable progress has been made in devising artificial photosynthetic systems, as described below in more detail, but Ciamician's dream has not yet been realized in a long-lived system that can be implemented on a wide scale. One of the biggest obstacles to artificial photosynthetic systems is the issue of stability and repair. The very energetic chemical processes that go on during either natural or artificial photosynthesis can easily lead to damaging conditions, so that the system is frequently damaged during operation. The two places where this seems to be most critical are the antenna system, where the damaging reactive oxygen species – such as singlet oxygen or superoxide – are produced if the excited state is not efficiently quenched by photochemistry, and in the highly oxidizing chemistry of the oxidation of water to form O_2. Natural photosynthetic systems are not immune from these damaging reactions, but they have evolved mechanisms that permit the system to repair damaged components. We encountered one of these systems in Chapter 10, the remarkable damage–repair cycle of Photosystem II, in which the D1 core reaction center protein is replaced frequently in oxygenic photosynthetic organisms. Artificial systems that are capable of repair have not yet been produced and this remains an important unrealized goal.

The basic physical principles that are employed in artificial photosynthetic systems are essentially the same as are found in natural systems. A light-driven electron transfer system uses photon absorption to create an excited state of a material or pigment, which then decays by an electron transfer reaction to form a charge-separated state. The initial charge-separated state is stabilized against recombination by a series of rapid secondary reactions that separate the charges before they have the chance to recombine. The primary difference between the natural and artificial photosynthetic systems is that the artificial system is not built by a living organism, but rather by synthetic chemists in a laboratory. Unlike the natural photosynthetic system, the components are usually not associated with a protein scaffold. Instead, they are often joined together either covalently or by molecular recognition forces to make a long-lived complex. In some systems, semiconductors are used as the basis for the device. For recent reviews see Gust *et al.* (2009, 2012), Nocera (2012), Tachibana *et al.* (2012), Barber and Tran (2013), Sherman *et al.* (2014), Zhang and Sun (2019), Zhou *et al.*, (2020) and Fukuzumi (2020).

13.7.1 Artificial photosynthetic systems that mimic reaction centers

Early efforts to produce an artificial photosynthetic complex, inspired by the chemistry that takes place in natural reaction center complexes, utilized what is called a dyad structure, in which two chemical groups are attached together, with light excitation of one molecular unit giving rise to an electron transfer process, resulting in a complex in which one part is oxidized and the other part is reduced. While this sort of system produces the charge-separated state with high quantum yield, it invariably decays rapidly by charge recombination, so that no long-lived states are produced. This early work has been reviewed by Gust and Moore (1989).

Dyad systems were an essential first step toward artificial photosynthetic systems, but were relatively quickly superseded by multistep complexes, such as the triad shown in Fig. 13.4 (Liddell *et al.*, 1997). In this complex, the central porphyrin is selectively

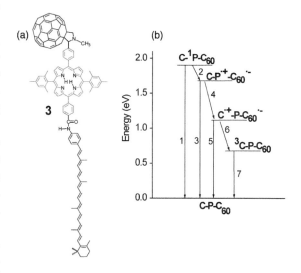

Figure 13.4 Molecular triad for artificial photosynthesis. (a) Structure of the triad, consisting of a fullerene C60 electron acceptor at the top colored blue, photoactive porphyrin in the center colored green, and carotenoid electron donor at the bottom colored red. (b) Energy level diagram describing the electron transfer processes in the triad. *Source:* Gust *et al.* (2012)/Royal Society of Chemistry.

excited and the excited state rapidly donates an electron to the fullerene C_{60} moiety, which acts as an electron acceptor. This results in the creation of the ion pair state $P^+ C_{60}^-$. This state is stabilized by a secondary electron transfer from the carotenoid, finally generating the state $C^+ P C_{60}^-$. This system operates using the same basic physical principles that are found in natural reaction centers, in that the initial charge-separated state is formed by excited-state electron transfer, and then rapidly reacts by a series of secondary processes to form a long-lived state, where recombination is prevented by spatial separation of the electron and hole.

Even more sophisticated systems with multiple components have been carefully designed to carry out multistep electron transfer processes. These more advanced systems include a pentad complex with five distinct portions, each of which serves a specific role in the overall activity of the complex (Gust *et al.*, 1990). The multistep electron transfer that goes on after photon absorption rapidly separates the positive and negative charges sufficiently to produce relatively long-lived charge-separated products, although they eventually decay by charge recombination. Systems that mimic proton-coupled electron transfer (PCET) processes in reaction centers have been reported (Mora *et al.*, 2018).

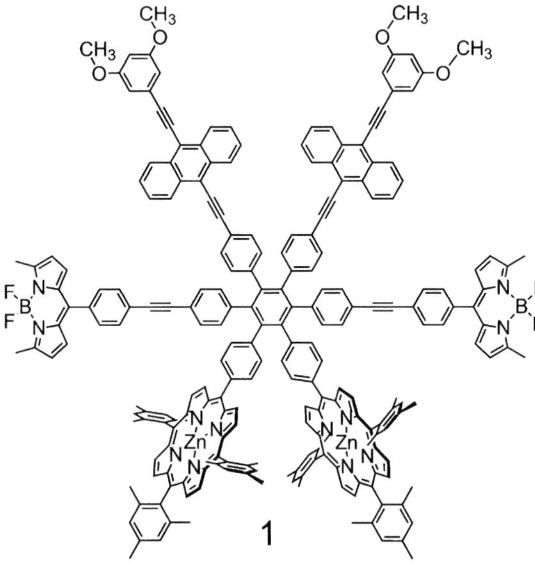

13.7.2 Artificial antenna systems

Other types of artificial photosynthetic systems have been built to mimic the action of antenna complexes. Here, multiple pigments are coupled into an array and energy transfer processes take place to deliver the excited states to a particular pigment, often at a lower energy, which can then initiate photochemistry in a multistep complex such as described above. An antenna array is shown in Fig. 13.5 (Gust *et al.*, 2012). Some artificial antennas have been made to include photoprotective functions similar to the nonphotochemical quenching found in Photosystem II.

Metal nanoparticles have been used as artificial antennas that work by a different physical principle

Figure 13.5 Artificial antenna complex, consisting of anthracene derivatives – which absorb in the blue spectral region, borondipyrromethenes – which absorb in the green spectral region, and Zn porphyrins – which absorb in the red spectral region, organized around a hexaphyl-benene core. *Source:* Gust *et al.* (2012)/Royal Society of Chemistry.

from any known natural antenna complex (Carmeli *et al.*, 2010). These devices utilize plasmonic principles instead of molecular absorption (Atwater and Polman, 2010).

13.7.3 Artificial water-splitting systems

A large effort has been expended to produce artificial complexes to oxidize water to form O_2 and H^+, similar to the activity of Photosystem II (Young *et al.*, 2012). Most of these systems utilize other sorts of cofactors instead of the Mn that is at the heart of the oxygen evolution center (OEC) in Photosystem II. The structure of the OEC is sufficiently complex and sensitive to the subtleties of the protein environment that it has resisted all efforts to produce a functional artificial version with the same basic structure. Instead, many researchers have utilized a variety of inorganic complexes, mostly in systems that are directly coupled to an electrode, which draws off the electrons (Zhao *et al.*, 2012). The long-term goal of these sorts of systems is to split water into H_2 and O_2 and recover the H_2 for use as a fuel in an internal combustion engine or a fuel cell. While these systems are promising, they have a number of shortcomings. These include the high cost of the materials, often including platinum and iridium, and often a requirement to use higher energy blue or UV photons. Many of these systems also utilize semiconductors (see below).

13.7.4 Semiconductor-based artificial photosynthetic systems

There is also a long history of attempting to mimic photosynthesis based on using semiconductors. One of the first of these attempts was the remarkable finding by Fujishima and Honda (1972) that TiO_2 can split H_2O into H_2 and O_2 upon illumination by ultraviolet light. The major problem with this was that the water splitting does not work when visible light is used, thereby making the system impractical for use as a solar energy conversion device. However, this finding spurred many researchers to attempt to find systems that will do the same overall chemistry using visible light (Meyer, 1989; Conception *et al.*, 2012). Many of these systems have utilized ruthenium complexes as critical components.

13.7.5 Multiple junction artificial photosynthetic systems

Recently, artificial photosynthetic systems have been described that employ dual or triple junction devices that are reminiscent of the two-photosystem tandem architecture of natural oxygenic photosynthetic organisms, in that one light-driven step oxidizes water and feeds electrons to the second light-driven step, which reduces H^+ to H_2 (Reece *et al.*, 2011; Liu *et al.*, 2013). They thus have essentially the same overall energy storage process as is

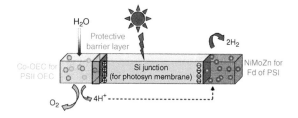

Figure 13.6 Schematic diagram of an artificial leaf. The water-splitting activity is performed by catalysts that are arranged on opposite sides of a Si junction, which absorbs photons and produces electrons and holes. The oxygen-evolving complex of Photosystem II is replaced by a Co complex and the ferredoxin electron acceptor of Photosystem I is replaced by a NiMoZn complex that produces H_2. *Source:* Nocera (2012)/American Chemical Society.

found in natural photosynthesis, as the free energy required is almost the same (Nocera, 2012). These sorts of systems, which are similar in many ways to the dye-sensitized solar cells described above in that they combine elements of photovoltaic cells with elements of photochemical systems, may represent the most efficient way for a manmade system to store solar energy in terms of chemical products. While they are structurally very far removed from the natural photosynthetic systems that are the main subject of this book, of course they must utilize the same physical principles that the natural systems do. A schematic diagram illustrating the concept of the "artificial leaf" is shown in Fig. 13.6 (Blankenship *et al.*, 2011; Nocera, 2012).

13.7.6 Concluding thoughts on artificial photosynthesis

Early attempts to mimic photosynthesis in artificial systems mostly utilized systems that had been designed to carry out the essential steps in photosynthetic energy conversion using similar components as occur in natural systems, including porphyrins as photosensitive pigments and quinones as electron acceptors. Later efforts implemented components such as fullerenes that are not found in natural systems. A parallel evolution of artificial photosynthetic systems that are based less on direct analogy with the natural system has utilized semiconductors and inorganic complexes or dyes. Recent efforts utilize some aspects of both approaches. However, the most promising of these artificial photosynthetic systems are looking less and less like the living organisms that inspired them. This is entirely appropriate and should not be cause for concern. A bird and an airplane are both devices that fly. They work in very different ways, yet they must obey the same laws of physics. Early concepts for manned flight, such as the drawings of Leonardo da Vinci, described devices with flapping wings. While none of these have ever proven to be practical, they represent an essential step in the progression towards workable flying machines (Spenser, 2008). If birds and other flying creatures did not exist, it is unlikely that people would have ever invented airplanes, as the idea would simply have been too far outside of the realm of possibility. The existence of natural photosynthetic systems provides the inspiration and the conceptual framework for artificial photosynthetic devices, even if the details of the implementation are entirely different in the two systems.

References

Atwater, H. A. and Polman, A. (2010) Plasmonics for improved photovoltaic devices. *Nature Materials* 9: 205–213.

BP Statistical Review of World Energy (2020) Statistical Review of World Energy. Can be downloaded at https://www.bp.com/content/dam/bp/business-sites/en/global/corporate/pdfs/energy-economics/statistical-review/bp-stats-review-2020-full-report.pdf.

Bar-Even, A., Noor, E., Lewis, N. E., and Milo, R. (2010) Design and analysis of synthetic carbon fixation pathways. *Proceedings of the National Academy of Sciences USA* 107: 8889–8894.

Barber, J. and Tran, P. D. (2013) From natural to artificial photosynthesis. *Journal of the Royal Society Interface* 10: 20120984.

Beckmann, J., Lehr, F., Finazzi, G., Hankamer, B., Posten, C., Wobbe, L., and Kruse, O. (2009) Improvement of light to biomass conversion by deregulation of light-harvesting protein translation in *Chlamydomonas reinhardtii*. *Journal of Biotechnology* 142: 70–77.

Blankenship, R. E. and Chen, M. (2013) Spectral expansion and antenna reduction can enhance photosynthesis for energy production. *Current Opinion on Chemical Biology* 17: 457–461.

Blankenship, R. E., Tiede, D. M., Barber, J., Brudvig, G. W., Fleming, G., Ghirardi, M., Gunner, M. R., Junge, W., Kramer, D. M., Melis, A., Moore, T. A., Moser, C. C., Nocera, D. G., Nozik, A. J., Ort, D. R., Parson, W. W., Prince, R. C., and Sayre, R. T. (2011) Comparing the efficiency of photosynthesis with photovoltaic devices and recognizing opportunities for improvement. *Science* 332: 805–809.

Bolton, J. R. and Hall, D. O. (1979) Photochemical conversion and storage of solar energy. *Annual Review of Energy* 4: 353–401.

Boschloo, G. (2019) Improving the performance of dye-sensitized solar cells. *Frontiers in Chemistry* 7: 77.

Carmeli, I., Lieberman, I., Kraversky, L., Fan, Z., Govorov, A., Markovich, G., and Richter, S. (2010) Broad band enhancement of light absorption in photosystem I by metal nanoparticle antennas. *Nano Letters* 10: 2065–2074.

Chen M. and Blankenship, R. E. (2011) Expanding the solar spectrum used by photosynthesis. *Trends in Plant Science* 16: 427–431.

Chen, M., Schliep, M., Willows, R. D., Cai, Z.-L., Neilan, B. A., and Scheer, H. (2010) A red-shifted chlorophyll. *Science* 329: 1318–1319.

Ciamician, G. (1912) The photochemistry of the future. *Science* 36: 385–394.

Concepcion, J. J., House, R. L., Papanikolas, J. M., and Meyer, T. J. (2012) Chemical approaches to artificial photosynthesis. *Proceedings of the National Academy of Sciences USA* 109: 15560–15564.

Dunlop, J. P. (2009) *Photovoltaic Systems*. Orland Park, IL: American Technical Publishers.

Fujishima, A. and Honda, K. (1972) Electrochemical photolysis of water at a semiconductor electrode. *Nature* 238: 37–38.

Fukuzumi, S. (2020) *Principles and Applications of Artificial Photosynthesis*. London: Royal Society of Chemistry.

Georgianna, D. R. and Mayfield, S. P. (2012) Exploiting diversity and synthetic biology for the production of algal biofuels. *Nature* 488: 329–335.

Gust, D. and Moore, T. A. (1989) Mimicking photosynthesis. *Science* 244: 35–41.

Gust, D., Moore, T. A., Moore, A. L., Lee, S. J., Bittersmann, E., Luttrull, D. K., Rehms, A. A., Degraziano, J. M., Ma, X. C., Gao, F., Belford, R. E., and Trier, T. T. (1990) Efficient multistep photoinitiated electron-transfer in a molecular pentad. *Science* 248: 199–201.

Gust, D., Kramer, D., Moore, A., and Moore, T. A. (2008) Engineered and artificial photosynthesis: Human ingenuity enters the game. *MRS Bulletin* 33: 383.

Gust, D., Moore, T. A., and Moore, A. L. (2009) Solar fuels via artificial photosynthesis. *Accounts of Chemical Research* 42: 1890–1898.

Gust, D., Moore, T. A., and Moore, A. L. (2012) Realizing artificial photosynthesis. *Faraday Discussions* 155: 9–26.

Hanna, M. C. and Nozik, A. J. (2006) Solar conversion efficiency of photovoltaic and photoelectrolysis cells with carrier multiplication absorbers. *Journal of Applied Physics* 100: 74510.

Kiang, N., Siefert, J., Govindjee, and Blankenship, R. E. (2007a) Spectral signatures of photosynthesis. I. Review of earth organisms. *Astrobiology* 7: 222–251.

Kiang, N., Segura, A., Tinetti, G., Govindjee, Blankenship, R. E., Cohen, M., Siefert, J., Crisp, D., and Meadows, V. S. (2007b) Spectral signatures of photosynthesis. II. Coevolution with other stars and the atmosphere on extrasolar worlds. *Astrobiology* 7: 252–274.

Knox, S. S. and Parson, W. W. (2007) Entropy production and the Second Law in photosynthesis. *Biochimica et Biophysica Acta* 1767: 1189–1193.

Kromdijk, J., Katarzyna Głowacka, K., Leonelli, L., Gabilly, S. T., Iwai,, M., Niyogi, K. K., and Long, S. P. (2016) Improving photosynthesis and crop productivity by accelerating recovery from photoprotection. *Science* 354: 857–861.

Liang, F., Lindberg, P., and Lindblad, P. (2018) Engineering photoautotrophic carbon fixation for enhanced growth and productivity. *Sustainable Energy and Fuels* 2: 2583.

Liddell, P. A., Kuciauskas, D., Sumida, J. P., Nash, B., Nguyen, D., Moore, A. L., Moore, T. A., and Gust, D. (1997) Photoinduced charge separation and charge recombination to a triplet state in a carotene-porphyrin fullerene triad. *Journal of the American Chemical Society* 119: 1400–1405.

Liu, C., Tang, J., Chen, H. M., Liu, B., and Yang, P. (2013) A fully integrated nanosystem of semiconductor nanowires for direct solar water splitting. *Nano Letters* 13: 2989–2992.

Melis, A. (2009) Solar energy conversion efficiencies in photosynthesis: Minimizing the chlorophyll antennae to maximize efficiency. *Plant Science* 177: 272–280.

Meyer, T. J. (1989) Chemical approaches to artificial photosynthesis. *Accounts of Chemical Research* 22: 163–170.

Mielke, S. P., Kiang, N. Y., Blankenship, R. E., and Mauzerall, D. (2013) Photosystem trap energies and spectrally-dependent energy-storage efficiencies in the Chl *d*-utilizing cyanobacterium, *Acaryochloris marina*. *Biochimica et Biophysica Acta* 1827: 255–265.

Miyashita, H., Ikemoto, H., Kurano, N., Adachi, K., Chihara, M., and Miyachi, S (1996) Chlorophyll *d* as a major pigment. *Nature* 383: 402.

Mora, S. J., Odella, E., Moore, G. F., Gust, D., Moore, T. A., and Moore, A. L. (2018) Proton-coupled electron transfer in artificial photosynthetic systems. *Accounts of Chemical Research* 51: 445–453.

Nakajima, Y. and Ueda, R. (1997) Improvement of photosynthesis in dense microalgal suspension by reduction of light harvesting pigments. *Journal of Applied Phycology* 9: 503–510.

Negi, S., Perrine, Z., Friedland, N., Kumar, A., Tokutsu, R., Minagawa, J., Berg, H., Barry, A. N., Govindjee, G., and Sayre, R. (2020) Light regulation of light-harvesting antenna size substantially enhances photosynthetic efficiency and biomass yield in green algae. *The Plant Journal* 103: 584–603.

Nocera, D. C. (2012) The artificial leaf. *Accounts of Chemical Research* 45: 767–776.

Ort, D. R., Zhu, X., and Melis, A. (2010) Optimizing antenna size to maximize photosynthetic efficiency. *Plant Physiology* 155: 79–85.

Ort, D. R., Merchant, S. S., Alric, J., Barkan, A., Blankenship, R. E., Bock, R., Croce, R., Hanson, M. R., Hibberd, J., Lindstrom, D. L., Long, S. P., Moore, T. A., Moroney, J., Niyogi, K. K., Parry, M., Peralta- Yahya, P., Prince, R., Redding, K., Spalding, M., van Wijk, K., Vermaas, W. F.J., von Caemmerer, S., Weber, W., Yeates, T., Yuan, J., Zhu, X. (2015) Redesigning photosynthesis to sustainably meet global food and bioenergy demand. *Proceedings of the National Academy of Sciences USA* 112: 8529–8536.

Oregan, B. and Grätzel, M. (1991) A low-cost, high efficiency solar-cell based on dye-sensitized colloidal TiO2 films. *Nature* 353: 737–740.

Page, L. E., Liberton, M., and Pakrasi, H. B. (2012) Reduction of photoautotrophic productivity in the Cyanobacterium *Synechocystis* sp., Strain PCC 6803 by phycobilisome antenna truncation. *Applied and Environmental Microbiology* 78: 6349–6351.

Perrine, Z., Negi, S., and Sayre, R. T. (2012) Optimization of photosynthetic light energy utilization by microalgae. *Algal Research* 1: 134–142.

Peterhansel, C. C. and Offermann, S. S. (2012) Reengineering of carbon fixation in plants – challenges for plantbiotechnology to improve yields inahigh-CO_2 world. *Current Opinion in Biotechnology* 23: 204–208.

Razeghifard, R. (2013) Algal biofuels. *Photosynthesis Research* 117: 207–219.

Reece, S. Y., Hamel, J. A., Sung, K., Jarvi, T. D., Esswein, A. J., Pijpers, Joep J. H., and Nocera, D. G. (2011) Wireless solar water splitting using silicon-based semiconductors and earth-abundant catalysts. *Science* 334: 645–648.

Schwander, T., von Borzyskowski, L. S., Burgener, S., et al. (2016) A synthetic pathway for the fixation of carbon dioxide in vitro;. *Science* 354: 900–904.

Sherman, B. D., Vaughn, M. D., Bergkamp, J. J., Gust, D., Moore, A. L., Moore, T. A. (2014) Evolution of reaction center mimics to systems capable of generating solar fuel. *Photosynthesis Research* 120: 59–70.

Shockley, W. and Queisser, H. J. (1961) Detailed balance limit of efficiency of p-n junction solar cells. *Journal of Applied Physics* 32: 510–519.

Skillman, J. B. (2008) Quantum yield variation across the three pathways of photosynthesis: not yet out of the dark. *Journal of Experimental Botany* 59: 1647–1661.

Spenser, J. (2008) *The Airplane: How Ideas Gave us Wings.* New York: Harper/Smithsonian Books.

Tachibana, Y., Vayssieres, L., and Durrant, J. R. (2012) Artificial photosynthesis for solar water-splitting. *Nature Photonics* 6: 511–518.

Tasleem, S. and Tahir, M. (2020) Current trends in strategies to improve photocatalytic performance of perovskites materials for solar to hydrogen production. *Renewable & Sustainable Energy Reviews* 132: 110073.

Work, V. H., D'Adamo, S., Radakovits, R., Jinkerson, R. E., and Posewitz, M. C. (2012) Improving photosynthesis and metabolic networks for the competitive production of phototroph-derived biofuels. *Current Opinion in Biotechnology* 23: 290–297.

World Energy Council (2010) *2010 Survey of Energy Resources.* London: World Energy Council. ISBN: 978 0 946121 021. Available for free download from: https://www.worldenergy.org/assets/downloads/ser_2010_report_1.pdf.

Yella, A., Lee, H.-W., Tsao, H., Yi, C. Y., and Chandiran, A. K. (2011) Porphyrin sensitized solar cells with cobalt (II/III)-based redox electrolyte exceed 12 percent efficiency. *Science* 334: 629–634.

Young, K. J., Martini, L. A., Milot, R. L., Snoeberger, R. C., Batista, V. S., Schmuttenmaer, C. A., Crabtree, R. H,, and Brudvig, G. W. (2012) Light-driven water oxidation for solar fuels. *Coordination Chemistry Reviews* 256: 2503–2520.

Zhang, B. and Sun, L. (2019) Artificial photosynthesis: opportunities and challenges of molecular catalysts. *Chemical Society Reviews* 48: 2216–2264.

Zhao, H., (ed.) (2013) *Synthetic Biology.* Amsterdam: Elsevier.

Zhao, Y., Swierk, J. R., Megiatto, J. D. Jr., Sherman, B., Youngblood, W. J., Qin, D., Lentz, D. M., Moore, A. L., Moore, T. A., Gust, D., and Mallouk, T. E. (2012) Improving the efficiency of water splitting in dye-sensitized solar cells by using a biomimetic electron transfer mediator. *Proceedings of the National Academy of Sciences USA* 39: 15612–15616.

Zhou, H., Xiao, C., Yang, Z., and Du, Y. (2020) 3D structured materials and devices for artificial photosynthesis. *Nanotechnology* 31: 282001.

Zhu, X.-G., Long, S. P., and Ort, D. R. (2010) Improving photosynthetic efficiency for greater yield. *Annual Review of Plant Biology* 61: 235–261.

Appendix 1
Light, energy, and kinetics

Photosynthesis is a biological process in which light energy is converted into chemical energy. In order to understand how photosynthesis works at the molecular level, it is necessary to have an understanding of some of the basic chemical and physical principles that operate. These involve concepts of thermodynamics, kinetics, electrochemistry, photochemistry, spectroscopy, and the nature of light. This Appendix 1 is designed as a brief introduction to these topics. The level is at what might be called "elementary physical chemistry." Because of the breadth of the subjects, the treatment given here is necessarily very abbreviated. Some more advanced concepts are also introduced in the chapters at the places where they are relevant. Textbooks that include more a more extensive treatment of these topics and other more advanced treatments of many of these concepts can be found in the reference section.

A.1 Light

Light is an electromagnetic disturbance that propagates through space as a wave. The properties of **wavelength** (λ), **frequency** (ν), and **speed** (c) that apply to other types of waves, such as sound waves and water waves, also apply to light waves. The wavelength is the distance between successive crests of the wave, the frequency is the number of crests that passes an observer in a second, and the speed is the rate of propagation of a crest as it moves through the medium. The relationship among these three parameters is given by Eq. (A1.1) and illustrated in Fig. A1.1.

$$c = \lambda / \nu \qquad (A1.1)$$

The speed of light in vacuum is $3 \times 10^8 \, \mathrm{ms}^{-1}$. Because their product is a constant, the frequency and wavelength are interdependent quantities, so that if one is specified, the other is determined. Only a very small portion of the electromagnetic spectrum is visible to our eyes.

The regions of the electromagnetic spectrum that are most important in photosynthesis are the visible and near infrared portions, as they are strongly absorbed by chlorophylls and other photosynthetic pigments. The visible portion of the spectrum extends from 400 nm (10^{-9} m, abbreviated nm) to 700 nm, and the near infrared region extends from 700 nm to about 1000 nm.

The other important concept involving light is the idea of the **photon**. A photon is a particle of

Molecular Mechanisms of Photosynthesis, Third Edition. Robert E. Blankenship.
© 2021 Robert E. Blankenship 2021 by John Wiley & Sons Ltd.
Companion website: https://www.wiley.com/go/blankenship/molecularphotosynthesis3e

(a)

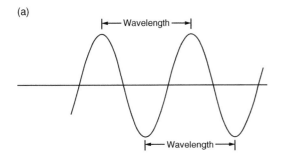

(b)

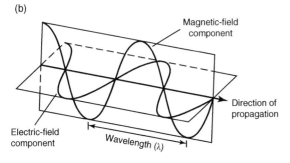

Figure A1.1　Properties of light waves. (a) The wavelength is the distance between crests, while the frequency is the number of crests that pass an observer per second. (b) Light is a combination of electrical and magnetic field waves that oscillate perpendicularly to each other and propagate through space.

light, much like a raindrop is a particle of rain. A photon has a definite energy, which depends on the frequency through the Planck relation, Eq. (A1.2).

$$E = h\nu \qquad (A1.2)$$

or combining with Eq. (A1.1)

$$E = \frac{hc}{\lambda} \qquad (A1.3)$$

In Eqs. (A1.2) and (A1.3), E is the energy of a photon of light (in Joules), h is Planck's constant, 6.63×10^{-34} J s, λ is the wavelength (in m), and c is the speed of light given above. A single photon of red light of wavelength 700 nm has energy 2.84×10^{-19} J. This energy can also be expressed in other more convenient units. One of these is the electron volt (eV). The conversion factor between J and eV is 1.60×10^{-19} J eV^{-1}, so that one photon of wavelength 700 nm has an energy content of 1.78 eV.

We have considered light first as a wave and then as a particle. Which one is correct? In our everyday experience, these are very different properties and are not found simultaneously in a single system. But light is a curious and sometimes surprising mixture of particle and wave. The fact that light simultaneously has both the properties of a wave and a particle is one of the remarkable results of quantum mechanics. This wave/particle dual nature also applies to matter such as electrons in atoms, as will be discussed in more detail below.

The wave that is associated with a photon has both electric and magnetic properties, which are at right angles to each other, as shown in Fig. A1.1. If the electric vectors have all possible orientations in space, the light is unpolarized. If the electric vectors are all oriented in a single plane, the light is known as plane polarized, while if the electric vectors rotate around the direction of propagation, the light is known as circularly polarized. This rotation can be either right- or left-handed. Normal sunlight is unpolarized, but can become plane polarized if it reflects off a shiny surface such as a lake. In many cases, important information about the molecules in photosynthetic systems can be learned by observing their interaction with polarized light.

The spectrum of the solar output is shown in Fig. 1.1. This spectrum is a complex curve that depends primarily on the temperature of the sun, but it is also affected by atomic species near the solar surface. The dependence of the solar spectrum on temperature is the same effect as we observe when something becomes so hot that it emits red light and even white light if it is even hotter. This effect is called **black-body radiation**, in that the effect was first explained with reference to a black body that was heated until it glowed. The black body radiation output curve, especially the falloff in the ultraviolet region, was explained by the German physicist Max Planck and formed the basis for the concept of quantization of light. The spectral output p of a sample heated to a temperature T is given by Eq. (A1.4).

$$\rho = \frac{8\pi hc}{\lambda^5}\left(\frac{1}{e^{hc/\lambda kT}-1}\right) \qquad (A1.4)$$

where k is Boltzmann's constant, and the other symbols are as defined earlier. Figure A1.2 shows output curves from black body radiators of various temperatures. As the temperature increases, the output increases and the maximum shifts to shorter wavelengths. The effective temperature of the sun is 5700 K.

It is instructive to consider the Planck photon relation (Eq. (A1.2)) and the black-body spectra of Fig. A1.2 simultaneously (also see the solar output curve Fig. 1.1). They tell us that the light in the shorter wavelength region of the spectrum has a high intrinsic energy content per photon but is not very intense, while the light at longer wavelengths has a low energy content per photon, although it is relatively intense.

A.2   Thermodynamics

Thermodynamics is the study of energetics. Since photosynthesis is an energy storage process, it is essential to have an understanding of the basic principles of thermodynamics in order to appreciate the way in which energy is converted from one form to another by photosynthetic systems.

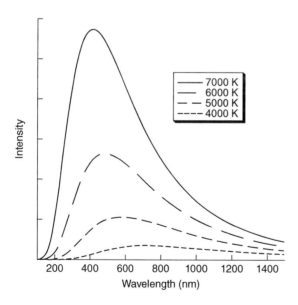

Figure A1.2   Black body emission curves, predicted by Eq. (A1.4). Curves are shown at temperatures as indicated. These curves should be compared to the solar output shown in Fig. 1.1.

The structure of thermodynamics is a bit like geometry, in that a small number of postulates or laws are formulated and then these are logically expanded into a remarkably complex and sophisticated subject. The laws of thermodynamics cannot be derived. Their validity rests in the fact that they do a remarkably successful job of describing how matter actually behaves. No documented exceptions are known. Occasionally, a claim is made that somehow the laws of thermodynamics can be circumvented by some clever device or process. A common one in the modern age is that an automobile can be modified to run without using any fuel other than water. Such claims are all variations on perpetual motion machines, in which a device either produces more energy than is put in or somehow extracts energy from a source in a way that violates the laws of thermodynamics. These devices and processes often initially attract considerable attention and many seem promising at first, but every one of them has ultimately turned out to be without merit. So the laws of thermodynamics provide a framework with which all processes and transformations in any branch of science must conform.

Our discussion of thermodynamics will mostly emphasize the macroscopic or classical formulation. Thermodynamics can also be formulated in a microscopic fashion, where it relies on statistical analysis. The two formulations are ultimately equivalent in almost every respect, and we will use elements of each in our discussion.

A.2.1   Thermodynamic functions: work, heat, internal energy, enthalpy, entropy, and free energy

Thermodynamic information is represented in terms of a number of mathematical functions. Some of these, such as **work** and **heat**, are familiar to us from everyday experience, while others, such as **entropy** or **free energy**, are more abstract and are formulated in terms of the simpler functions. Work is done when a mass is moved against a force. Heat is energy in the form of random thermal motion of molecules.

Most of the thermodynamic functions are called **state functions** and are represented by upper case letters. The values of state functions do not depend on the path that is used to cause the change. Internal energy is a state function, as are temperature, pressure, volume, and the derived functions enthalpy, entropy, and free energy that we will consider below. Heat and work are not state functions, so that a given change in heat or work may leave the system in different states, depending on how the changes are carried out. Intuitively, the idea that some changes are path-dependent while others are path-independent makes sense in terms of a mechanical analogy. We may desire to move a large rock from a lower to a higher elevation. The amount of work we expend getting the rock to the top very much depends on the way we do the job; for example, dragging the rock over rough or smooth territory, or rolling it up a specially constructed ramp. In contrast, once it is on top of the hill the rock's potential energy (which is a state function) is entirely independent of the path we took to get it to the top.

A.2.2 The first law of thermodynamics

The **internal energy** (U) is the total energy of the system under consideration. The first law of thermodynamics states that the change in internal energy of a system is the net of the **work** done on a system and the **heat** added to the system. A quantitative expression of the first law is as follows:

$$\Delta U = q - w \qquad (A1.5)$$

where ΔU is the change in internal energy, q is the heat added to the system, and w is the work done on the system. The essence of the first law is conservation of energy. Heat and work may appear to be different but are in reality similar and often interconvertible forms of energy. The energy of the system can be changed by either doing work on it or by adding heat to it. The important result is that the internal energy change is independent of the relative contributions of heat and work; instead, it depends only on the sum of the two. Interestingly, while heat and work are not state functions, the sum of the two, the internal energy, is a state function. As might be expected, the units of heat and work are the same, Joules (J) after James Joule, the nineteenth-century Scottish physicist who established the equivalence of heat and work.

The first law of thermodynamics is like a balance sheet, such as might be kept in a bank that does currency exchange. You can withdraw or deposit money in either of two different currencies: heat or work. However, the value of your bank account only depends on the net amount of money left in it after these transactions, not on what currency you used.

While the internal energy is perhaps the most fundamental of the thermodynamic functions, it is not the one that is the most convenient for most biochemical systems. If we make some restrictions on the conditions that apply and use a different function, it turns out to make calculations much simpler.

The **enthalpy** (H) of the system is a new thermodynamic function defined as:

$$H = U + PV \qquad (A1.6)$$

where P and V are the pressure and volume of the system. The utility of the enthalpy function is that under conditions of constant pressure and no other forms of work involved (the usual situation for most biological systems), the change in enthalpy of a system is numerically equal to the change in heat:

$$\Delta H = q_p \qquad (A1.7)$$

where q_p is the heat absorbed by the system under conditions of constant pressure. Many biochemical processes involve heat changes that take place at constant pressure, so the enthalpy is a natural function to use to describe them. If the enthalpy change is negative, so that heat is given out by the system, the process is called **exothermic**, while if the process takes up heat, it is called **endothermic**.

A.2.3 Entropy and the second law of thermodynamics

The first law of thermodynamics tells us what energy changes accompany a change in state of a system. Unfortunately, it cannot tell us whether or not that change is likely to actually take place. To understand why one process happens and another doesn't, we need to introduce a second law of thermodynamics. The second law is usually formulated in abstract ways that are sometimes difficult to understand. However, we all know the underlying principle of the second law of thermodynamics from everyday experience. If we pour a cup of hot coffee and leave it, after a time it will get cold. An ice cube placed on the kitchen counter will melt. We never have to worry about a cold cup of coffee getting hot again all by itself, or a puddle of water at room temperature spontaneously freezing; these things just don't happen! The knowledge of what are and what aren't spontaneous processes is just common sense based on our experiences with the world, but the first law permits all sorts of processes like this that we know are impossible. Another example that also introduces the idea of probability involves taking a new deck of cards in sequential order and shuffling it several times. If we then look at the order of the cards, it will be very different from what we started with, and no amount of shuffling will ever restore the initial order.

The second law of thermodynamics helps us understand why these examples work the way they do. The second law is often stated in terms of the impossibility of converting heat completely into work. While the first law tells us that a given amount of heat and work may have the same energy content, the second law tells us that they differ in terms of how available that energy is to us. A more intuitive formulation of the second law is based on probabilities. To understand this, we must first define a new thermodynamic function called the **entropy** (S). The entropy of a system is defined in statistical terms as:

$$S = k_B \ln W \qquad (A1.8)$$

where k_B is a fundamental physical constant known as Boltzmann's constant ($1.3807 \times 10^{-23}\,\mathrm{J\,K^{-1}}$), ln is the natural logarithm function, and W is the probability of finding the system in a given state or, alternatively, the number of distinct states that it can adopt. This equation defining the entropy in terms of probabilities can actually be found chiseled into the Austrian physicist Ludwig Boltzmann's gravestone in Vienna. Returning to our example of a deck of cards, initially the deck was completely ordered, and there is only one way to arrange the cards in order. The entropy of the ordered deck of cards is therefore zero, because the ln of one is zero. There are, however, a very large number of possible orders for the shuffled deck to adopt, so the act of shuffling the deck increases the entropy of the system.

Another statement of the second law of thermodynamics is that the entropy of the universe must increase when any spontaneous process takes place. It may seem inconvenient to have to keep track of the entire universe to know if a process is spontaneous, but the situation is not quite that difficult. In many cases, the system may be isolated, so that no heat or work changes are involved with the rest of the universe, called the surroundings. In this case, the entropy changes in the system are the same as those in the universe. In other cases, it is possible to rather simply keep track of the surroundings, as we will see in the next section. In any case, just knowing the sign of the entropy change for the universe tells us whether the process can occur. If a process is accompanied by a decrease of entropy in the universe, then it will not happen. In fact, the opposite process will be spontaneous, because it is accompanied by an increase of entropy in the universe. So the coffee cools off because by doing so the entropy of the universe increases, and the coffee doesn't spontaneously warm up above room temperature because this process would take place with a decrease in the entropy of the universe. Of course, entropy changes in the system can be either positive or negative; when the coffee cools off its entropy actually goes down, but this is more than offset by an increase in the entropy of the surroundings, so that the entropy change in the universe is positive.

A.2.4   Gibbs free energy

It is possible to define a new thermodynamic function of the system that under certain conditions tells us whether the process being described is spontaneous or not. In essence, we are including the entropy changes of the surroundings into the new function, so that it is simply a more convenient way of representing the predictions of the second law. This new function is called the **Gibbs free energy** after J. Willard Gibbs, the American chemist who was instrumental in establishing the modern formulation of thermodynamics. Usually, we will simply call it the free energy. The free energy is defined in terms of the enthalpy, entropy, and absolute temperature. (The absolute or Kelvin temperature (C+273) is always used in thermodynamic analysis. Use of the Celsius temperature will in almost all cases lead to wrong results).

$$G = H - TS \qquad (A1.9)$$

Given this definition of the free energy, it is straightforward to derive the result (although we won't do it here) that under the very important restrictions of constant pressure and temperature, the change in free energy of the system must be negative for a spontaneous change. If the free energy change is positive, then the reverse process is spontaneous, and if it is zero then the process is at equilibrium and there is no net change. The free energy change ΔG can be thought of as the difference of an enthalpic term and an entropic term, the latter weighted by the absolute temperature.

$$\Delta G = \Delta H - T\Delta S \qquad (A1.10)$$

This is something of an oversimplification, but it is true that exothermic processes with negative enthalpy changes ($\Delta H < 0$) will always be spontaneous if the entropy of the system also increases ($\Delta S > 0$), because both the ΔH term and the $-T\Delta S$ term are then negative and ΔG must be negative. An endothermic process ($\Delta H > 0$) can be either spontaneous or not depending on the signs and magnitudes of the two terms in Eq. (A1.10). A process in

which ΔG is negative is called **exergonic,** and one in which ΔG is positive is called **endergonic**.

The ability to focus on just the system is a great convenience, because we no longer have to explicitly calculate what happens in the surroundings. We will see that it is possible to predict all sorts of things about a process such as a chemical or biochemical reaction by simply calculating the free energy change of the process. Tables of free energies have been tabulated for many molecules, so it is a simple matter to determine whether a given chemical reaction is spontaneous by simply calculating the free energy change for the reaction. We must always keep in mind the restrictions of constant temperature and pressure, because the sign of the free energy change only indicates whether or not a process is spontaneous under those conditions. However, this is not too much of an inconvenience, especially for biochemical processes, which typically take place at ambient temperatures and pressures.

A.2.5   Chemical potentials and equilibrium constants

We are now almost ready to make a fundamental connection between the free energy change of a chemical reaction and the equilibrium constant of the reaction. To do this, we need to first introduce a new function called the **chemical potential**, μ. The chemical potential of species i is defined as:

$$\mu_i = \mu_i^o + RT \ln a_i \qquad (A1.11)$$

where μ_i^o is the molar free energy of species i at 1 atm pressure. R is the gas constant ($8.314\,\mathrm{JK^{-1}mol^{-1}}$), T is the absolute temperature, and a_i is the thermodynamic activity of species i. The activity is a quantity that is very close to the concentration for many compounds in dilute solution. Important exceptions are ionic species, where activities differ from concentrations even in quite dilute solutions. For our discussion, we will assume that concentrations and activities are the same, but you should always keep in mind that this is an approximation. The main advantage of the chemical potential is that it gives us a mechanism for representing the free energy of a species as a function of concentration in a solution and provides a route towards relating free energy changes and equilibrium constants of chemical reactions.

For the general chemical reaction

$$aA + bB \rightleftharpoons cC + dD \qquad (A1.12)$$

we can write the chemical potentials of each of the species involved, as given by the example of species A

$$\mu_a = \mu_a^o + RT \ln[A] \qquad (A1.13)$$

where the quantity in brackets is the concentration of that species in solution. The quantities μ_a, μ_b, μ_c, and μ_d are defined analogously. The chemical potential of each species is thus a standard value μ_i^o which represents the free energy of the compound at $1\,M$ concentration, plus a correction term that adjusts the free energy for concentrations other than $1\,M$. The free energy change of the overall reaction is just the free energy of the products minus that of the reactants

$$\Delta G = \sum \left(G_{products} - G_{reactants} \right)$$
$$= c\mu_c + d\mu_d - a\mu_a - b\mu_b \qquad (A1.14)$$

where the individual chemical potentials are weighted by their stoichiometric coefficients in the reaction. If we then collect terms and do a little algebra, we obtain the fundamental relationship between free energy changes in reactions and a measure of the progress of the reaction, which is often called the reaction quotient, Q.

$$\Delta G = \Delta G^\circ + RT \ln Q \qquad (A1.15)$$

where

$$\Delta G^\circ = c\mu_c^o + d\mu_d^o - a\mu_a^o - b\mu_b^o \qquad (A1.16)$$

ΔG° is the **standard state free energy change** of the reaction under standard conditions of $1\,M$ concentration, $1\,atm$ pressure and $298\,K$ and the reaction quotient Q is given by:

$$Q = \frac{[C]^c [D]^d}{[A]^a [B]^b} \qquad (A1.17)$$

If the reaction has come to equilibrium, the free energy change is zero by definition and Q becomes the **equilibrium constant** K

$$0 = \Delta G = \Delta G^\circ + RT \ln K \qquad (A1.18)$$

Rearranging, we find the simple but very powerful quantitative relationship between the standard state free energy change and the equilibrium constant:

$$\Delta G^\circ = -RT \ln K \qquad (A1.19)$$

If ΔG° is less than zero, K is greater than 1 and the reaction is spontaneous as written, while if ΔG° is greater than zero, then K is less than 1 and the reverse reaction is spontaneous. To determine the numerical value of the equilibrium constant, all we need to do is find the standard state free energy change ΔG° for the reaction. In most cases, the easiest way to do this is to use tabulated values of the free energies of formation, ΔG_f° of compounds from their elemental constituents. The overall standard state free energy change is then

$$\Delta G^\circ = \sum \left(\Delta G_{fproduct}^\circ - \Delta G_{freactants}^\circ \right) \qquad (A1.20)$$

Table A1.1 gives free energies of formation for some chemical compounds that are important in photosynthesis. The free energies of formation of elements in their most stable form are defined to be zero. This choice is acceptable because energy (unlike entropy) has no natural zero and is always compared to an arbitrary reference point. Since the reference point is used consistently and only differences in energy are considered, no errors are introduced.

A.3 Boltzmann distribution

We will introduce one important result from statistical thermodynamics, which is how a system with more than one energy level is described as a function

Table A1.1 Free energies of formation of some chemical compounds important in photosynthesis.

Species	ΔG_f° (kJ mol^{-1})[a]
H_2O (liq)	−237.19
O_2 (g)	0
CO_2 (g)	−394.38
Glucose $C_6H_{12}O_6$ (aq)	−914.54
H^+ (aq)	0

[a] ALL values are in units of kJ mol^{-1} and refer to the standard state of 298 K and 1 atm pressure.

of temperature. Consider a system with only two energy levels. The spacing between the two energy levels is ΔE, and the absolute temperature is T. The populations of the two levels at equilibrium are given by Eq. (A1.21), which is known as the **Boltzmann distribution**:

$$\frac{P_2}{P_1} = \exp\left(-\Delta E / k_B T\right) \qquad \text{(A1.21)}$$

where k_B is Boltzmann's constant. This relation shows that the relative populations of the two levels depend on both the energy gap and the temperature. At low temperatures or large energy gaps, the population is almost all in the lower level, while at very high temperatures and small energy gaps, the two levels are almost equally populated. The energy that permits the population of the higher levels comes from random thermal energy, hence the critical role of the temperature in the expression. Eq. (A1.21) describes a system at equilibrium. When the system is out of equilibrium, which can happen for a variety of reasons, it will always relax toward the energy distribution given by the Boltzmann distribution. Systems that have more than two levels can be described with a modified version of the equation, but the basic principle is maintained, that the lower energy levels are the most highly populated, with the population of higher levels falling off exponentially.

A.4 Electrochemistry: reduction–oxidation reactions

Many of the early chemical reactions that take place in photosynthesis are pure electron transfer reactions, in which an electron is added to a species (**reduction**) or taken away from it (**oxidation**). It is important for us to have an understanding of these reduction–oxidation (often shortened to **redox**) reactions. Fortunately, we have already developed most of the formalism for describing them in our discussion of thermodynamics, so that only a few additional concepts are needed.

A general redox reaction can be written as:

$$A_{ox} + e^- \rightleftharpoons A_{red} \qquad \text{(A1.22)}$$

where A_{ox} and A_{red} represent the oxidized and reduced form of species A. For each species that is reduced, another must become oxidized

$$B_{red} \rightleftharpoons B_{ox} + e^- \qquad \text{(A1.23)}$$

The overall reaction is

$$A_{ox} + B_{red} \rightleftharpoons A_{red} + B_{ox} \qquad \text{(A1.24)}$$

If we use Eq. (A1.15) for the free energy change associated with a chemical reaction, we obtain for the reaction given in Eq. (A1.24)

$$\Delta G = \Delta G^{\circ} + RT \ln \frac{[A]_{red}[B]_{ox}}{[A]_{cox}[B]_{red}} \qquad \text{(A1.25)}$$

The free energies of electrochemical reactions are most conveniently measured using electrical measurements. There is a simple relationship between the free energies, measured in Joules, and the free energies measured in volts. The proportionality between the two forms of free energy is given by

$$\Delta G = -nF\Delta E \qquad \text{(A1.26)}$$

where n is the number of electrons involved in the redox reaction (one in our example), F is the

Faraday constant (96,485 Coulombs mol^{-1}), and ΔE is the electrical potential measured in volts. Combining Eqs. (A1.25) and (A1.26), we obtain the **Nernst equation**, which is the fundamental equation describing electrochemical processes:

$$\Delta E = \Delta E^{\circ} - \frac{RT}{nF} \ln \frac{[A]_{red}[B]_{ox}}{[A]_{ox}[B]_{red}} \quad (A1.27)$$

If we convert to base 10 logs and substitute in the numerical values for the constants, we obtain an alternate version of the Nernst equation

$$\Delta E = \Delta E^{\circ} - \frac{0.059}{n} \log \frac{[A]_{red}[B]_{ox}}{[A]_{ox}[B]_{red}} \quad (A1.28)$$

In most cases, it is convenient to measure the potentials (and thereby the free energies) of electrochemical reactions with reference to a standard reaction. The standard that has been chosen for electrochemical reactions is the hydrogen electrode

$$2H^{+} + 2e^{-} \rightleftharpoons H_2 \quad (A1.29)$$

Electrical free energy, like energy in general, has no absolute value and is always set relative to a reference state. For the Normal Hydrogen Electrode (NHE), the standard state redox potential E° is defined as 0 V (at 298 K and 1 M H^{+} concentration). This primary reference electrode is not convenient for biochemical systems, so secondary reference electrodes are utilized for measurements, and then, the values obtained are converted to the hydrogen electrode reference. Common secondary reference electrodes are the calomel and the silver/silver chloride electrodes. This method permits the determination of redox potentials for a large number of electrochemical "half reactions" of the form shown in Eq. (A1.22). Table A1.2 gives values of standard redox potentials for a number of electrochemical reactions that are important in photosynthesis. Note that these reactions are by convention written as reductions. If the reaction that takes place is an

Table A1.2 Midpoint redox potentials for reactions of interest in photosynthesis.

Redox reaction	E'_m (V)
NADP^{+} + H^{+} + 2e^{-} $\rightleftharpoons$ NADPH	−0.324
O$_2$ + 2H^{+} + 4e^{-} $\rightleftharpoons$ 2H$_2$O	+0.816
P700^{+} + e^{-} $\rightleftharpoons$ P700	+0.49
P870^{+} + e^{-} $\rightleftharpoons$ P870	+0.45
P680^{+} + e^{-} $\rightleftharpoons$ P680	~1.1
2H^{+} + 2e^{-} $\rightleftharpoons$ H$_2$ (g)	−0.414
UQ + 2e^{-} + 2H^{+} $\rightleftharpoons$ UQH$_2$	+0.060
Chl^{+} + e^{-} $\rightleftharpoons$ Chl	+0.78
BChl^{+} + e^{-} $\rightleftharpoons$ BChl	+0.64

All values are in units of V and refer to the standard state of 298 K, and pH 7.

oxidation, the potential is the negative of the value for the reaction as a reduction. Potentials for biochemical redox reactions are by convention measured at pH 7; this is indicated by a superscript prime on the symbol.

We also need to understand one other distinction between the standard state redox potentials and experimentally determined values, which are called midpoint potentials. This is discussed in the next section.

A.4.1 Measurement of midpoint potentials

We will now consider how one goes about measuring the redox potential of a species that may be either a small molecule in solution or a redox-active protein such as a cytochrome. As discussed above, the measurement is made relative to a reference electrode, usually a secondary reference such as the saturated calomel electrode. The measured potential is then given by a modified form of the Nernst equation

$$\begin{aligned} E_{meas} &= E_A - E_{ref} \\ &= E_{mA} - \frac{0.059}{n} \log \frac{[A]_{red}}{[A]_{ox}} - E_{ref} \end{aligned} \quad (A1.30)$$

where E_{meas} is the electrical potential actually measured as shown in Fig. A1.3, E_{ref} indicates the

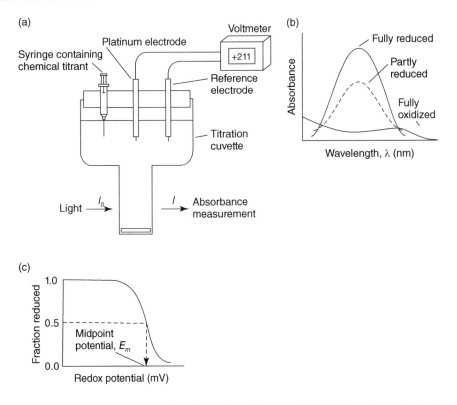

Figure A1.3   Redox measurements to determine electrochemical potentials. (a) Experimental setup for redox titrations, including platinum and reference electrodes, a voltmeter, and a light beam to detect absorbance. (b) Spectral absorbance band of a redox-active species. (c) Redox titration of a molecule. The midpoint potential is the potential where the species being investigated is half oxidized and half reduced.

potential of the reference electrode, E_A is the potential of the species under investigation, and E_{mA} is the **midpoint redox potential** of species A. Typically, the reference electrode and an inert electrode, usually platinum, are inserted into a solution containing the species of interest. In order to determine values for E_A°, the standard state potential, it is necessary to adjust the concentrations (activities) of A_{red} and A_{ox} to 1 M. Usually, it is not possible to obtain such high concentrations of biomolecules, so an alternative approach is used. The concentrations of A_{red} and A_{ox} are adjusted to be the same, whereupon the log-term in Eq. (A1.30) vanishes. The parameter that is obtained under these conditions, the midpoint potential, E_m, is close to but not exactly the same as a standard state potential. To preserve this distinction, it is given the name

midpoint potential, to indicate that it is the redox potential measured at the midpoint of the conversion of the species from fully oxidized to fully reduced. Most of the standard potentials tabulated in Table A1.2 are midpoint potentials. Midpoint potential values are an essential starting point to understand the workings of electron transport chains such as are found in photosynthetic or respiratory complexes.

Experimentally, midpoint potentials are determined by making a series of measurements of the concentration of the oxidized and reduced forms of the species of interest, while at the same time measuring the ambient electrical potential as shown in Fig. A1.3. The change in the redox state of the species can be monitored in any number of ways, as long as the method distinguishes oxidized from reduced molecules. The most common method is to use absorbance measurements, in which the UV/Vis absorption of the species is monitored as a function of redox potential. Most molecules have significantly different absorbance properties in the oxidized and reduced forms. A series of spectra at different potentials is recorded, and the fraction reduced *vs.* measured potential is tabulated. The midpoint potential is then found as the potential at which the species is half reduced and half oxidized. This is shown schematically in Fig. A1.3b. The form of the Nernst equation (Eq. (A1.30)) requires that the system is at thermodynamic equilibrium when the actual measurement is made. This is essential because redox measurements are thermodynamic measurements and are meaningless if the system is not at equilibrium. An additional electrode can be included in the system to inject or withdraw electrons directly and therefore avoid the addition of the chemical titrants. However, the principle of the method is the same regardless of how the oxidation or reduction is carried out.

A.5   Chemical kinetics

In addition to a basic understanding of thermodynamics, it is essential to have an appreciation of the basic principles of **kinetics** in order to fully appreciate the mechanisms of photosynthesis. In this section, we will explore a few basic concepts of kinetics that apply directly to photosynthetic processes. In contrast to thermodynamics, which tells us what processes are possible energetically, but little or nothing about how quickly they take place, kinetics tells us about the time course of chemical processes, but little about the energetics. Actually, in a few special cases connections can be made between kinetics and thermodynamics, the most famous being the Marcus theory of electron transfer processes. We discuss this topic in more detail in Chapter 6.

An essential principle of chemical kinetics is the concept of the rate or **velocity** v of the reaction. This is an experimental quantity that represents the rate of change with time of the concentration of either one of the reactants R or one of the products P. The units of reaction velocity are $M\,s^{-1}$.

$$-\frac{d[R]}{dt} = \frac{d[P]}{dt} = v \qquad (A1.31)$$

The velocity of the reaction usually depends on the concentrations of one or more of the reactants or products. A series of experiments is run in which the concentrations of various species such as reactants, products, possible catalysts or inhibitors, and so on are systematically varied. From these experiments, a **rate law** is deduced, which represents quantitatively how the velocity of the reaction depends on concentrations of the various species.

$$v = [A]^n [B]^m \dots \qquad (A1.32)$$

In Eq. (A1.32), the species A and B affect the rate of the reaction. The **order** of the reaction with respect to each of the species is the exponent associated with that species. In Eq. (A1.32) the reaction is nth order with respect to A and *m*th order with respect to B, for an overall order of $m+n$. Orders can be either positive or negative, the latter in the case of an inhibitor, and need not in all cases be integers.

The order of the overall reaction is an empirical relationship and need not be the same as the stoichiometry of the reaction. The study of chemical kinetics is devoted to the establishment of the rate laws for chemical reactions and then trying to use this information to deduce the underlying molecular mechanism of the reaction. This is a difficult task, because there is no general one-to-one correspondence between observed rate laws and mechanisms. If a detailed mechanism is proposed, it will predict a particular rate law. If the rate law is not the experimentally observed one, then the mechanism is incorrect in at least some aspects. However, if the mechanism predicts the correct rate law, it does not necessarily mean that the correct mechanism has been found, because often several clearly distinct mechanisms predict the same rate law. So kinetic data can prove a mechanism wrong but can never prove it right.

There is one situation in which the observed order for a kinetic process must coincide with the reaction stoichiometry, and that is when an elementary step of the reaction sequence is being observed. An elementary step is the simplest possible process in the reaction mechanism, such as breaking a bond or transferring an electron from one species to another. In this case the rate law gives information about the underlying mechanism. Fortunately, a large number of kinetic processes that are observed in photosynthetic systems are in this category. It is also fortunate that the effects of diffusion of molecules in solution, which are often dominant for many chemical reactions, are significantly less important for most of the reactions involved in photosynthesis. This is because many of the processes take place on large complexes embedded in a membrane system. The system in many ways exhibits solid-state behavior rather than solution behavior. This simplifies many of the observed reactions and makes most processes kinetically first order. This is not true for the soluble reactions of the carbon fixation pathway, which exhibit kinetic behavior typical for enzymes. We introduce some aspects of enzyme kinetics in Chapter 9 when these processes are discussed.

A.5.1   *First-order kinetics*

A chemical reaction with general form $A \rightarrow B$ that is **first order** in species A obeys the following rate law:

$$v = -\frac{d[A]}{dt} = k[A] \qquad (A1.33)$$

The first-order rate constant k for this reaction has units s^{-1}. This rate law can be easily integrated to give the time dependence of the disappearance of A, where $[A]_t$ is the concentration of species A at an arbitrary time t, $[A]_0$ is the concentration of A at zero time

$$[A]_t = [A]_0 \, e^{-kt} \qquad (A1.34)$$

Figure A1.4 shows the time course of a first-order decay. The rate constant k is the reciprocal of the characteristic time τ. The parameter τ is the time required for A to decay to $1/e$ (0.37) of the initial concentration $[A]_0$. Sometimes first-order reactions are described in terms of the **half time** $t_{1/2}$, which is the time required for A to decay to $[A]_0/2$. The value of $t_{1/2}$ is 0.69τ. One of the characteristics of first-order decay is that the $t_{1/2}$ is the same throughout the entire decay process, so that the time required to decay from $[A]_0$ to $[A]_0/2$ is the same as that required to decay from $[A]_0/2$ to $[A]_0/4$.

One of the classic examples of a first-order process is radioactive decay, in which an unstable nucleus decays spontaneously to products. Many of the kinetic processes in photosynthetic systems follow first-order kinetics. First-order behavior is typically observed when a chemical species is somehow created in a high-energy state and then

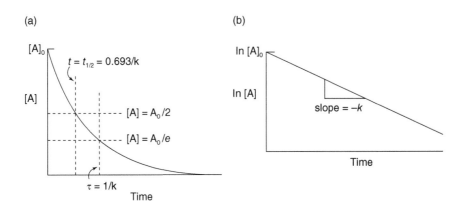

Figure A1.4   First-order kinetic decay. (a) Direct plot of concentration of species *A* versus time. The half life $t_{1/2}$ and characteristic time τ are the time required for *A* to decay to half or 1/e of its initial value, respectively. (b) Semilog plot of *A* vs time, with slope = $-k$.

relaxes spontaneously to a lower-energy state, without involvement of another reacting species. Many high-energy states are found in photosynthetic systems that result from photon absorption, and the subsequent relaxation of the system is often kinetically first order.

If the first-order reaction is reversible, the measured kinetics of the decay are still exponential, although the end point of the decay is not zero, but is instead the equilibrium distribution of the reactant and product. In this case, where the equilibrium state is being approached, the observed first-order decay constant is the sum of the intrinsic forward and reverse rate constants. This is true regardless of whether the initial concentration distribution is dominated by product or reactant. The equilibrium constant of the reaction is the ratio of the forward and reverse rate constants.

A.5.2   Parallel first-order reactions

In many cases, a species decays by a first-order kinetic process, but can follow any one of a number of decay pathways from the initial state. An example of this behavior relevant to photosynthesis is the decay of the chlorophyll excited states that are produced by photon absorption. This is therefore an important special case to master in order to obtain an understanding of photochemistry and photophysics of excited state processes.

A parallel first-order reaction has a common origin but two or more possible products, as shown in Fig. A1.5a. Each pathway has its own characteristic rate constant. The rate expressions for the decay of A and the formation of the products are as follows:

$$-\frac{d[A]}{dt} = k_1[A] + k_2[A] + k_3[A] \quad (A1.35a)$$

$$\frac{d[B]}{dt} = k_1[A] \quad (A1.35b)$$

$$\frac{d[C]}{dt} = k_2[A] \quad (A1.35c)$$

$$\frac{d[D]}{dt} = k_3[A] \quad (A1.35d)$$

Each of the rate constants k_i is a first-order rate constant with units of s^{-1}. The integrated form of the rate law for disappearance of A is as follows:

$$[A]_t = [A]_0 e^{-k_{obs}t} \quad (A1.36)$$

where

$$k_{obs} = k_1 + k_2 + k_3 = \sum_n k_n \quad (A1.37)$$

Equation (A1.35a) indicates that A decays exponentially but with an observed rate constant that is equal to the sum of the rate constants of the individual decay processes. Somewhat surprising at

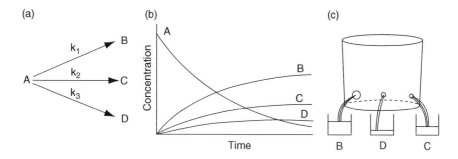

Figure A1.5   Parallel first-order decay processes. (a) Reaction scheme. (b) Time course of decay of precursor and buildup of products. (c) Bucket analogy to describe the amount of each species that forms via different paths.

first is the kinetic behavior for the appearance of the products. For species B, the time course is as follows:

$$[B]_t = \frac{k_1[A]_0}{k_{obs}}\left(1 - e^{-k_{obs}t}\right) \qquad (A1.38)$$

$[C]_t$ and $[D]_t$ have the same form as Eq. (A1.38) but the k_1 in the numerator is replaced with k_2 or k_3, respectively. The appearance of B is exponential, and the observed rate constant for its appearance is k_{obs}, the same as for the disappearance of A. This reflects the fact that the precursor state A for B formation is decaying away with rate constant k_{obs}, so that as A disappears, B formation must stop. The amount of B (C or D) formation does, however, depend on the values of the rate constants for those pathways. Consider the final fractional yield of B, which is the limit of Eq. (A1.38) at $t = \infty$:

$$Y_B = \frac{[B]_\infty}{[A]_0} = \frac{k_1}{k_{obs}} \qquad (A1.39)$$

The fractional yield (Y) of each product species is

$$Y_i = \frac{k_i}{\displaystyle\sum_n k_n} \qquad (A1.40)$$

The total yield of all product species is 1.

$$\sum_i Y_i = 1 \qquad (A1.41)$$

The time course of decay of A and formation of B, C, and D are shown in Fig. A1.5b. A useful analogy to aid in understanding parallel first-order decays is to visualize a bucket filled with water, which represents the total amount of species A (Fig. A1.5c). The bucket has a number of holes in it, through which water can leak. Smaller buckets are positioned beneath each of the holes to catch the water that leaks out of each hole. The overall rate constant with which the water leaks out of the bucket (k_{obs}) depends on the number and size of the holes. The amount of water that leaks out of each hole (Y_i) depends on the

size of that hole compared to the total size of all the holes combined (see Eq. (A1.40)).

An important variation on the theme of parallel first-order reactions is the case where one of the reactions is not first order but rather second order. Consider an additional reaction similar to those described by Eq. (A1.35).

$$\frac{d[E]}{dt} = k_4[A][Q] \qquad (A1.42)$$

where species Q is another reactant that is involved in the decay of A, but only via this new pathway. The rate constant k_4 is a second-order rate constant, with units $M^{-1}\ s^{-1}$. This additional pathway will have no influence on the decay of A if there is no Q present, but if the Q concentration is high, it will become the dominant decay pathway, because the rate of this process depends linearly on the concentration of Q present. The yield of pathway i (which is often not a pathway that involves Q) in the absence of $Q\left(Y_i^0\right)$ is given by Eq. (A1.40). However, in the presence of some amount of Q, the yield of pathway i will be

$$Y_i^Q = \frac{k_i}{\displaystyle\sum_n k_n + k_4[Q]} \qquad (A1.43)$$

where Y_i^Q is the yield of pathway i in the presence of Q, and the summation is over all the paths (1 through 3 in this example) that do not involve Q. If we simply take the ratio of the two yields

$$\frac{Y_i^0}{Y_i^Q} = \frac{\dfrac{k_i}{\displaystyle\sum_n k_n}}{\dfrac{k_i}{\displaystyle\sum_n k_n + k_4[Q]}} \qquad (A1.44)$$

This equation is easily simplified to give the **Stern–Volmer** equation:

$$\frac{Y_i^0}{Y_i^Q} = 1 + K_{SV}[Q] \qquad (A1.45)$$

where K_{SV}, known as the Stern–Volmer constant, is given by:

$$K_{SV} = \frac{k_4}{k_{obs}^0} \qquad (A1.46)$$

where k_{obs}^0 is the observed decay constant when $[Q] = 0$. K_{SV} has units of M^{-1}. The yield of pathway i is thus linearly dependent on the amount of Q, as shown in Fig. A1.6. This simple relationship is very frequently used to describe the yield of fluorescence from an excited state in the presence of varying amounts of a quencher molecule Q. The slope of the line in Fig. A1.6 is K_{SV}. Somewhat surprisingly, this result can be easily used to find k_4, the quenching rate constant, by simply substituting into Eq. (A1.46), provided the decay rate constant with no Q present has been measured. In this way, a single time-resolved measurement can be carried out to establish k_{obs}^0, and then, the quenching rate constants of any number of quenchers can be determined by simple steady-state fluorescence measurements at a number of quencher concentrations. This is convenient because quenching rate constants may be quite large, requiring special techniques to measure. If this has to be measured for each of a large number of quenchers being tested, it can be difficult. However, the Stern–Volmer equation permits the extraction of the needed kinetic information with only one time-resolved measurement and a number of steady-state ones.

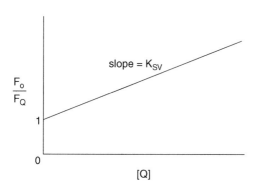

Figure A1.6 Stern–Volmer plot of fluorescence intensity *vs.* quencher concentration.

A.5.3   Sequential first-order reactions

Another kinetic mechanism that is very common in photosynthetic systems is a series of sequential first-order reactions. The reaction sequence is given by Eq. (A1.47):

$$A \xrightarrow{k_1} B \xrightarrow{k_2} C \qquad (A1.47)$$

The rate laws for the three species are as follows:

$$-\frac{d[A]}{dt} = k_1[A] \qquad (A1.48a)$$

$$-\frac{d[B]}{dt} = -k_1[A] + k_2[B] \qquad (A1.48b)$$

$$\frac{d[C]}{dt} = k_2[B] \qquad (A1.48c)$$

This can be solved to give the time course for each of the three species:

$$[A]_t = [A]_0\, e^{-k_1 t} \qquad (A1.49a)$$

$$[B]_t = \frac{k_1[A]_0}{k_2 - k_1}\left(e^{-k_1 t} - e^{-k_2 t}\right) \qquad (A1.49b)$$

$$[C]_t = [A]_0\left\{1 - \frac{1}{k_2 - k_1}\left(k_2 e^{-k_1 t} - k_1 e^{-k_2 t}\right)\right\} \qquad (A1.49c)$$

The intermediate species B begins with zero concentration, reaches a maximum, and then declines as the final product C builds up. The amount of B that transiently accumulates depends on the relative magnitudes of the rate constants k_1 and k_2 for creation and destruction of B. Figure A1.7 shows some examples of sequential first-order reactions with different values of the rate constants.

A peculiarity of the series first-order kinetic scheme is that often it is difficult to distinguish experimentally

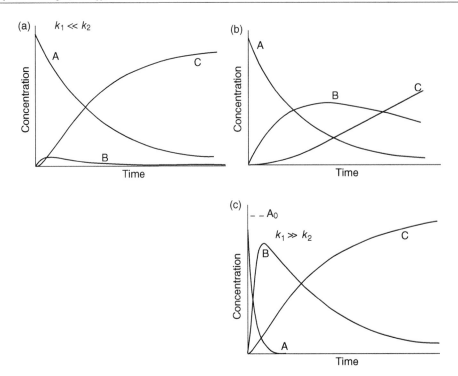

Figure A1.7 Sequential first-order reactions under three different relative magnitudes of the rate constants of the first and second reactions: (a) $k_1 \ll k_2$; (b) $k_1 = k_2$, (c) $k_1 \gg k_2$.

between the case where $k_1 \gg k_2$ and the opposite case where $k_1 \ll k_2$. Note that the shape of the curve for species B is similar in the two cases, although the amplitude is different. If only the curve for species B is observed, then it is natural to assume that the fast-rising kinetic phase represents the formation of B from A and the slow phase the decay of B to C (Fig. A1.7a). However, just the opposite is true. This can be a source of confusion if the concentration profiles for all the species are not available, as is often the case.

A.5.4 Temperature dependence of reaction rates

Reaction rates are almost always observed to increase as the temperature is increased, at least over the range of stability of the species involved. This temperature dependence of the rate constant is described empirically by the **Arrhenius equation**:

$$k(T) = Ae^{-\frac{E_a}{RT}} \qquad (A1.50)$$

where R is the universal gas constant, A is known as the pre-exponential factor, and E_a is called the **activation energy**. The activation energy is represented as the energy barrier that must be overcome if the reaction is to take place, as shown in Fig. A1.8a. The activation energy of a reaction is usually determined by measuring the rate constant at a number of temperatures and then graphing $\ln k$ vs. $1/T$, as shown in Fig. A1.8b. The slope of the straight line is $-E_a/R$. Some of the electron transfer reactions that take place in photosynthetic complexes exhibit a remarkable lack of temperature dependence, and in some cases actually become slower as the temperature is increased rather than faster. This behavior is not easily explained using the simple Arrhenius concept of an activation barrier, but is described by a more sophisticated theory of electron transfer

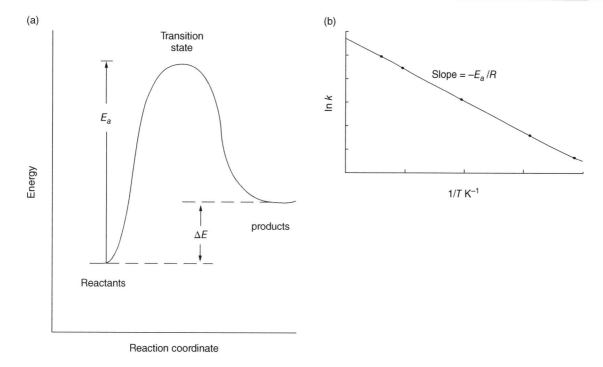

Figure A1.8   (a) Activation energy barrier for a chemical reaction. (b) Arrhenius plot of ln k vs. $1/T$.

known as the Marcus theory. We discuss this theory in more detail in Chapter 6.

A.6   Quantum versus classical mechanics

A major scientific revolution took place in the early twentieth century. It involved the recognition that the motion and energies of very small particles do not follow the same rules as those followed by large particles. The science of mechanics was formulated in the 1600s by Isaac Newton. It does a remarkable job of describing and predicting the properties of macroscopic objects like planets and baseballs. However, science was now pushing into a new realm, the world of ultrasmall atoms and molecules. At first, it was simply assumed that the same laws that worked for planets would work for atoms. However, it slowly became clear that this wasn't valid and that a new mechanics were needed for

such small particles with tiny energies. Beginning with Bohr, Planck, and Einstein and continuing with Pauli, Heisenberg, and Schrödinger, the period from 1900 to 1930 witnessed the birth of this new **quantum mechanics**. It didn't replace Newtonian, now called **classical**, mechanics. Rather, it explained the limits of applicability of classical mechanics and described the behavior of systems on the atomic and molecular scale.

A.7   Quantum mechanics – the basic ideas

Quantum mechanics is a little like thermodynamics in that a small number of laws or postulates are formulated and then elaborated into a formalism that accurately describes the behavior of real systems. The basic results of quantum mechanics of interest to us are that in the world of atoms and molecules only certain select energies are valid states of the system and that the position of the particles like electrons cannot be described with infinite precision. Atoms and molecules do not have a continuous distribution of energy states. Instead, they have quantized energy levels that are the only ones that the system is permitted to adopt. Energies intermediate between any two quantized levels are not possible. We also must give up the idea of a well-defined trajectory for a quantum particle. In classical mechanics, an example of the trajectory is the path a bullet has when fired from a rifle. In quantum systems, we talk instead about the probability of the particle being in a certain region of space. This information is carried by the wave function, as discussed below. Similarly, the energy of a system has an uncertainty associated with it that depends on the lifetime of the state.

The basic equation of quantum mechanics is the **Schrödinger equation**, which describes the energies of a quantum system. The Schrödinger equation was formulated by the Austrian physicist Erwin Schrödinger during a skiing trip over the Christmas holidays in 1925.

Some aspects of the quantum world are extremely difficult, if not impossible, to appreciate on an intuitive level. The reason is that we live our lives in the classical world. The quantum world is just too small for us to experience on a personal level. In this respect, quantum mechanics is more difficult than thermodynamics. In thermodynamics, while the formulations may be abstract, at least the phenomena such as energy and even entropy are things that we experience in our lives and therefore understand almost unconsciously. Quantum mechanics presents us with any number of situations in which everyone's natural reaction is "How can that possibly be!" An interesting book that explores what the world might be like if the values of the physical constants were such that quantum effects were observable in daily life is *George Gamow's Classic Mr. Tompkins in Paperback* (Gamow and Stannard, 1999). Quantum mechanics also challenges ingrained concepts such as causality. The philosophical implications of quantum mechanics are vigorously debated even today. Perhaps the most counterintuitive aspect of quantum mechanics is

the idea of the wave function. Quantum particles, including light (as discussed above) have a dual nature, with some aspects of both particle-like and wave-like behavior. The wave function is a mathematical way to represent this dual nature. The generally accepted interpretation of the physical reality of the wave function is that the square of this mathematical function represents the probability of finding a particle in a certain region of space.

Unfortunately, relatively sophisticated mathematics is needed to develop quantum mechanics in a proper manner. We will not attempt to go into the details of this complex subject in such a brief introduction. Instead, we will introduce some of the important results and emphasize how they apply to the processes that take place in photosynthesis. One of the most powerful ways to learn about the properties of molecules is to probe their interactions with light. This interaction of light with matter is known as spectroscopy.

A.7.1 Quantum mechanical formalism

The Schrödinger equation is an **operator equation**, in that it uses a mathematical operator for the energy, acting on a wave function to give as a result the energy levels of the system. An operator is simply a set of instructions for how to operate on a mathematical function. An example is the differential operator d/dx, which instructs us to take the derivative of a function with respect to the variable x. An operator must have a function to operate on; an example of an operator equation involving the differential operator is given by Eq. (A1.51)

$$\frac{d}{dx}x^2 = 2x \qquad (A1.51)$$

The particular details of each type of quantum system are found in the operator and in the wave functions. The Schrödinger equation appropriate for most of our purposes is given by Eq. (A1.52):

$$\widehat{H}\Psi = E\Psi \qquad (A1.52)$$

in which $\widehat{H}$ is the Hamiltonian energy operator, Ψ is the wave function, and E is the energy associated with a particular quantum level. The Hamiltonian energy operator consists of two terms, one for potential energy and another for kinetic energy, as given by Eq. (A1.53):

$$\widehat{H} = \widehat{T} + \widehat{V} \qquad (A1.53)$$

where $\widehat{T}$ is the kinetic energy operator, and $\widehat{V}$ is the potential energy operator.

The potential energy term depends on what system is being considered and is different for different systems. For example, the potential energy acting on an electron in an atom is different from the potential energy that is found in a spring such as is used for the harmonic oscillator discussed below. The kinetic energy operator is always the same for all quantum systems, although it may take different forms in different coordinate systems. The kinetic energy operator appears to be different from the $1/2\,mv^2$ that is encountered in elementary physics, although the result is the same. The kinetic energy operator for motion of a particle in one dimension is given by Eq. (A1.54):

$$\widehat{T} = -\frac{h^2}{2m}\frac{d^2}{dx^2} \qquad (A1.54)$$

where h is Planck's constant, m is the mass of the particle, and $\dfrac{d^2}{dx^2}$ is the second derivative operator.

To solve the Schrödinger equation, one needs to first define the problem, so that the potential function can be properly chosen. Then, all that is necessary is to find a wave function Ψ that is unchanged when operated on by the Hamiltonian operator. The operation gives back the wave function multiplied by the energy of the system, which is just a numeric quantity. Note that this is not the case for the example operator and function that we used in Eq. (A1.51). The sorts of mathematical functions that are unchanged after the second derivative is taken of them are oscillatory or wavelike functions such as sin and cos, hence the term wave function.

Depending on the quantum system being analyzed, wave functions can be simple, such as just a sin function, or very complicated, such as the orbital functions that describe the electron density in atoms and molecules, but the basic principle is the same. One of the postulates of quantum mechanics is that the square of the wave function gives the probability of finding the particle at a given position in space. This has the consequence that if several measurements are made of the position of the particle, then one time we may find it in one position and another time it may be somewhere else. If we do enough experiments and plot the positions of the particles found, the distribution will follow the prediction of the wave function. However, we cannot predict with certainty where the particle will be for any given experiment. Hence, the idea of a well-defined trajectory for a quantum particle must be abandoned.

One other important aspect of quantum mechanics is that there is always more than one possible solution to the Schrödinger equation. The different solutions have different energies and different wave functions, but are all equally valid. The energies of the various quantum levels increase from the lowest energy level, called the ground state, to a number of excited states of higher energy. Each state has its characteristic energy obtained by solving the Schrödinger equation and probability distribution given by the square of the wave function. The system cannot occupy states with energies that are not solutions of the Schrödinger equation. However, it can jump from one quantum state to another if energy is either added to or taken away from the system. Often the energy difference is in the form of light either absorbed or emitted, which gives rise to spectroscopic transitions. A more complex form of the Schrödinger equation, called the time-dependent Schrödinger equation, describes these transitions in detail. We will not explore the time dependence of quantum mechanics as it is beyond the scope of this discussion. However, the time-independent solutions to the Schrödinger equation are usually sufficient to give the main ideas of the quantum results we are after.

A.7.2 *The Heisenberg uncertainty principle*

Another consequence of the quantum world is known as the Heisenberg uncertainty principle, after the German physicist, Werner Heisenberg, who first formulated it. Earlier, we indicated that quantum mechanics forces us to give up the concept of a well-defined trajectory of a particle. An alternative way of thinking about this is to examine the limits that quantum behavior imposes on where we can localize a particle. The uncertainty principle states that the product of the uncertainty of the position and momentum of a quantum particle must be greater than a certain value:

$$\Delta x \Delta p \geq \frac{h}{4\pi} \qquad (A1.55)$$

where Δx is the uncertainty in position of a particle, Δp is the uncertainty in its momentum, and h is Planck's constant. Another formulation of the Heisenberg uncertainty principle concerns the energy and time, as given by Eq. (A1.56):

$$\Delta E \Delta t \geq \frac{h}{4\pi} \qquad (A1.56)$$

where ΔE is the uncertainty in the energy of the system, and Δt is the lifetime of the state. The main consequence of this lies in very short-lived transient states. The uncertainty principle states that the shorter the lifetime of the state, then the less well-defined the energy of that state. So we should think of the energies that we obtain from solving the Schrödinger equation as not precise values, but rather as somewhat fuzzy depending on how long the state lasts. In practical terms, the uncertainty of an electronic state of a molecule is not uncertain by a significant amount until the lifetime of the state is in the femtosecond time regime.

A.8   Molecular energy levels and spectroscopy

A.8.1 *Absorption and fluorescence*

Molecules are formed from atoms according to the laws of quantum mechanics. These same laws also apply to the atoms themselves, although we will not specifically consider atomic energy levels here. Light absorption gives substances their characteristic colors. The light that is not absorbed by the substance can cause a sensation by being absorbed by the pigments in our eyes. So when we see the characteristic green color of a plant, it is due to the green light that was not absorbed. Light absorption is at the heart of photosynthesis, so we will explore this phenomenon in a little more depth.

If a system, such as a molecule, is irradiated with light with photon energy equal to the energy spacing between two quantum levels, then the molecule may absorb the energy and make a transition to the higher energy state. The light that causes the transition must have just the correct energy content; if it is either too low or too high, it will simply not be absorbed. This is described in Fig. A1.9a.

This effect is called **resonance absorption** and gives rise to an absorption spectrum, in which the absorption of light is plotted versus the wavelength, as shown in Fig. A1.9b. We learned in the beginning of this Appendix 1 about the quantum nature of light, so that the energy content of a photon is directly proportional to the frequency of the light (Eq. (A1.2)), or inversely proportional to the wavelength (Eq. (A1.3)). We can express this resonance condition in a simple equation:

$$M + h\nu \rightarrow M^* \qquad (A1.57)$$

where M is the unexcited molecule, M^* is the excited molecule, and $h\nu$ represents a photon of frequency ν and energy $h\nu$ The energy change in the molecule is given by:

$$\Delta E = h\nu \qquad (A1.58)$$

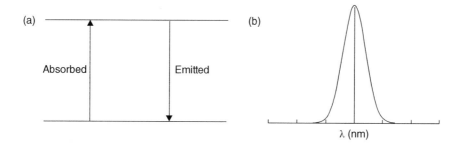

Figure A1.9   (a) Energy level diagram for a simple two-level system. (b) Stick spectrum predicted by this energy level diagram and the absorption spectrum of a typical molecule in condensed phase.

This equation appears very similar to Eq. (A1.2). The only difference is that the ΔE refers to the change in energy of the molecule, which is induced by the absorption of light. Conservation of energy requires that the sum of the energy of the photon and the ground state of the molecule equals the energy of the excited molecule.

In addition to the upward absorption transition, a downward transition is also possible:

$$M^* \rightarrow M + h\nu \qquad (A1.59)$$

This transition deactivates the excited state and is called **fluorescence**.

Figure A1.9a is an oversimplified version of an energy level diagram for a real molecule. It leads to the stick spectrum in Fig. A1.9b, in which only a single wavelength of light is absorbed. However, real molecules always have a broader and usually more complex spectrum. The reasons for these differences require a more in depth look at molecular energy levels.

A.8.2 Molecular potential energy curves

The simple two-level energy system is a drastic simplification from the multitude of energy levels that are found in real molecules, but is sufficient for many purposes. Now we will explore the details of the types of molecular energy levels in more detail.

Molecules have a variety of quantized energy levels that stem from different physical origins. We can divide the classes of energy levels into four types: electronic, vibrational, rotational, and translational, with decreasing energy spacing. The rotational and translational energy levels for molecules in condensed phase (liquid or solid) are very closely spaced and generally do not affect the characteristic spectra of molecular systems, except to broaden the vibrational and electronic transitions. We will therefore focus on the electronic and vibrational energy levels and transitions between them.

To begin the study of molecular energy levels and spectroscopy, we first need to examine the **potential energy curve** for the ground state of a typical molecule, shown in Fig. A1.10. The characteristic shape of this graph of potential energy vs. internuclear separation is generally applicable to most molecules. However, this graph is really only applicable to a diatomic molecule; more complicated molecules will have much more complex multidimensional potential energy diagrams that are impossible to visualize. At large values of internuclear separation, the energy of the system is just that of the isolated atoms (recall that energy has no natural zero, so that we are free to choose this as a reference point). As the atoms come closer, the electron distributions of each atom are attracted to both nuclei, and partly because of this extra electron-nuclear attraction, a chemical bond is formed. However, if the nuclei approach too close, then the positively charged nuclei strongly repel each other and the total energy goes up. The compromise of the attractive and repulsive tendencies of the system is at the equilibrium nuclear separation, r_e.

The solid line in Fig. A1.10 is the potential energy function that the nuclei move in. If they stray from the equilibrium position, then the potential energy increases and a restoring force tends to pull them back to the equilibrium separation, which is the state of lowest energy. The electrons and nuclei have an unusual interdependent relationship. The electrons quickly follow the positions of the nuclei, as the electrons are much less massive and can move much more quickly. So one might think the nuclei are in charge and the electrons are just followers. However, ultimately, the distribution of electrons is what determines the most stable positions for the nuclei, so in the end they have an essential role in determining the properties of the bond. A useful analogy is to visualize some campers in an outdoor area that is infested with mosquitoes. The light, fast mosquitoes follow the slow, massive campers wherever they go. However, the campers will finally retreat to their tent under the influence of the constant pestering of the mosquitoes, so their movements are in the final analysis determined by the mosquitoes.

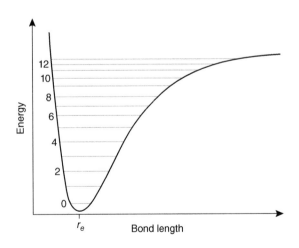

Figure A1.10 Potential energy curve for a diatomic molecule. The energy of the molecule as a function of internuclear separation is shown, with the energies of vibrational levels superimposed.

A.8.3 Harmonic oscillator

Next, we will consider molecular vibrations. We can make a useful approximation to the potential energy function appropriate to the atoms in a molecule by considering a somewhat simpler system, known as a **harmonic oscillator**. It consists of two masses held together by a spring. We will first analyze this system classically and then indicate the changes necessary to include quantum effects. The atoms that are bonded can be considered to be like masses connected to a spring with a spring constant, k_s, as shown in Fig. A1.11a. The spring constant is a measure of the stiffness of the spring, and determines the steepness of the potential energy *vs.* position curve, as shown in Fig. A1.11b. The potential energy curve is given by Eq. (A1.60):

$$V = \frac{1}{2} k_s \left(r - r_e\right)^2 \qquad (A1.60)$$

This potential curve defines the harmonic oscillator and is the same for both the classical and quantum oscillators. The total energy of the harmonic oscillator is the sum of the kinetic and potential energies. The spring goes back and forth between the two extremes of position. The total energy is constant, but alternates between being purely potential energy when the spring is either extended or compressed the maximum amount, to being purely kinetic energy when the spring is at the equilibrium position. The probability of finding the classical oscillator at any given position is shown in Fig. A1.11c. The spring spends most of its time near the two ends of its motion, where its velocity goes through zero as it changes direction. These end points of the motion are called the classical turning points. It spends little time near the equilibrium position, as its velocity is maximum here. The classical oscillator can have any energy, which depends on the amplitude of the motion, the masses of the objects, and the spring constant. The harmonic oscillator has a natural frequency, V_0, of its back and forth oscillation, given by:

$$v_0 = \frac{1}{2\pi} \left(\frac{k_s}{\mu}\right)^{\frac{1}{2}} \qquad (A1.61)$$

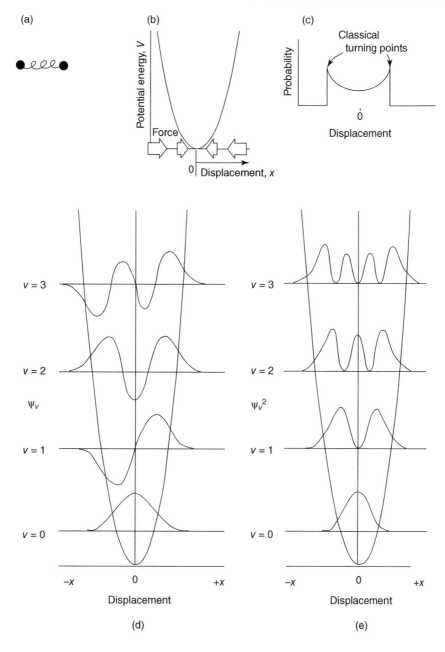

(a)   (b)   (c)

(d)   (e)

Figure A1.11 Classical and quantum mechanical harmonic oscillator. (a) Diagram of masses and spring. (b) Potential energy curve. (c) Probability distribution of classical oscillator. (d) Energy level diagram and wave functions of the lowest energy states of a quantum mechanical harmonic oscillator. (e) Energy level diagram and probability distribution (wave functions squared) of the lowest energy states of a quantum mechanical harmonic oscillator.

In Eq. (A1.61), k_s is the spring constant discussed above and μ is the reduced mass, defined as:

$$\frac{1}{\mu} = \frac{1}{m_1} + \frac{1}{m_2} \qquad \text{(A1.62)}$$

The reduced mass is a kind of average mass that is weighted in favor of the less massive body. According to Eq. (A1.61), the frequency of the vibration will be higher for lighter masses. On the molecular scale, light atoms such as H, bonded to a large molecule (which if it is much higher in mass than the H can be considered to have infinite mass, so that μ equals the mass of the light atom) have higher frequency vibrations than do heavy atoms. The frequency also depends on the value of k_s, the spring constant. A very strong spring has a high frequency of vibration. Again on the molecular scale, a strong bond, such as a double or triple bond, will have a large value of k_s compared to a single bond and therefore a higher vibrational frequency. The harmonic oscillator potential function is a reasonably good approximation to the actual potential energy curve experienced in molecules, especially at the lower energies where the system is most likely to be found.

There are some important similarities and differences between the classical and the **quantum mechanical harmonic oscillator**. The potential energy function and the frequency of oscillation given by Eqs. (A1.60) and (A1.61) are the same. The most important of the differences is the fact that only some of the possible energies of the system are allowed. These are drawn in on Fig. A1.11d, which shows a quantum mechanical harmonic oscillator. We won't go through the details of the solution of the quantum mechanical harmonic oscillator here, but will just quote the results. The energy levels allowed for the harmonic oscillator are as follows:

$$E_{ho} = \left(v + \frac{1}{2}\right)h\nu_0 \qquad \text{(A1.63)}$$

where v is the **vibrational quantum number**, which must be an integer beginning with zero. The quantum mechanical harmonic oscillator has a ladder of equally spaced energy levels, as shown in Fig. A1.11d. The lowest of these energy states is called the zero point energy and has an energy of one half the spacing between the levels. Transitions between any two adjacent energy levels can take place, with the interesting result that the frequency of the photon that is either absorbed or emitted is equal to the frequency of the molecular vibration. The analysis of vibrational spectra can tell a great deal about the chemical identity and even the environment of the molecule being detected. Vibrational frequencies are typically in the infrared region of the electromagnetic spectrum. In order to provide a convenient unit for the reporting of vibrational frequencies, it is customary to use the reciprocal of the wavelength in centimeters of the light required to cause a transition. This frequency unit, which is called wavenumbers (cm^{-1}) and has the symbol $\bar{\nu}$, does not appear to have the proper units to be a frequency, which is normally s^{-1}. However, because the frequency and wavelength of any wave are reciprocally related (Eq. (A1.1)), the wavenumber unit is equal to the actual frequency divided by the speed of light and is therefore directly proportional to the frequency. Typical values for molecular vibrational frequencies range from about $3000\ cm^{-1}$ for an H atom vibration in a large molecule down to a few tens of cm^{-1} for an intermolecular vibration of two molecules that are not covalently attached, but are just very weakly associated.

Vibrational spectra can also be obtained by using a scattering technique known as Raman spectroscopy, in which a photon incident on a molecule is scattered, but the scattered photon has a frequency that is slightly shifted from the incident photon. The difference in frequencies between the incident and scattered photons is the frequency of the molecular vibration. A variation on Raman spectroscopy, in which the incident photon has a frequency that corresponds to an electronic transition, is known as resonance Raman. It is useful because the intensity of vibrations of atoms that are in the region of the electronic transition are dramatically enhanced. This allows the vibrations of strongly absorbing molecules, such as chlorophylls or carotenoids, to be selectively observed over the background of the nonabsorbing protein.

Perhaps the most surprising aspect of the quantum mechanical harmonic oscillator is the probability distribution of the positions of the masses. Recall that the square of the quantum mechanical wave function gives information about the positions of the particles. Each of the levels of the harmonic oscillator has its own characteristic probability distribution, a few of which are shown superimposed on the potential energy curve in Fig. A1.11d. This probability is found by taking the square of the vibrational wave functions that result from the solution of the quantum mechanical harmonic oscillator. The lowest energy state has highest probability towards the middle, with only a very small probability of being found near the turning points, where the classical oscillator spends most of its time. Higher energy states have distributions that include nodes, or regions where there is zero probability of finding the masses. As the quantum level gets higher in energy, the probability distributions begin to look more like the classical result that is shown in Fig. A1.11c. At very high values of the quantum numbers, the quantum results merge smoothly with the classical results. This is a necessary result, because there is no definite line between the quantum and the classical worlds. The smooth transition from quantum to classical behavior is called the **correspondence principle**.

One feature of the quantum mechanical harmonic oscillator that is very different from the classical result is the extension of the probability distributions beyond the classical turning points. In this region, the particle has a higher potential energy than it has total energy, so it is classically forbidden. However, the quantum mechanical oscillator can have a small probability of occupying this forbidden region. This effect is the origin of the quantum mechanical tunneling effect that is observed in some electron transfer reactions in photosynthetic systems at low temperatures.

A.8.4 Molecular electronic energy levels: singlets and triplets

So far we have considered only the lowest energy **electronic state** of a molecule, with its associated vibrational energy levels. Molecules also can exist in a number of different electronic states, each of which has its own potential energy function for the nuclear motions. The lowest in energy of these electronic states is called the **ground state**, while the more energetic ones are called **excited states**. In the ground state, the electrons that make up the molecule fill the lowest energy molecular orbitals of the molecule, in accordance with the **Pauli exclusion principle**, which states that each orbital can hold a maximum of two electrons and that if two electrons are present in an orbital that their spins must be opposite or paired. The electron spin is a purely quantum mechanical concept that has no classical analog. When a molecule has two electrons occupying each of the filled orbitals, they must be paired, which is represented as one electron with spin up and the other with spin down. The electrons in the molecule fill up the orbitals beginning with the lowest energy and up to the highest occupied molecular orbital, or **HOMO** (Fig. A1.12). There is only a single way that this can happen, so the ground electronic state with all paired electrons is called a singlet state. Oxygen is a very rare exception in that the ground electronic state is a triplet instead of a singlet. The properties of triplet states are discussed below.

To a crude first approximation, the excitation of a molecule from the ground to an excited state involves movement of an electron to one of the higher energy molecular orbitals that is not occupied in the ground state. The lowest of these higher energy molecular orbitals is called the lowest unoccupied molecular orbital or **LUMO**. The lowest energy excited states thus involve one electron in the HOMO and one in the LUMO (Fig. A1.12). Higher energy excited states usually involve electron population in excited orbitals higher in energy than the LUMO. The orbital occupation diagram is an important factor but not the only factor in determining the energies of the ground and excited states. Another factor concerns the spin properties of the electrons.

In an excited state, it is possible for the two electrons that occupy the two different orbitals to have their spins either paired as in the ground state, or unpaired, with the same spin orientation. The unpaired arrangement does not violate the

(a)   (b)

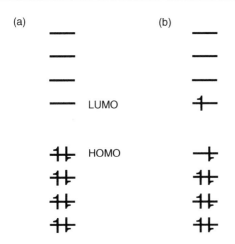

Figure A1.12 Molecular orbital diagram for a typical organic molecule. (a) Ground state. (b) Excited state.

Pauli exclusion principle, because the electrons occupy different orbitals. There are three ways for the unpaired electrons to be arranged, so this is called a triplet state. Two of these are easy to visualize, with the two electrons either both with spin up or both with spin down. The third way is more subtle, and involves the two electrons with opposite spins, seemingly similar to the singlet state. However, there is an important difference, in that the two electrons that are paired in the singlet state have the property that if the labels on the electrons are interchanged, the wave function that describes them changes sign. This is called antisymmetric behavior and is required for singlet states. The two electrons in a triplet state have a wave function that does not change sign when the labels are interchanged and therefore has a symmetric behavior. A vector diagram that illustrates the singlet and three triplet states and their symmetric and antisymmetric natures is shown in Fig. A1.13.

The energy of the triplet state is invariably lower than the energy of the singlet state with the same orbital occupation diagram. This is a result of a quantum mechanical effect called electron correlation. While it is beyond the scope of our treatment to go into the details of this effect, the result is that in triplet states the electron motions are correlated to stay further apart on average than they are in singlet states. The triplet state energy is therefore somewhat lower than the corresponding singlet state because of reduced electron–electron repulsion. The three triplet state energies are almost the same. They can, however, be split using a magnetic field and observed, usually by employing magnetic resonance techniques.

A.8.5 Electronic transitions

When a molecule makes a jump from the ground electronic state to an excited state by absorbing a photon, or a downward transition by emitting a photon, the distribution of electrons changes essentially instantaneously. The transitions and resulting absorption and fluorescence spectra are shown in Fig. A1.14. The distribution of electrons in the excited state defines a new potential energy function for the excited state of the molecule. This new potential function typically has a displacement toward longer equilibrium internuclear separation. This happens because when an electron is promoted to an antibonding orbital, the bond strength almost always weakens, which is in turn manifested as a longer bond length.

The nuclei do not adjust their positions instantaneously to this new potential function, because they move much more slowly than the electrons. Because of this difference in timescales of motion, the transition is often described as "vertical" with the electronic change taking place while the nuclei are in effect "frozen" during the transition. Because of the horizontal displacement of the two electronic states and the vertical nature of the electronic transition, the vibrational state populated by an absorption is usually not the lowest energy vibrational level of the excited state. In a collection of molecules, there will be a distribution of populations of the final vibrational states that essentially reflect the probability of the nuclei being in that position according to the distributions in Fig. A1.11. Because the transition involves changes of both vibrational and electronic states of the molecule, it is often called a **vibronic** transition. The probabilities of transition to each vibrational state of the

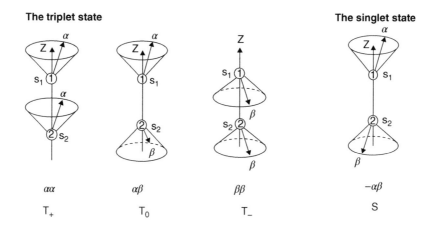

The triplet state   The singlet state

$\alpha\alpha$ $\alpha\beta$ $\beta\beta$ $-\alpha\beta$

T_+ T_0 T_- S

Figure A1.13   Singlet–triplet vector diagram. The singlet state has spins opposed and out of phase. Two of the triplet states have spins aligned while one has spins opposed. All three triplet states have spins in phase.

excited state are called **Franck–Condon factors**. These factors are the squares of the overlap of the vibrational wave functions. They thus depend on the amplitude (and also the sign, which we have not explicitly considered) of the vibrational wave functions in both the ground and excited states. An analogous set of probabilities are critical for explaining the rates of electron transfer reactions and are considered in detail in Chapter 6.

After an equilibration period, which typically is on the order of picoseconds or less in solution, the molecule will relax back toward the equilibrium separation, and the vibrationally excited molecule will return to the ground vibrational state of the excited electronic state. The intrinsic lifetime of this vibrationally relaxed but electronically excited singlet state is typically several nanoseconds and is the initial state for fluorescence emission (Fig. A1.14b). Absorption and emission spectra typically have a mirror image relationship because of the overall similarity of the ground and excited state potential energy surfaces and their vibrational modes (Fig. A1.14b).

Energy level diagrams such as those shown in Fig. A1.14 are often further simplified into energy level diagrams in which each electronic state is a single horizontal line and transitions between them are indicated by arrows. Usually, straight arrows indicate radiative transitions in which a photon is either absorbed or emitted and wavy arrows indicate nonradiative transitions in which no photon is involved.

The excited state can decay by any of several possible pathways (Fig. A1.14). The excited state decay processes include emission of a photon to return to the ground state, a radiative process called fluorescence. The fluorescence emission is also a "vertical" process that takes place instantaneously compared to movement of the nuclei. In this case, the final state that is initially populated is usually an excited vibrational state of the ground electronic state. The result of the processes described above is that the fluorescence emission is usually at somewhat longer wavelengths than the absorption, as illustrated in Fig. A1.14. This shift is called the **Stokes shift**. The magnitude of the Stokes shift is a measure of how much the excited state potential energy curve is shifted relative to the ground state potential energy curve.

In many experimental situations, the excitation may be into a higher excited state than the lowest energy excited state. The molecule relaxes nonradiatively to the lowest energy excited state which then emits. Higher excited states are usually very short lived, so that the only state that lives long enough to emit is the lowest excited state. This effect, that only the lowest energy excited state emits fluorescence, is called **Kasha's rule**, after

(a)

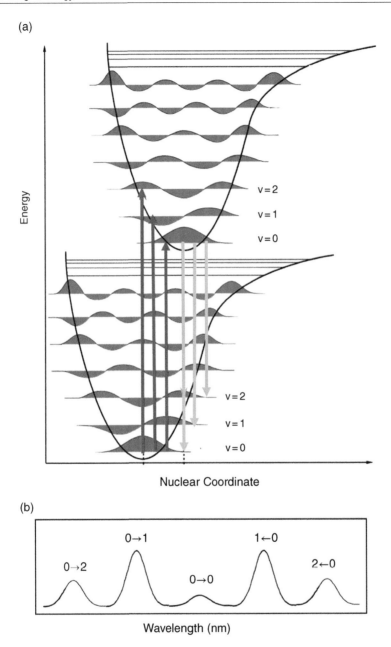

(b)

Figure A1.14   Potential energy diagrams and spectra for absorption and fluorescence electronic transitions in organic molecules. (a) The initial state for absorption is usually the ground vibrational state of the ground electronic state, while the final state is usually an excited vibrational state of the excited electronic state. The initial state for fluorescence is the ground vibrational state of the first excited state, while the final state is an excited vibrational state of the ground electronic state. (b) Absorption and fluorescence spectra that result from the transitions shown in (a). The absorption and fluorescence spectra are mirror images of each other if the ground and excited state potential energy curves have the same shape.

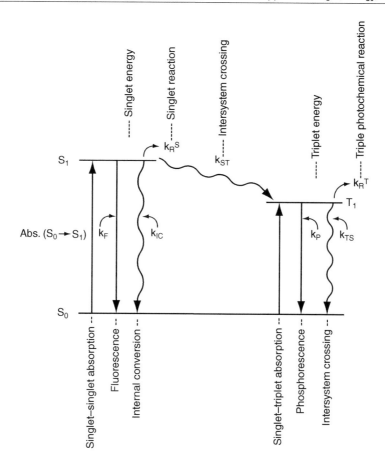

Figure A1.15 Excited state decay pathways. S_0 is the ground electronic state, while S_1 is the first excited singlet and T_1 is the first excited triplet state. Radiative transitions are represented by straight lines, and nonradiative transitions are represented by wavy lines.

Michael Kasha, the spectroscopist who first described this effect. A common example of this effect is when the Soret band of a chlorophyll is excited in the blue spectral region and the emission is observed in the red spectral region. Chlorophyll spectra are discussed in more detail in Chapter 4.

Other possibilities for excited state decay are one of several nonradiative processes including conversion to the triplet state, called **intersystem crossing**; direct conversion to a very highly vibrationally excited level of the ground electronic state, a process called **internal conversion**; **energy transfer** to another molecule; or finally **photochemistry**. The relative amounts of excited state decay that go via

each pathway are given by the equations describing parallel first-order decay processes, Eq. (A1.33)–(A1.44). For maximum efficiency of energy storage in photosynthesis, it is necessary for photochemistry to be the dominant excited state decay process.

A.8.6 Absorption intensity, oscillator strength, and transition dipole moments

Not all light absorption transitions have equal probabilities. In some cases, the absorption is very intense, while in others it is so weak that it is undetectable. The intensities of transitions are determined by quantum mechanical selection rules, which are based on symmetry properties of the molecules and the type of transition involved. Transitions that are very intense are called "allowed," while those that are very weak are called "forbidden." Intermediate cases are also possible in complex molecules. Singlet to triplet transitions are forbidden and are therefore very weak.

There are several quantitative measures of the intensity of an absorption transition. One is called the **oscillator strength** (f) and is proportional to the area under the absorption band. The oscillator strength (which is unitless) is based on an analogy to a classical oscillator and ranges from 0 for a forbidden transition to 1 for a fully allowed one. Quantitatively, f is given by:

$$f = 1.44 \times 10^{-19} n \int \varepsilon(v) dv \qquad (A1.64)$$

where n is the refractive index of the medium, and the integral of the molar extinction coefficient ε is evaluated over the entire absorption band plotted as a function of frequency. The extinction coefficient is defined below. A rough estimate of the value of the integral in Eq. (A1.64) can be obtained assuming the absorption band is triangular in shape, and multiplying the extinction coefficient at the peak maximum by the full width of the peak at half maximum (measured in appropriate frequency units).

A second measure of the transition probability is called the **transition dipole moment**, μ. This property is particularly useful because it gives an indication of the polarization properties of the transition and can be used to determine orientations of molecules. The transition dipole moment arises from the fact that the absorption process involves the interaction of light, which contains an oscillating electric field, with the molecule. The effect of the light is to couple to the electrons in the molecule and cause them to change their spatial distribution. This coupling takes place by the coupling of the electric field of the light to the dipole moment of the molecule, so that the absorption of light causes the dipole moment of the molecule to change. This is described using the formalism of quantum mechanics, but an important result that we can use in a qualitative fashion is that the transition has a directionality within the molecule. This directionality can be determined by theoretical calculations but also by experiment. If plane polarized light is used to cause the transition, then those molecules that are oriented with their transition dipole moments parallel to the electric vector of the light will absorb a photon with highest probability, while those with their transition dipole moments perpendicular to the electric vector of the light will not absorb a photon. The intensity of the absorption is proportional to the square of the transition dipole moment and is measured in units of Debyes[2], where $1 \text{ Debye} = 10^{-18} \text{ esu cm}$.

A.8.7   Einstein coefficients

Although they may seem to be unrelated quantities, the absorption intensity and the fluorescence lifetime are intimately tied to each other. This relationship was first described by Einstein, so the quantities are called the **Einstein coefficients**. Consider a simple system with two energy levels, such as illustrated in Fig. A1.9a. The probability of absorption of a photon to cause a transition from the lower state to the upper state (P_{iu}) is given by the population of the lower level (N_l) times the intensity of light of frequency such that it can cause the transition ($I(v)$) times a coefficient B:

$$P_{l \to u} = N_l I(v) B \qquad (A1.65)$$

The Einstein coefficient B is equal to a collection of constants times the square of the transition dipole moment, so it is directly related to the intensity of the absorption. Quantum mechanics requires that the intrinsic probability P for the photon-induced downward emission transition be the same as that of the upward absorption, so the rate of downward stimulated emission transitions is as follows:

$$P_{u \to l} = N_u I(v) B \qquad (A1.66)$$

where N_u is the population of the upper state. **Stimulated emission** is a process in which a photon interacts with the excited state, causing emission of a second photon of the same frequency. It is the physical basis for laser action.

If these were the only factors involved, the populations of the two levels would soon become equal and there would be no net absorption or emission of photons. This clearly doesn't happen, so it is necessary to include another process that returns the system to the lower state, so that at equilibrium the lower state has a higher population than the upper state. Here we can make a connection with thermodynamics. The Boltzmann distribution tells us that at equilibrium the population difference of the two levels must be given by (Eq. (A1.21)). The new process is called **spontaneous emission** and has a probability given by the Einstein A coefficient. Spontaneous emission is the source of the normal fluorescence effect usually measured in biochemistry. Spontaneous emission and therefore also the A coefficient do not involve incident photons, so this decay process is independent of light intensity, in contrast to stimulated emission which depends on light. Incorporating that result into the expression given above gives:

$$P_{u \to l} = N_u \left[I(\nu) B + A \right] \qquad (A1.67)$$

After some additional manipulations which are beyond the scope of our treatment, we arrive at the result that is important for our purposes:

$$A = (const) \nu^3 B \qquad (A1.68)$$

This result tells us that the probability of spontaneous emission is directly proportional to the intensity of the absorption. The Einstein coefficient A is just the intrinsic rate constant for fluorescence, k_f. The important result is that an intense absorption will result in an excited state that has a short intrinsic lifetime, while a weak absorption will result in an excited state with a long intrinsic lifetime. The reciprocal of the fluorescence rate constant k_f is called the natural radiative lifetime τ_0 and is the lifetime of the excited state if fluorescence is the only decay pathway. Values of τ_0 are in the few tens of nanoseconds range for chlorophylls. This intrinsic fluorescence lifetime represents an inevitable loss process that competes for the decay of the

excited state. For a major fraction of the excited states to decay by some other process such as photochemistry, it is necessary that they be significantly faster than fluorescence. This then requires that the photochemical processes take place on the subnanosecond timescale, which is indeed the case. Thus, we can see that there is a trade-off between absorption strength and excited state lifetime. Light absorption is more efficient if the molecule has an intense absorption band, but the excited state produced by an intense absorption is intrinsically short lived, so the processes that store the energy must then be ultrafast in order to be effective.

This relationship between the absorption strength and excited state lifetimes expressed in the Einstein coefficients has profound consequences in photosynthetic systems. Because the excited state lifetimes of photosynthetic pigments are so short, the entire system has to be constructed to function at an extremely rapid pace. This puts large constraints on the organizational principles of the system that can be utilized. For example, it is not sufficient to employ diffusional collision of the photoactive pigments and their reaction partners. The diffusion process is limited by how fast molecules move about in solution and such a system would inevitably be inefficient and prone to damaging side reactions. Instead, photosynthetic reaction centers all incorporate the species that interact with the excited state directly in the reaction center complex, so diffusion is not required prior to reaction. In addition, these species are optimally positioned so as to have just the right distance and orientation to enhance reactivity without creating wasteful side reactions. This leads to efficient photochemistry with minimal undesirable effects.

A.8.8 Fluorescence quantum yield and lifetime

We are now in a position to combine several of the concepts that we have already established separately. We have already discussed qualitatively the processes that deactivate the excited state. In addition, we have quantitatively treated the case of parallel first-order decay processes. Excited state decay is an excellent example of this kinetic mechanism. The major kinetic processes and their intrinsic rate constants are as follows: fluorescence, k_f, intersystem crossing, k_{isc}, internal conversion, k_{ic}, energy transfer, k_{en}, and photochemistry by electron transfer, k_{et}. The yield of each pathway is given by Eq. (A1.40). These yields are called **quantum yields**, because the processes result from quantum absorption to form the excited state. The symbol for quantum yield is ϕ. The fluorescence quantum yield is given by a slightly modified version of Eq. (A1.40):

$$\phi_f = \frac{k_f}{k_f + k_{isc} + k_{ic} + k_{en} + k_{et}} \quad \text{(A1.69)}$$

Recall from Eq. (A1.37) that the observed rate constant for decay is the sum of all the individual rate constants. Here we will use the intrinsic time, τ, which is the reciprocal of the sum of these rate constants. The **fluorescence quantum yield** can then be written simply as:

$$\phi_f = \frac{\tau}{\tau_0} \quad \text{(A1.70)}$$

where τ_0, the intrinsic fluorescence lifetime, was defined above. The fluorescence quantum yield can also be measured by quantitatively measuring the number of fluorescent photons emitted by a sample and dividing by the total number of photons absorbed. This can either be done by making an absolute measurement or by comparison to the amount of fluorescence of a reference compound whose fluorescence quantum number is known.

A.9 Practical aspects of spectroscopy

A.9.1 Beer–Lambert law of absorption

One of the most useful properties of the absorption of light is that it can be used to make quantitative determinations of the amount of material present in a sample. The relationship between the light absorbed and the sample concentration is called the **Beer–Lambert law** or often just Beer's law. The equation that relates these two quantities is given by Eq. (A1.71)

$$A = \varepsilon c \ell \quad \text{(A1.71)}$$

where A is the **absorbance** (a unitless quantity also sometimes called the optical density), ε is a proportionality constant called the **molar extinction coefficient**, l is the pathlength of light, usually 1 cm and c is the concentration of the absorbing substance in M. The units of ε are $M^{-1}\,cm^{-1}$. Typically, ε has a different value at every wavelength, so that a graph of ε vs. λ defines the absorption spectrum of a sample. The value of ε at the wavelength of maximum absorbance, λ_{max}, is often tabulated. This extinction coefficient is a measure of the intrinsic intensity of the absorption for a given substance. It can range from near zero for a transition that is forbidden to upwards of $10^5\,M^{-1}\,cm^{-1}$ for a fully allowed transition in a molecule like chlorophyll.

We will next derive Beer's law from simple considerations. Consider a very thin sample of thickness Δx containing c moles per liter of a substance that absorbs light (Fig. A1.16a). A beam of light of intensity I_0 passes through the sample and some of the light is absorbed by the substance. The fraction of light that is transmitted by the sample is called the **Transmission** (T):

$$T = \frac{I}{I_0} \quad \text{(A1.72)}$$

The T can vary from 0 when all the light is absorbed to 1 when none of it is absorbed. Sometimes $\%T$ is used, which is equal to $100T$ and therefore ranges from 0 to 100. T is not directly proportional to the concentration of absorbing material, so a related but slightly different formulation is usually used instead, which establishes a linear relationship between concentration and light absorption. This is called the absorbance, A, which is equal to $\log \frac{I_0}{I}$.

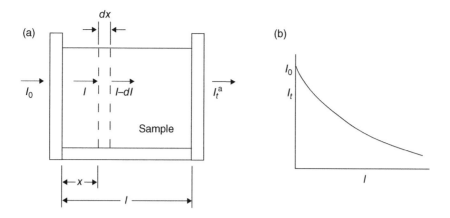

Figure A1.16   Absorption spectroscopy. (a) Derivation of Beer–Lambert law. (b) Intensity of light as it traverses an absorbing sample.

To justify why A is proportional to concentration, consider the following analysis. The fractional change in intensity of the light passing through the thin layer of Fig. A1.16a, $\frac{\Delta I}{I}$, is proportional to Δx, the thickness of the layer, the concentration of absorbing species c, and a proportionality constant, k:

$$\frac{\Delta I}{I} = -kc\Delta x \qquad (A1.73)$$

The negative sign indicates that the light intensity is decreasing as the beam traverses the sample. If a large number of such thin layers are added together to form a sample of macroscopic size, the sum of the effects of the layers on the light intensity is as follows:

$$\sum_{i=1}^{n} \frac{\Delta I}{I} = -kc \sum_{i=1}^{n} \Delta x \qquad (A1.74)$$

In the limit of very thin layers, the sum can be converted to the integral:

$$\int_{I_0}^{I} \frac{dI}{I} = -kc \int_{0}^{\ell} dx \qquad (A1.75)$$

where ℓ is the total thickness of the sample. If we solve this integral:

$$\ln \frac{I}{I_0} = -kc\ell \qquad (A1.76)$$

or taking the exponential of both sides:

$$I = I_0 e^{-kc\ell} \qquad (A1.77)$$

If we convert to base 10 logs and substitute $\varepsilon = k/2.303$, then (after some minor algebraic substitutions) we get:

$$\log \frac{I_0}{I} = \varepsilon c\ell \qquad (A1.78)$$

The absorbance is defined as $\log \frac{I_0}{I}$, so that:

$$A = \varepsilon c\ell \qquad (A1.79)$$

which is the Beer–Lambert law as given earlier.

The advantage of using the absorbance over the transmission for quantitative studies, where the amount of a substance is determined using absorption spectroscopy, is that A is directly proportional to concentration, while T is not. This is perhaps the single most common measurement in all of biochemistry. Figure A1.16b shows how the light intensity decreases exponentially as it traverses from one side of the sample to the other, as predicted by Eq. (A1.77). Because of the logarithmic nature of the absorbance, an A of 1 has $\frac{I_0}{I} = 10$, A of 2 has $\frac{I_0}{I} = 100$, A of 3 has $\frac{I_0}{I} = 1000$, and so on. At the other end of the scale, an A of 10^{-3} has $\frac{I_0}{I} = 1.002$. Practically, the ability to make accurate measurements of $\frac{I_0}{I}$ to very high accuracy limits the range of absorbance values that can be measured with confidence.

A.9.2   Absorption measurements

Absorption spectra are measured using a **spectrophotometer**. The most common type measures absorption in the ultraviolet and visible regions of

the electromagnetic spectrum, and is called a UV/Vis spectrophotometer. A block diagram of a UV/Vis spectrophotometer is shown in Fig. A1.17a. The essential parts are a light source (usually a tungsten lamp for the visible region and a deuterium lamp for the UV region, a wavelength selection device (usually a monochromator, but simple systems can utilize bandpass filters), a sample holder (often a quartz cell or cuvette of pathlength 1 cm), a detector (usually either a photomultiplier tube or photodiode), and a readout device (either a chart recorder or a computer in most modern machines).

As given by Eq. (A1.78), the instrument must measure both I_0, the light intensity incident on the sample, and I, the light intensity after the light traverses the sample. This is usually done by having a matched set of sample cells, one of which contains only the solvent while the other contains both

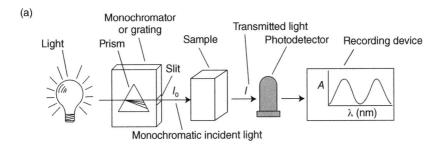

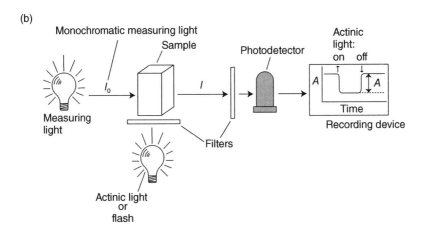

Figure A1.17   (a) Block diagram of a UV/V is absorption spectrophotometer. (b) Experimental setup for difference spectroscopy, either continuous or flash photolysis.

solvent and sample material. The incident light is sent through both cells and the intensity measured by the detector. The intensity detected for the reference beam is I_0, while that of the sample beam is I. The log of the ratio of the two intensities is then taken electronically to give the absorbance at a given wavelength. The wavelength can usually also be scanned and a spectrum recorded. This arrangement is called a double beam spectrophotometer and is usually found in higher quality instruments. Alternatively, a measurement can be recorded with only solvent in the cell to establish the I_0 value at a given wavelength, and then, the sample is placed in the cell and the measurement recorded a second time to establish the I value. The two values are then divided and the log taken to establish the absorbance. This process can also be automated to record an entire I_0 spectrum and store it in a computer. Spectra of samples can then be taken from this stored I_0 spectrum. This is called a single beam spectrophotometer. It is usually used in less expensive instruments, because fewer optical components are required. A single beam spectrophotometer is adequate for most measurements, but when accurate values of either very high or very low absorbance values are required, a double beam spectrophotometer is usually superior. This is because it measures I_0 and I at nearly the same time and is therefore less subject to drift.

A.9.3   Difference spectroscopy and flash photolysis

Measurements of absorbance changes as a function of both time and wavelength in response to a stimulus are indispensable in photosynthesis research. These are usually carried out using a single beam spectrophotometer set to a single wavelength, with the absorbance change initiated by a perturbation, usually light. The overall arrangement is very similar to that of a UV/Vis spectrophotometer. The technique can be done on a continuous basis, producing **light-induced difference spectra**. This technique was pioneered by Louis Duysens from the Netherlands in the early 1950s. The changes

can also be initiated by a flash of light, often from a laser. This technique is called **flash photolysis** and was invented by the English chemist George Porter in the late 1950s. Experimental setups for continuous difference measurements and flash photolysis are shown in Fig. A1.17b. The flash of light that causes the photochemical change is called the actinic light, while the light that probes the sample is called the measuring light. The intensity of measuring light transmitted through the sample before the flash is defined as I_0, while I is the intensity measured vs. time after the flash. The change in intensity of the transmitted light can be either positive, if the absorbance decreases through the action of the flash, or negative, if the absorbance increases. Recall that the higher the absorbance, the lower the intensity of transmitted light. These changes can either be plotted *vs.* time at a single wavelength to give a kinetic curve, or the spectrum of the changes at a given time after the flash can be plotted to give a difference spectrum, as shown in Fig. A1.18. A positive change in the difference spectrum indicates the appearance of a new species, while a negative change indicates the disappearance of a species that was present before the flash. Both the kinetic curve and the difference spectrum are essential for a complete picture of the changes induced by the flash of light. Some sophisticated instruments can measure both the time and spectral dependence of the absorbance changes simultaneously, so that a three-dimensional "flying carpet" is produced, with time along the x axis, wavelength along the y axis, and ΔA along the z axis, as shown in Fig. A1.18c. A vertical cut parallel to the y axis then gives a difference spectrum at a given time, while a vertical cut parallel to the x axis gives a kinetic trace at a given wavelength. Another way of representing these flash photolysis experiments is shown in Fig. A1.19, where a "heat map" color codes the three dimensions of data into two dimensions, with the magnitude of absorption changes represented by different color (Niedzwiedzki *et al.*, 2019).

Difference spectroscopy plays an important role in many investigations in photosynthesis research and highly sophisticated techniques for measuring difference spectra have been developed. Some of

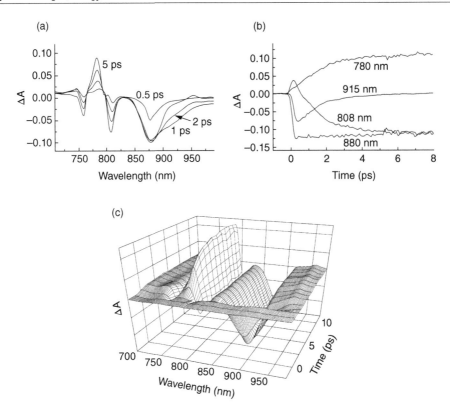

Figure A1.18   Results from a flash photolysis experiment. (a) Difference spectrum at one time after the flash. (b) Kinetic trace at one wavelength. (c) "Flying carpet" representation of both spectral and time dependence of flash photolysis data. *Source*: Figure courtesy of Su Lin.

these techniques involve the use of light flashes on the picosecond or femtosecond timescale. These methods are able to probe the ultrafast processes that take place in photosynthetic systems immediately after photon absorption. Other systems that operate on much slower timescales are better suited to probe later processes in the photosynthetic process, so faster is not always better. It is important to match the instrument to the question being asked in a particular study.

A.9.4   *Fluorescence measurements*

Fluorescence measurements provide an important probe of the excited states of a system following photon absorption. A variety of types of fluorescence measurements are possible, each of which gives different information. Both steady-state and

time-resolved measurements are used extensively in photosynthesis research. A fluorescence spectrophotometer is diagrammed schematically in Fig. A1.20. The essential elements are a light source, excitation and emission monochromators, a sample holder, a photodetector, and a readout device. Most of these components are very similar to those described for an absorption spectrophotometer. The major difference is that in contrast to an absorption spectrophotometer, where the incident and transmitted light are at the same wavelength, in a fluorescence measurement the incident light is usually at a different wavelength than the emitted light. The reasons for this wavelength shift between absorbed and emitted light are discussed above.

Two basic types of fluorescence spectra are usually measured. The most common is the emission spectrum, in which the excitation monochromator

(a)

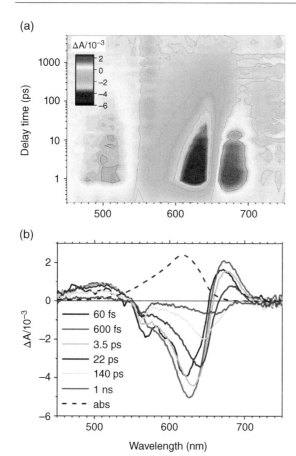

(b)

is set to a wavelength that excites the molecule of interest, and the emission monochromator is scanned as a function of wavelength. The other type of spectrum is the excitation spectrum, in which the emission monochromator is set to detect emission from the molecule of interest, and the excitation monochromator is scanned as a function of wavelength. The fluorescence excitation spectrum is usually very similar to the absorption spectrum of the emitting molecule. Excitation spectra can be instrumental in analyzing energy transfer between pigments with different absorption spectra. Photons absorbed by one molecule may be emitted by another after energy transfer. Both pigments will contribute to the excitation spectrum. This energy transfer process is considered in more detail in Chapter 5.

A.9.5 Fluorescence lifetime measurements

In addition to the spectra discussed above, measurements of the fluorescence lifetime are commonly carried out. This technique measures the excited state lifetime τ directly and is often very informative. Because the excited state lifetimes of chlorophylls and other photosynthetic pigments are usually in the nanosecond or picosecond time range, special techniques are required to measure these rapid events. A number of ultrafast fluorescence techniques have been developed, including single-photon counting methods, phase shift methods, and upconversion methods.

Figure A1.19 Transient absorption (TA) results of the phycobiliproteins from the cyanobacterium *Acaryochloris marina*. The sample was excited at 570 nm. (a) Pseudo-color TA profile, (b) Exemplary TA spectra taken at various delay times after excitation. The steady-state absorption spectrum of the sample is shown as a dashed line. *Source*: Niedzwiedzki *et al.* (2019). Reproduced with permission from Elsevier.

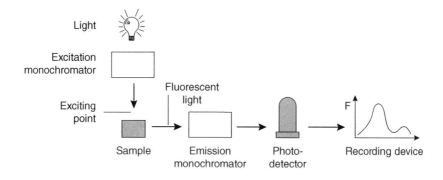

Figure A1.20 Block diagram of a fluorescence spectrophotometer.

A.9.6   *Fluorescence polarization and anisotropy*

A third type of fluorescence measurement uses polarized light. The excitation light is made plane polarized through use of a polarizing filter or prism. The emission spectrum is then measured through a polarizer arranged either parallel or perpendicular to the excitation light. The polarized light selectively excites those molecules whose transition dipoles are oriented parallel to the plane of the excitation polarization. The emission polarizer preferentially transmits emitted light that is polarized parallel to it. The maximum polarization is observed when the transition dipole that is responsible for absorbing a photon is also the one that emits. The polarization will be decreased either if the excited molecule rotates during the lifetime of the excited state or if energy transfer takes place to a molecule with a different orientation. The **fluorescence polarization** (P) is defined as:

$$P = \frac{I_{//} - I_{\perp}}{I_{//} + I_{\pm}} \qquad (A1.80)$$

where $I_{//}$ is the intensity of light emitted parallel to the plane of polarization of the exciting light, while $I_{\pm}$ is the intensity of light emitted perpendicular to the polarization of the exciting light. In a randomly oriented sample, statistical averaging means that the maximum values of P are +0.5 and −0.33. These values are observed when the emitting transition dipole moment is parallel or perpendicular to the absorbing one. A related quantity is the **fluorescence anisotropy** (r), which is defined slightly differently from the polarization:

$$r = \frac{I_{//} - I_{\perp}}{I_{//} + 2I_{\perp}} \qquad (A1.81)$$

The anisotropy varies from +0.4 to −0.2. In both measurements, rotation of the molecule or energy transfer to another molecule at a different orientation reduces the magnitude of the polarization or anisotropy. Both these measurements can be made in a time-dependent manner on the ultrafast time-scale. Modern work usually involves the anisotropy, as the denominator is equal to the total emission intensity and is therefore more convenient experimentally. This type of measurement can give information about the kinetics of energy transfer between pigments with identical absorption spectra that is difficult to obtain any other way.

A.10   Photochemistry

Photosynthesis is a biological process that begins with **photochemistry**. All the principles of photochemistry that have been elucidated in purely chemical systems must also apply to photosynthesis. In this section, we will briefly introduce some of the basic concepts of photochemistry, some of which we have already discussed in earlier sections.

The basic principle of photochemistry is **Einstein's law of photochemistry**, which states that the primary photochemical process is caused by action of one photon absorbed by a molecule. This law emphasizes the important fact that only absorbed photons can do photochemistry. The other important concept is that of the **photochemical quantum yield**, ϕ, which is defined as:

$$\phi = \frac{\text{number of photochemical products}}{\text{number of absorbed photons}} \qquad (A1.82)$$

The quantum yield is a measure of the efficiency of the photochemical process. It is important to distinguish between the primary quantum yield and the overall quantum yield, which can involve a large number of secondary processes. The primary quantum yield ranges from 1 for a process in which every absorbed photon leads to products to 0 when no products are formed, usually because of efficient loss processes. Some chemical chain reactions have overall quantum yields of more than a thousand. This results from the fact that a single primary photochemical event initiates a chain reaction in which many products are formed. In photosynthetic systems, the primary quantum yields are often close to 1 under optimal conditions, indicating that almost all absorbed photons are effective in producing initial products. However, more than one photon is usually needed to produce stable products, so that the overall quantum yields are significantly less than 1. For example, the overall quantum yield for O_2 production by oxygen-evolving photosynthetic organisms is typically about 0.1, so that ten photons must be absorbed for each O_2 produced. We explore the controversial history of these measurements in Chapter 3 and the mechanistic reasons for the observed value in later chapters. The reciprocal of the quantum yield is called the quantum requirement and is an often-cited quantity. For O_2 production the quantum requirement is therefore about 10.

Another important property of a photochemical process is whether the excited state that initiates the chemistry is a singlet state or a triplet state. When an excited singlet state initiates photochemistry, photon absorption leads to an excited singlet state. If this excited state is the lowest excited singlet, photochemistry occurs directly from this initially populated state. If it is a higher excited singlet state, photochemistry almost always takes place from the lowest excited state, to which the higher excited state rapidly relaxes. When a triplet state initiates photochemistry, photon absorption leads first to an excited singlet state, which then undergoes intersystem crossing to form an excited triplet state. The excited triplet then reacts photochemically.

Most organic photochemical processes in solution proceed from the excited triplet state, because it is usually much longer lived than the excited singlet, and therefore has more time to react with other molecules that must diffuse to it before the reaction can take place. Also, the recombination processes that can deplete the primary products once formed are slowed because of spin restrictions if the reactive state is a triplet. However, a variety of lines of evidence clearly shows that the primary photochemistry of photosynthesis takes place from the first excited singlet state. Triplet states are important in some photoprotective processes in photosynthetic systems but are not involved in the primary events.

A.10.1 Excited state redox potentials

The essence of photosynthetic energy storage is the conversion of photon energy to chemical redox energy. How this happens is a complex process that occupies much of this entire book. However, a concept that greatly aids in visualizing the energetics and mechanism of the initial conversion of electronic excitation to charge-separated states is the idea of the **excited state redox potential**.

The excited state of a molecule has many chemical properties that are quite different from the same molecule in the ground electronic state. One example is acid–base behavior, which may be very different in the ground *vs.* the excited state. The most important of these properties for understanding photosynthesis is the redox potential of the excited state. Surprisingly, the excited state can be both a stronger oxidant and a stronger reductant than the ground state. A simple orbital occupation diagram illustrates the reason for this seemingly contradictory behavior. Figure A1.21 shows the orbital diagram for the ground and first excited state of a molecule such as a chlorophyll. In the ground state, an electron that is lost must come from the low energy HOMO, while an electron that is gained must go into the high energy LUMO. The result is that the molecule is neither a very strong oxidant nor a very powerful reductant. In the excited state, the electron in the LUMO is now rather loosely bound and is easily lost, making the molecule a powerful reducing agent. In addition, the hole in the HOMO has a high affinity for another electron, making the molecule a powerful oxidizing agent. Both types of behavior are known in various photochemical systems, but in all known photosynthetic organisms the excited state acts as a powerful reductant.

It is even possible to estimate the midpoint redox potential of the excited state quantitatively, although the excited state is much too transient a species to carry out a standard redox titration such as described above. The redox potential of the excited state depends on the redox potential of the ground

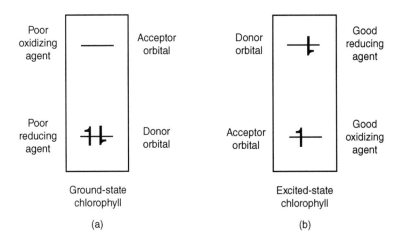

Figure A1.21 Excited state redox behavior. (a) Ground state. (b) Excited state. In (a), the ground state is both a poor oxidizing agent and a poor reducing agent, while in (b) the excited state is both a strong oxidant and a strong reductant.

state, as well as the energy of the excited state above the ground state. This is essentially the difference in the energies of the $v = 0$ vibrational levels of the potential energy curves of the ground and reactive excited state, and is therefore called the 0–0 energy. The excited state redox potential is then given by the simple equation:

$$E_m\left(P^+/P^*\right)=E_m\left(P^+/P\right)-E\left(P/P^*\right) \qquad (A1.83)$$

where $E_m(P^+/P^*)$ is the midpoint redox potential of the excited state reaction:

$$P^+ + e^- \rightleftharpoons P^* \qquad (A1.84)$$

$E_m(P^+/P)$ is the midpoint potential of the ground state reaction:

$$P^+ + e^- \rightleftharpoons P \qquad (A1.85)$$

and $E(P/P^*)$ is the 0–0 spectroscopic energy of the excited state above the ground state (in units of electron volts):

$$P + h\nu \rightleftharpoons P^* \qquad (A1.86)$$

In Eqs. (A1.83) through (A1.86), P stands for the photoactive pigment. All the reactions shown are

written as reductions, according to custom, although the important excited state redox process is really the excited pigment acting as a strong reductant, which leads to its oxidation. Excited state redox potentials of some of the pigments important in photosynthesis are given in Table A1.3. The excited state redox potentials are in most cases more negative than $-1\,V$ (*vs.* NHE), which means that they are extremely powerful reductants. When viewed in this way, the basic chemical mechanism of the photosynthetic energy storage is easily visualized. Light redistributes the electrons within the photoactive pigment by producing the highly reducing and reactive excited singlet state. All subsequent processes, including the primary charge

Table A1.3 Excited-state redox potentials and excitation energies.

Redox or excitation process	$E_m\,(P^*/P^+)$ (V)	$E_{00}(P/P^*)$ (eV)
P700$^+$+e$^-$ $\rightleftharpoons$ P700*	−1.26	1.75
P870$^+$+e$^-$ $\rightleftharpoons$ P870*	−0.94	1.39
P680$^+$+e$^-$ $\rightleftharpoons$ P680*	∼−0.7	1.80
Chl$^+$+e$^-$ $\rightleftharpoons$ Chl*	−1.07	1.85
BChl$^+$+e$^-$ $\rightleftharpoons$ BChl*	−0.94	1.58

Redox potentials were calculated using Eq. (A1.83). Excitation energies were calculated using Eq. (A1.3). See Table A1.2 for ground state redox potentials of reaction center oxidation. * signifies excited state. Data from Blankenship and Prince (1985).

separation, are simply a series of downhill electron transfer processes. Chapters 6 and 7 discuss the mechanisms and pathways of both the primary photochemical process and the secondary processes that follow.

References

Blankenship, R. E., and Prince, R. C. (1985) Excited state redox potentials and the Z scheme of photosynthesis. Trends in Biochemical Sciences 10: 382-383.

Gamow, G. and Stannard, R. (1999) *The New World of Mr Tompkins: George Gamow's Classic Mr. Tompkins in Paperback.* Cambridge: Cambridge University Press.

Niedzwiedzki, D. M., Bar-Zvi, S., Blankenship, R. E., and Adir, N. (2019) Excitation energy migration in phycobilisomes from the cyanobacterium *Acaryochloris marina. Biochimica et Biophysica Acta* 1860: 286–296.

Further reading

Allen, J. P. (2008) *Biophysical Chemistry.* Oxford: Wiley-Blackwell.

Atkins, P. and De Paula, J. (2015) *Physical Chemistry for the Life Sciences*, 2nd Edn. San Francisco: W. H. Freeman.

Chang, R. (2005) *Physical Chemistry for the Biosciences.* Sausalito, CA: University Science Books.

Creighton, T. E. (2011) *The Physical and Chemical Basis of Molecular Biology: Fundamentals.* Helvetian Press.

Tinoco, I. Jr., Sauer, K., Wang, J. C., Puglisi, J., Harbison, G., and Rovnyak, D. (2013) *Physical Chemistry. Principles and Applications in Biological Sciences*, 5th Edn. Englewood Cliffs, NJ: Prentice Hall.

van Holde, K. E., Johnson, W. C., and Ho, P. S. (1998) *Principles of Physical Biochemistry*, 2nd Edn. Upper Saddle River, NJ: Prentice Hall.

Further advanced reading

Amesz, J. and Hoff, A. J., (eds.) (1996) *Biophysical Techniques in Photosynthesis.* Dordrecht: Kluwer Academic Publishing.

Aartsma, T. J. and Matysik, J., (eds.) (2008) *Biophysical Techniques in Photosynthesis*, Vol. II. Dordrecht: Springer.

Atkins, P. W. and Friedman, R. S. (1997) *Molecular Quantum Mechanics*, 3rd Edn. New York: Oxford University Press.

Cantor, C. R. and Schimmel, P. R. (1980) *Biophysical Chemistry.* San Francisco: W. H. Freeman.

Cramer, W. A. and Knaff, D. B. (1990) *Energy Transduction in Biological Membranes: A Textbook of Bioenergetics.* New York: Springer-Verlag.

Klostermeier, D and Rudolph, M. G. (2017) *Biophysical Chemistry.* Boca Raton: CRC Press.

Turro, N. J. (1991) *Modern Molecular Photochemistry.* Mill Valley, CA: University Science Books.

Index

Molecular Mechanisms of Photosynthesis, Third Edition. Robert E. Blankenship.
© 2021 Robert E. Blankenship 2021 by John Wiley & Sons Ltd.
Companion website: https://www.wiley.com/go/blankenship/molecularphotosynthesis3e

Printed and bound by CPI Group (UK) Ltd, Croydon, CR0 4YY

27/10/2024

14580303-0001